MW00760385

Practical Multiservice LANs: ATM and RF Broadband

For a complete listing of the *Artech House Telecommunications Library*, turn to the back of this book.

Practical Multiservice LANs: ATM and RF Broadband

Ernest O. Tunmann

Artech House
Boston • London

Library of Congress Cataloging-in-Publication Data
Tunmann, Ernest O.
    Practical multiservice LANs : ATM and RF broadband / Ernest O. Tunmann.
        p.   cm. — (Artech House telecommunications library)
    Includes bibliographical references and index.
    ISBN 0-89006-408-3 (alk. paper)
        1. Local area networks (Computer networks).   2. Asynchronous transfer mode.
    3. Broadband communication systems.   I. Title.   II. Series.
    TK5105.7.T85   1999
    004.6'8—dc21   99-17793
                                                                                          CIP

British Library Cataloguing in Publication Data
Tunmann, Ernest O.
    Practical multiservice LANs : ATM and RF broadband — (Artech House
    telecommunications library)
    1. Broadband communication systems   2. Telephone switching systems,
    Electronic   3. Local area networks (Computer networks)
    I. Title
    621.3'82'16

    ISBN 0-89006-408-3

Cover design by Lynda Fishbourne

© 1999 ARTECH HOUSE, INC.
685 Canton Street
Norwood, MA 02062

International Standard Book Number: 0-89006-408-3
Library of Congress Catalog Card Number: 99-17793

10 9 8 7 6 5 4 3 2 1

To my wife Margret,
and my daughters Jeannine and Linda

Special thanks go to Linda
for spending long hours creating the figures

Contents

	Preface	*xix*

1	**The Ubiquitous LAN—ATM and RF Broadband**	**1**
1.1	Megahertz Versus Megabits	2
1.2	RF Broadband Networking	4
1.3	The Integrated Voice/Data and Video Facility	5
1.4	Digital and Analog Services—Transparency	6
1.5	Asynchronous Transfer Mode—Interoperability	6
1.6	No More Moves and Changes	7
1.7	Nonobsolescence	8
1.8	Scalability	9
1.9	The Reasons for Obscurity	9
1.9.1	RF Broadband Is a Multivendor Network	10
1.10	Convergence	11

2	**The Evolution of RF Broadband**	**13**
2.1	The Delivery of Cable Television	14
2.1.1	Early Systems	15
2.2	The Many Deficiencies	16
2.2.1	Cascades of Noise	17

2.2.2   Composite Triple Beat   19

2.2.3   Ingress and Egress   19

2.2.4   Multipath Problems   20

2.2.5   Customer Service   21

2.3   The Regulatory Environment   22

2.3.1   Federal Regulations   22

2.3.2   State and Local Regulations   23

2.4   The Telecommunications Act of 1996   24

2.5   Upgrading RF Broadband to Hybrid Fiber/Coax   25

2.5.1   Single-Mode Fiber Trunk and Coaxial
        Distribution   27

2.5.2   Sizing the Fiber Node   28

2.5.3   Downstream and Upstream Spectral Capacities   31

2.6   The Operations Center—Gateway/Headend   32

2.6.1   ATM Interoperability   34

2.7   RF Broadband Services   34

**3   The Evolution and Future of Corporate
        Communications   37**

3.1   The Development of Telephony and the PBX   38

3.1.1   Analog and Digital Technology   38

3.1.2   Voice Networking in the WAN   39

3.1.3   The Future of Telephony in the WAN   40

3.1.4   The Future of Telephony in the Corporate
        Network   41

3.1.5   The Alternative RF Broadband Solution   41

3.2   The Evolution of Data Networking   42

3.2.1   Data Network Wiring Systems   43

3.2.2   Routers—Switches and IP in the LAN   44

3.2.3   ATM in the LAN   45

3.2.4   RF Broadband in the LAN   47

3.3   Connectivity With the Outside World   53

3.3.1   Alternative Local Loop Carriers   53

3.4   The Future of Enterprise Communications   55

3.4.1   Virtual Private Networks   56

4   RF Broadband—The Spectral Capacity   61

4.1   The Bandwidth of Fiber-Optic Cables   62

4.1.1   Multimode Fiber   62

4.1.2   Single-Mode Fiber   65

4.2   The Bandwidth of Unshielded Twisted-Pair Cables   67

4.3   The Bandwidth of Coaxial Cables   70

4.3.1   Coaxial Service Drop Cables   71

4.3.2   Coaxial Backbone Cables   72

4.4   Bandwidth Limitations   73

4.4.1   Infrastructure—The Bandwidth Conduit   74

4.4.2   Fiber-Optic Transmission Equipment   75

4.4.3   Coaxial Transmission Equipment   76

4.5   RF Broadband—Architecture Limitations   77

4.5.1   The Subsplit Architecture   77

4.5.2   The High-Split Architecture   79

4.5.3   The Dual System   80

4.6   The Future Corporate LAN   81

5   The Importance of SONET and ATM   83

5.1   The Synchronous Optical Network   84

5.1.1   The SONET/SDH Hierarchies   84

5.1.2   The SONET Advantages   86

5.1.3   The SONET Frame Structure   88

5.1.4   The Virtual Tributary Frame Structure   92

5.1.5   The SONET Transport System   92

5.2   Asynchronous Transfer Mode   95

5.2.1   The ATM Cell Structure   96

5.2.2   The ATM Adaptation Layers   98

5.2.3   ATM-UNI Signaling   100

5.2.4   ATM Switching   101

5.2.5   Network Operation and Routing—Private
        Network-to-Network Interface (PNNI) and
        Network Management   102

5.2.6   Traffic Management and Congestion Control   104

5.2.7   Service Classes and QoS   105

5.2.8   LAN Emulation   110

5.2.9   Multiprotocol Over ATM (MPOA)   112

5.3   ATM—Interoperability and Multiservice
        Technology   113

5.3.1   ATM and Frame Relay   114

5.3.2   ATM and TCP/IP and the Internet   114

5.3.3   IPv6 and ATM   116

5.3.4   System Network Architecture in ATM   116

5.3.5   ATM and Ethernet: Fast Ethernet, FDDI,
        and Gigabit Ethernet   118

5.3.6   ATM: Video and Telephony   119

5.4   ATM Forum Specifications   121

5.5   Cable Modem Standards and ATM   125

5.5.1   The Residential Broadband Working Group   125

5.5.2   Institute of Electrical and Electronic
        Engineers—IEEE 802.14   125

5.5.3   The Digital Audiovisual Council   126

5.5.4   The Society of Cable Telecommunications
        Engineers   128

5.5.5   Cable Labs   128

5.6   ATM Applications for RF Broadband LANs   129

6   ATM-Based Network Architectures and
        RF Broadband   131

6.1   The Global Multiservice Network   132

6.2	The Enterprise Network	132
6.3	The Campus Network	136
6.3.1	Migration Toward the Multiservice Network	137
6.3.2	The RF-ATM Corporate Campus Network	139
6.4	Versatility of the Campus Network Architecture	142
6.4.1	The Fiber Backbone Star Architecture	142
6.4.2	The Fiber Backbone Single-Ring Architecture	144
6.4.3	The Fiber Backbone Dual-Ring Architecture	145
6.4.4	The RF Broadband Star Architecture in the Dual Ring	146
6.5	Versatility of the Building Network Architecture	147
6.5.1	RF Broadband in the Small Building	147
6.5.2	RF Broadband in the High-Rise Building	148
6.5.3	RF Broadband in the Horizontal Wiring	153
6.6	Flexibility in RF Broadband Networking	155
6.7	Performance Considerations of the RF Broadband Network	158
6.7.1	The Carrier-to-Noise Ratio	159
6.7.2	The Composite Triple Beat	160
6.7.3	Reliability of Service	161
6.8	Selecting the Desired RF Broadband Network Architecture	162
6.8.1	Optimizing the Spectral Capacity	162
6.8.2	Optimizing the Availability	164
6.8.3	Economical Considerations	164
	Reference	164
7	**ATM-Based Network Applications and RF Broadband**	**165**
7.1	RF Broadband—Versatility	165
7.2	The SOHO and Small Office LAN	166
7.2.1	Conventional Choices	166

7.2.2   The ATM-RF Broadband Alternative   168

7.3   The Intelligent High-Rise Building   169

7.4   Apartment Complexes   172

7.4.1   Conventional Wiring Methods   172

7.4.2   Multiservice Infrastructure Alternatives   173

7.5   The Campus Network   175

7.5.1   Conventional Wiring Systems   176

7.5.2   Data Networking Alternatives   177

7.5.3   Computer Telephony Integration   178

7.5.4   Multiservice ATM and RF Broadband   179

7.6   Ancillary Applications Using RF Broadband   181

7.6.1   Cable Television   181

7.6.2   Video on Demand   181

7.6.3   Videoconferencing   182

7.6.4   Cable Data   183

7.6.5   Cable Telephony   183

7.6.6   Personal Communication Services   184

8   Planning the Multiservice Broadband LAN   187

8.1   Existing Plant Records   188

8.1.1   Outside Plant   188

8.1.2   Inside Plant   190

8.1.3   Computer Center and Server Farms   193

8.1.4   PBX Equipment   195

8.1.5   Data Network Center   195

8.2   Conformance to Structured Cabling Standards
        (BICSI)   195

8.2.1   Single-Mode Fiber   197

8.2.2   UTP Wiring   198

8.2.3   Coaxial Wiring   202

8.3   Network Architecture Preferences   203

8.3.1   Preplanning Considerations   204

8.4   The Planning Document   207
8.4.1 The Single-Mode Fiber Interconnect Facility   207
8.4.2 Outside Plant Requirements   208
8.4.3 Power and Space Requirements   209
8.4.4 IDF and Vertical Riser Requirements   210
8.4.5 Outlet and User Requirements   210
8.4.6 Horizontal Wiring Requirements   211
8.5   The Multiservice LAN Project Plan   211
8.5.1 Facility Construction Requirements   212
8.5.2 Implementation of the Multiservice LAN   213

9   The Status of Cable Modem Development   215

9.1   The Evolution of the Cable Modem   215
9.1.1 Analog Video Cable Modulators and
       Demodulators   216
9.1.2 Digital Video Modems and Set-Top Converters   216
9.1.3 Digital Cable Modems   216
9.2   Cable Modem Manufacturers and Market   217
9.2.1 The Cable Modem Market   217
9.2.2 Symmetric and Asymmetric Cable Modem
       Providers   219
9.3   Cable Modem Technology   220
9.3.1 The Physical Layer   221
9.3.2 The MAC Layer   222
9.3.3 MAC Management   224
9.3.4 QoS Considerations   225
9.3.5 Security Considerations   226
9.4   Present Cable Modem Standards   228
9.4.1 The MCNS Interoperable Cable Modem   228
9.4.2 ATM-Based Cable Modems   233
9.5   Cable Modem Applications   233
9.5.1 Internet Access for Residences and Businesses   234

9.5.2 Applications in the RF Broadband LAN   235

10 The Status of Cable Telephony   237

10.1 Cable Telephony in the Cable Industry   237

10.1.1 Synchronous Transfer Method   238

10.1.2 Voice Over IP and Over ATM   238

10.1.3 Business Telephone Systems   240

10.1.4 Network Management   242

10.2 Telephony in the Corporate LAN   243

10.2.1 Computer Telephony Integration   243

10.2.2 ATM-Based and RF Broadband Cable
Telephony   244

11 Videoconferencing and Video-on-Demand   249

11.1 Analog Videoconferencing (NTSC)   250

11.1.1 Automation Equipment   250

11.2 Digital Videoconferencing   251

11.2.1 Videoconferencing Standards   251

11.2.2 A Comparison of JPEG and MPEG-2   253

11.2.3 JPEG Over ATM   254

11.2.4 MPEG-2 Over ATM   255

11.2.5 ATM Networking and RF Broadband   256

11.2.6 Videoconferencing in the WAN   257

11.3 Video-on-Demand in the Corporate Network   257

11.3.1 MPEG-2 and the Digital Video Disk   258

11.3.2 VOD Technology   258

11.3.3 VOD Network Architectures   260

12 Fiber-Optic Transmission Equipment   263

12.1 Single-Mode Fiber-Optic Cables   263

12.1.1 Loose Tube Cables   264

12.1.2 Tight Buffer Cables   265

12.1.3 Break-Out Cables 267

12.2 Fiber-Optic Cable Termination Equipment 268

12.2.1 Fiber Distribution Frames 269

12.2.2 Rack Mounting and Cable Management 270

12.2.3 Single-Mode Fiber Connectors 270

12.2.4 Preassembled Patchcords 271

12.3 Connectorization Considerations 271

12.4 Fiber-Optic Transmission Equipment 272

12.4.1 High-Power DFB Laser Transmitters 273

12.4.2 Low-Power DFB Laser Transmitters 274

12.4.3 Receiving Equipment 275

12.5 Fiber-Optic Couplers 276

13 Coaxial Transmission Equipment 277

13.1 Coaxial Cables 277

13.1.1 Outside Plant, Riser, and Plenum Cables 278

13.1.2 Service Drop Cables—Riser and Plenum 280

13.2 Passive Devices 282

13.2.1 Splitters and Directional Couplers 282

13.2.2 Multitaps 285

13.2.3 Connectors 288

13.3 Amplification Equipment 289

13.3.1 Distribution Amplifiers 290

14 Implementation Considerations for the HFC-RF Broadband Network 293

14.1 Implementing the Planning Document 293

14.1.1 Implementation Categories 294

14.2 Outside Plant Installation Considerations 295

14.2.1 Underground Duct Installations 295

14.2.2 Single-Mode Fiber-Optic Cable Pulling 297

14.3 Inside Plant Considerations 297

14.3.1   Conduit Installations   297
14.3.2   Equipment Rooms   298
14.3.3   Vertical Risers and Horizontal Wiring   300
14.4   Vendor Capability Assessments   302
14.5   Preparation of RFPs and Bid Specifications   305
14.5.1   RFP Preparation   305
14.5.2   Technical Specifications   307
14.5.3   Terms and Conditions   308
14.6   Implementation Supervision   310

15   The Design of the RF Broadband Network   311

15.1   Overview of the Design   311
15.1.1   Fiber Design Criteria   311
15.1.2   Coaxial Design Criteria   312
15.2   The Design of the Fiber Backbone   314
15.2.1   Forward and Return Considerations   315
15.2.2   Optical Design Budgets   317
15.2.3   Fiber Couplers   318
15.3   The Design of Coaxial Segments   319
15.3.1   The Service Drop Segment   320
15.3.2   The Riser Distribution Segment   324
15.3.3   The Headend Interconnect Segment   329

16   Cost Considerations   333

16.1   General Assumptions   333
16.1.1   Standard Network Components   333
16.1.2   RF Broadband Applications   334
16.2   A Comparison of IP and RF Broadband Costs   334
16.3   A Cost Comparison Between ATM and
        RF Broadband   336
16.4   Costing an ATM-Based Videoconferencing
        System   338

16.5 Costing an HFC-Based VOD Network   338
16.6 Costing an ATM-Based Telephone Overlay   340
16.7 Price Forwarding   340

17   Outlook   341

17.1 The Changing Telecommunications
      Environment   341
17.2 The Multiservice Public Network   342
17.2.1 The Long-Distance Segment   342
17.2.2 The ISPs and the Internet   343
17.3 The Multiservice Local Loop   343
17.3.1 CLECs   344
17.3.2 RF Broadband Cable Networks   345
17.4 The Multiservice WAN   346
17.4.1 ATM and IP Choices   347
17.4.2 VPNs   348
17.5 The Multiservice Corporate LAN   349
17.5.1 ATM and IP Choices   349
17.5.2 The Integration of RF Broadband   350
17.6 Electronic Commerce   351
17.7 Thoughts in Closing   351

List of Acronyms and Abbreviations   353

Bibliography   367
Books   367
Standards (Samples)   368
Standards Sources   369

About the Author   371

Index   373

Preface

This book is of interest to anyone who would like to better understand the basic telecommunications technologies of today's transmission networks. It provides the reader with an overview of the evolution of telecommunications technologies and predicts future trends in the long-distance segment, the local loop, the wide area network (WAN), and the local area network (LAN).

The book more specifically addresses telecommunications professionals, communications information officers, network managers, facility managers, telephone network specialists, information technology experts, and the many individuals who are experts in their own fields and wish to expand their understanding of the new interrelated technologies. It discusses the integration of voice, data, and video in the long-distance segment, in the enterprise WAN, and in the corporate LAN; brings the numerous trends in these technologies into context; and offers guidance for an economical transition to multiservice networking. It also highlights the application of the RF broadband technology in the corporate LAN as a forgotten but useful alternative for future networking and the for the integration of voice, data, and video in a single wiring system.

While telephone was the major form of communications for half a century, the last decades have seen enormous technological advances in the way we communicate and conduct business. The result is the new information age and the ability to instantly acquire and share knowledge with anyone on this fast-shrinking planet. This book aims to provide insight into the evolution and explosive present-day development of new technologies in all areas of telecommunications—whether in telephony, the transfer of data, the many uses of video, the evolution of RF broadband-based cable television, or personal communications services (PCS)—and discusses how they have become daily necessities and will develop in the future.

Furthermore, the book touches on the competitive developments in the local loop; the transition of the cable TV system into a multiservice voice, data, and video service provider; the competitive local exchange carriers (CLECs) and wireless local multipoint distribution services (LMDS) offerings; the resulting corporate transition to more economical enterprise services; and the need for readiness to benefit from the asynchronous transfer mode (ATM)-based multiservice outside world. Also, the long-distance segment and the new global optical information superhighways are briefly discussed to give the reader a basic understanding of synchronous optical network (SONET), Internet, and ATM technologies that form the basis for the rapid development of global multimedia communications. In addition, new competitive developments in the local loop are discussed, along with the resulting transition of the business community to more economical virtual private network (VPN) enterprise services and the need to adapt the evolving ATM-based technology into the corporate network for end-to-end multimedia services.

An overview of present and future WAN and LAN architectures explains the advantages of SONET/ATM technology, addressing the need for integrated end-to-end multiservice networking and discussing the importance of ATM in the campus backbone, the uses of RF broadband and hybrid fiber-optic/coaxial (HFC) in the campus LAN, and the step-by-step integration of voice, data, and video into a single multiservice wiring system. In addition, the benefits of using single-mode fiber in the backbone of the LAN meeting the higher speed ranges of the future and establishing alternative RF broadband services are examined as are the importance of good planning practices and the advantages of using a Building Industry Consulting Services, International (BICSI)-compliant structured wiring system. Finally, the book discusses the need for a vendor-independent infrastructure and provides a guide for network planning, design, implementation, and cost issues.

1

The Ubiquitous LAN—ATM and RF Broadband

The RF broadband network has become the favorite platform of the cable industry and of many CLECs. RF broadband technology uses a combination of single-mode fiber-optic strands and coaxial cables operating to RF broadband transmission standards. RF broadband technology permits the transmission of multiple services over discrete RF frequencies. Because of the mixture of single-mode fiber and coax, RF broadband technology is often also called HFC. HFC has been proven in many field trials and will change the competitive picture in the local loop before the end of this century. Over $14 billion has already been invested by the major communications delivery companies to upgrade the outside plant, meet HFC transmission parameters, and ensure the delivery of integrated voice, data, and video over a single cabling system to residencies and businesses. The timeline of these upgrade activities will extend, in some areas, to 2000 and beyond and, together with wireless offerings, will change the landscape of telecommunications.

This book deals with the implications of the change toward HFC-based RF broadband technology in the outside world and the great advantages that it can bring to campus-wide networking, premises wiring, and the integration of voice, data, and video into a single multiservice infrastructure.

1.1 Megahertz Versus Megabits

Present wiring solutions in corporate networks are the result of an evolutionary development. Wiring for voice communications was the first requirement in every business establishment. From key systems to large private branch exchange (PBX) systems, more wiring was required to satisfy companies' internal and external communications needs. Then came the computer revolution and the beginning of the information age. The transfer of an ever-increasing information requirement led to the development of LAN technologies and associated wiring requirements. For the first time, the terms *bits, bytes, kilobits per second* (Kbps), *megabits per second* (Mbps), and *gigabits per second* (Gbps) entered the vocabulary of the user, who, at first, only wanted to interconnect a few computer workstations. The term *intelligent islands* was often used to refer to the difficulties that were experienced in attempts to interconnect various data streams and protocols. The advent of bridges, routers, and switches enabled the interconnection of these intelligent islands and formed the basis for a new and separate wiring technology—the data network.

Subsequently, wiring standards changed. New shielded and unshielded twisted-pair cabling was developed to permit faster transfer of data. Multimode fiber-optic cables and category-5 (CAT-5) twisted-pair cabling established a new standard for data transmission systems—first for 10-Mbps Ethernet and later for 100-Mbps fast Ethernet. Lately, we hear about ATM at 155 Mbps, a cell switching system using 53 byte cells and even Gbps transmissions to the desktop, only to be reminded that higher speeds may require different wiring solutions.

The terms *Kbps, Mbps,* and *Gbps* mean that one thousand, one million, or one billion bits per second can be transported over a particular wiring system. These terms apply to the transport of zeros and ones over a wire at baseband frequencies. The higher the speed of the zeros and ones that are going through the wire, the more bandwidth or spectrum is required to accommodate the faster modulation methods that are being applied. Higher transmission speeds require shorter distances between devices or a change of the wiring. As a result, the infrastructure of corporate LANs has been rebuilt several times.

In contrast to the baseband transmission using units in bits per second, there is the term *megahertz* (MHz), which also is used to describe bandwidth. Megahertz has been used to describe the bandwidth requirement of any analog transmission. A video channel, which can include a number of audio carriers, utilizes a 6-MHz spectrum of frequencies.

When the Federal Communications Commission (FCC) assigned a frequency spectrum to the television broadcast industry, the 6-MHz spectrum segments were assembled in a sequential manner. Channel 2 was assigned the 54–60-MHz band, channel 3 used the 60–66 MHz-band, and so on, and television was born.

The beginning of the television era also established the need for cable television. In the 1950s, the first cable television systems operated in the low band channels. Channels 2–6 were received off the air and distributed on coaxial cable.

Thus, RF broadband technology was born. Since then, the ability to stack 6-MHz channels, one on top of the other, has increased the channel capacity to over 100 channels. Today, cable distribution systems can deliver a 1-GHz spectrum over considerable distances using HFC networks. It is helpful to compare Mbps bandwidth with MHz bandwidth to understand the capacity and limitations of these wiring systems. A video channel complying to the National Television Standards Committee (NTSC) specifications occupies a 6-MHz spectrum segment when transmitted in analog form. Digitization of a full-motion video sequence, without the use of compression, would require 145 Mbps to properly present all the artifacts. Compression technology has reduced this high bit rate requirement using various compression algorithms. There is the DS-3 compression system that uses the standard common carrier T-3 speed of 45 Mbps. An 80-channel video delivery system, if digitized to T-3 standards, would require eighty T-3 connections but could be carried over a single HFC cable. As T-3 compression is only used for long distance and sometimes for distance learning applications, this example is somewhat academic. Nevertheless, it points to the enormous carrying capacity of RF technology.

The term *cable telephony* is used to describe the transmission of voice channels over HFC networks. A total of 240 voice channels or 10 T-1 assignments can be carried in one 6-MHz frequency assignment. Referred to as *cable data*, current cable modem technology permits the transmission of a 40-Mbps Ethernet connection or a 25.6 ATM-based bit stream to be carried within a 6-MHz bandwidth slot. New modulation techniques will expand this carrying capacity to the ATM speed of 51.2 Mbps or beyond.

While most video channels are transmitted in the old and economical analog format, it will not take too long before cable television is fully digital. The delay in digitization has been caused by standardization problems and economic considerations. The benefits will outweigh these problems over time. Every 6-MHz channel can be subdivided into five to eight video channels using the digital Motion Picture Experts Group (MPEG-2) format or even 20 video channels operating in the MPEG-1 format.

1.2 RF Broadband Networking

Since the term *broadband* has frequently been used to market higher Mbps speeds, the terminology *RF broadband* will be used to describe the enormous spectral capabilities of this networking technology. An RF broadband network consists of single-mode fiber-optic cables for longer distances and a coaxial tail for the distribution of services. Hence, the term *HFC* has become quite common.

In the service area of the cable television company, single-mode fiber is used to connect the headend or gateway of the serving cable company with fiber node locations. The cable television industry spent years analyzing traffic patterns, user habits, and service reliability and reached the conclusion that each fiber node location can serve between 200 to 500 users with a single bidirectional coaxial distribution system. The important fact is that coaxial cable distribution is far more economical than bringing the fiber to the home.

In a premise wiring system, a comparable scenario is to use the single-mode fiber-optic segment in the backbone to interconnect the various buildings in a campus setting. The coaxial delivery system can then be deployed to distribute the transmissions throughout a single building and to each individual user. Either a single coaxial cable can be used for all bidirectional services or two coaxial cables—one for downstream and one for upstream traffic—can be used. It is also possible to eliminate the coaxial distribution altogether and use CAT-5 wiring for desktop connectivity. In this scenario, the use of multicarrier RF broadband technology is restricted to the single-mode fiber segment of the network.

The spectral capacity of present-day HFC-RF broadband systems occupies a bandwidth between 50 and 1,000 MHz with distance limitations only in the coaxial segment. Current electronic equipment limits the upper frequency to 860 MHz, but passive components such as cable, connectors, and multitaps are available for 1,000-MHz operation. Even using the lowest 750-MHz limitation of present fiber-optic transmission equipment, a bandwidth of 50 to 750 MHz is available with a capacity of over one hundred 6-MHz channel assignments. Such a system can handle future simultaneous voice, data, and video transmission requirements in either analog or digital formats without ever changing its architecture. Here are some examples:

- Forty analog NTSC quality video/audio channels plus 30 channels for voice (30 × 240 = 7,200 voice channels), plus 30 channels for 25.6-Mbps ATM data with a combined data speed equivalent to 786 Mbps;

- Two hundred digital MPEG-2 video/audio channels plus 7,200 voice and 30 ATM data channels at 25.6 Mbps;
- One hundred ATM data channels at 25.6 Mbps for any desired mix of ATM-based voice, data, and video in MPEG-1 or MPEG-2 formats for an equivalent capacity of 2.56 Gbps.

Even the new Gbps technology cannot measure up to these speed ranges and would require single-mode fiber strands to cover the distances that can be covered by the RF broadband network. Using a single coaxial tail, the HFC network does not lose its bidirectional capabilities.

The subsplit system with only a return bandwidth of 5 to 42 MHz is the choice of the cable industry. A high-split system with a downstream bandwidth of 222–750 MHz can provide eighty-eight 6-MHz channels and thirty 6-MHz channels in the upstream direction. These topologies are preferred when asymmetric requirements—such as cable television, video on demand, or Internet access—exist and can also be applied to premises wiring systems.

1.3   The Integrated Voice/Data and Video Facility

The HFC-RF broadband network is a cabling infrastructure that can replace existing telephony, data, and video wiring networks. It can also be used to add new services that would otherwise require expensive wiring changes. When properly designed, RF broadband offers almost unlimited expansion capabilities for voice, data, and video services throughout a campus of multiple buildings, throughout a single building, or even for a single floor. An HFC-RF broadband network consists of three major components: the headend or operations center, the fiber segment, and the coaxial distribution network. The main point of presence (MPOP) provides the interconnectivity with the public network. When interconnected through an ATM-based backbone, the RF broadband network can become a workable and economical adjunct to campus networking. The various architectures and applications of the RF broadband network are discussed in Chapters 6 and 7 of this book.

Since video studio, PBX, and computer center are seldom in close proximity, it is wise to consider their integration into the network and assess the added costs before proceeding with system design and implementation. The most effective use of the HFC-RF broadband technology is in campus backbones with restricted fiber counts, in the elimination of numerous workgroup switches, and in the provision of alternate networking for high-speed Internet access, multiple video-on-demand, and multiple bidirectional video requirements.

1.4 Digital and Analog Services—Transparency

An HFC-RF broadband infrastructure is ubiquitous and fully transparent. Digital and analog services can live side by side. Each 6-MHz spectrum can be used for different traffic. Special cutoff filters, called SAW filters, at the edge of each 6-MHz channel provide for separation and prevent adjacent channel interference. Telephony, data, and video can use digital modulation methods and transmit within the confines of the 6-MHz channel assignments. Cable modems are used to convert the digital transmission to the desired RF frequency and can be used for high-speed Internet access, intranet data requirements, videoconferencing, and telephony. Commonly used Internet protocol (IP) networks can easily interface with the cable modem and transport the digital information as IP or IP over ATM to interconnect with the local exchange carrier (LEC). VPN strategies can be used to extend the connectivity throughout the global enterprise.

If broadcast quality video is required in a campus, it can be left in the analog format for a few years. The established broadcast NTSC standards have served the delivery of full-motion video well and have kept the cost of television sets low. The transmission of video and audio in the comparable MPEG-2 format is still expensive and requires the availability of economical digital television sets before it can be used. While it is possible to integrate MPEG-2 into the ATM bit stream, the transmission of video at bit rates of 6–15 Mbps per channel can reduce the throughput of the network. Video represents a heavy load even when variable bit rate parameters are limited to 8 Mbps. While the ragged pictures of video presentations that have been encoded for integrated services digital network (ISDN) or MPEG-1 are fine for desktop (talking head) transmission, they cannot compare with standard analog television. The RF broadband infrastructure enables the optimization of economy, quality, reliability, and throughput. The bandwidth capabilities of the RF broadband architecture can accommodate side-by-side analog/digital video and multiple high-speed data transmissions without reducing the throughput of the data traffic—a benefit that no other infrastructure can claim.

1.5 Asynchronous Transfer Mode—Interoperability

While the RF broadband technology in the LAN can be used to combine multiple Ethernet and voice over IP (VOIP) transmissions, it can also be used to concentrate workgroup switches at the backbone network. A good example is an ATM-based backbone and IP-based campus zones. It is no longer necessary to locate workgroup switches in each campus building, since multiple services can use different RF frequencies. The result is a resilient network with legacy equipment in the outlying areas and with a robust ATM-based core. ATM is a technology that permits the transport of constant and variable bit rate (VBR) services in a fully transparent manner and performance assurance through multiple quality of service (QoS) levels. Any existing data format or protocol can be transported within the cell structure of ATM using LAN emulation (LANE) protocols. While LANE can support voice, data, and video, it also has the ability to finally close the gap between enterprise-wide LANs and the public telecommunications carriers. The multiprotocol over ATM (MPOA) is used to interconnect the connectionless IP technology at high speeds, using the SONET physical layer as the conduit. ATM can be transmitted over twisted-pair, fiber, or coaxial cables. It conforms to globally accepted standards and soon will serve the majority of all end-to-end global information transfers. In contrast, present-day transport technologies such as Ethernet, fast Ethernet, fiber distributed data interface (FDDI), gigabit, and token ring are isolated to LANs and require adaptation through ATM to transform a "best-effort" service into committed QoS levels for global interconnectivity.

The advent of ATM technology promises that existing data networks and protocols can interface with the outside world without any translation. The integration of ATM technology with the RF broadband infrastructure in the backbone of any corporate LAN combines the advantages of both technologies and provides a multiservice facility with excess capacity for the twenty-first century. The ATM structure defines a small 53-byte cell with a 5-byte header and 48 bytes of information. Addressing can be connection-oriented or multicast and define the destination of these cells in any legacy format, whether they are Ethernet, token ring, voice, or video. An ATM-compatible cable modem can transmit 40-Mbps ATM over the RF broadband system in a 6-MHz spectrum assignment at a wide range of frequencies. The capacity of the HFC-RF network, together with ATM cell transport technology, permits an almost unlimited expansion of internal and external multiservice traffic. Chapter 5 provides a more detailed discussion of ATM technology.

1.6 No More Moves and Changes

The RF broadband network carries the entire frequency spectrum to all outlet locations. Good planning and a saturation of the outlet locations can preclude any moves and changes. A change of location of one or more employees only requires a different carrier frequency assignment and authentication. The move of multiple offices to a different building only requires bandwidth reallocation and possibly the setting of a few cable modems to a different RF frequency.

Good spectrum management is necessary to efficiently utilize the enormous bandwidth, but wiring changes, equipment additions, and router port changes can be eliminated. Should it be impossible to select every outlet location that may be used in the future, the system can be designed to permit future expansion. Multitaps have four or eight ports for service drop connections. Empty ports can be terminated and activated when needed. Additional service drops and outlets can then be installed whenever required. When using existing CAT-5 wiring, only a single RF channel is delivered to a group of workstations. All required frequency assignment changes can be performed at the horizontal cross-connect (HC) for the workgroup or used to assemble different compositions of workgroups.

1.7 Nonobsolescence

An HFC-RF network can be designed with ample spare capacity to meet any traffic requirement that may be encountered in the future. Initial assignment of frequencies and a continuous control of the spectrum is important to safeguard future growth.

Should there be a need for more RF spectrum at any time in the future, it will be possible to replace or upgrade both fiber and coaxial electronics to 1,000 MHz without changing any part of the wiring. Major additional requirements are easily met by illuminating additional fiber strands in the backbone and by a further subdivision of workgroups. Many examples of obsolescence-reducing planning and design activities are mentioned in Chapters 8, 14, and 15.

The life expectancy of both single-mode fiber and coaxial plant is considered extremely high. Both fiber and coaxial cables are proven products, designed for outdoor use, and they do not disintegrate unless they are cut. The life expectancy of all RF-HFC components exceeds 20 years, especially in the premises environment. The cable industry has studied the longevity of all HFC components for many years and would not have made the enormous investment in HFC upgrade projects if it believed that the systems would require any rebuilding in the foreseeable future. The HFC-RF system components are all designed to withstand the rigors of the outdoor environment. We have heard of frequent outages of cable television systems because these systems require pole-mounted amplifiers in large cascade numbers, a condition that does not apply to the corporate network. Premises networks are also not exposed to the elements. All active equipment, including fiber-optic transmission equipment and coaxial amplifiers, is located inside equipment closets. Even the passive equipment in riser spaces enjoys dry and warm conditions. The cable industry claims that the service availability of its new HFC systems will exceed 99.999% and surpass the availability of the phone system. This means that a properly engineered and installed RF-HFC system within a campus or building complex can provide telephony, data, and video with an availability of better than 99.9999%.

1.8 Scalability

The spectral capacity of the RF broadband network is determined by the design of the infrastructure. All components of the RF broadband network are more economical than other wiring schemes. Once the infrastructure is implemented, the system can grow at a speed commensurate with yearly budgets and needs, adding peripheral equipment and services as required. Scalability applies to the utilization of the system for any new service, whether an improvement of Internet access, the integration of digital telephony, the addition of new telephones, the expansion of the data system to ATM technology, the downsizing of a number of workgroups, or the assignment of new RF frequencies.

1.9 The Reasons for Obscurity

If HFC-RF technology is so great, why isn't everyone installing it? There are a number of factors that may explain why the technology is not mentioned by the media, not highlighted in technical seminars, not featured in exhibits, and not known to CIOs and telecommunications managers.

Upon considering who the players are, it becomes evident that HFC-RF broadband in the corporate environment does not warrant any advertisement by anyone. On one side, there are the cable companies that are busy upgrading their networks to HFC. This upgrade is done with only one purpose in mind: the increase of revenue potential by offering new services to the public. There are also competitive factors between the large multiple system operators (MSOs) and the Regional Bell Operating Companies (RBOCs). These competitive factors prevent disclosure of construction plans and promotion of technological features. In the battle for the lucrative local loop, secrecy is an important adjunct to the all-out marketing strategy to secure the largest possible customer base at the time of a new service offering. In addition, there are other newcomers entering the market. Electric utility companies have shown an increasing interest in entering the local loop communication business. These factors and the desire to sell profitable interconnect agreements to the residential and business community—when the service offerings become available—preclude any publicity about utilizing HFC-RF broadband technology in the corporate environment.

On the other side are the data networking companies. As equipment designers and manufacturers, these companies are working hard to sell the next level fast backplane switch to the business community. The data networking equipment providers are in competition with each other and have well-paid marketing personnel working to convince the CIO of an enterprise that their way of upgrading to ATM is the best solution. The data networking companies are just beginning to promote voice and data integration within the hierarchy of their existing product lines. The move to OC-12 in the backbone and the interconnection of workgroups at OC-3 are certainly of more interest than the utilization of cable modems.

ATM or Gbps switches and fast backplane motherboards are expensive items when they are needed on a per-building or per-floor basis. Accordingly, the marketing effort is hard-hitting and directed toward getting the highest return from the business community, which is only calling for help and assistance to increase the speed of its data network. Thanks to the initial cable modem development conducted by LanCity (recently incorporated into Bay Networks and then into Nortel), all data network vendors are concentrating on the cable modem market and hope to enter the HFC-RF broadband arena. Competition in pursuit of the lucrative cable industry, and the cable industry's efforts toward integrated residential communication services, will bring economical price levels to RF frequency-based modem equipment and advance the development of RF broadband infrastructures as an economical alternative to multiservice premises LANs.

1.9.1 RF Broadband Is a Multivendor Network

Data networking companies also have the ability to market their LAN technology as a sole source procurement. Except for the installation of fiber cable and CAT-5 service wiring, all equipment can be provided by one supplier. For HFC-RF equipment there are various fiber equipment vendors, many coaxial equipment supply houses, and a number of cable modem manufacturers. Although there are a few turnkey implementors that have experience with both single-mode fiber and coaxial installations, technical expertise is limited. An HFC-RF system needs to be planned, designed, and engineered for optimal routing and performance, and system integration is required to integrate cable modem management and voice/data network management. These negative factors, however, do not outweigh the many advantages of RF broadband infrastructures, which can easily meet the most stringent telecommunications requirements of the twenty-first century.

1.10 Convergence

Excited by the many HFC field trials of the cable industry, data network equipment providers have decided that RF broadband technology is an interesting area for future sales. Knowing that the cable companies have to interface HFC-based, high-speed data and telephony with LECs or CLECs, there is a market for IP- and ATM-based switches and SONET-based transmission equipment. The missing link between the cable subscriber and the ATM switch locations was the cable modem. The public demand for faster Internet access also aided the development of the cable modem.

Now, convergence has set in. Most data networking companies are either developing cable modems or have forged joint venture agreements with cable modem manufacturers: Cabletron is in field trials with its own cable modem design; 3Com has a licensing agreement with Com21 for the supply of cable modems; and Fore has announced that it will develop an ATM-based cable modem system jointly with General Instruments. In addition, computer telephony integration (CTI) is being integrated into ATM-based data networks and the integration of voice, data, and video on a single multiservice facility is becoming more feasible every day.

ATM interoperability in the long-distance and WAN segments as well as the application of dense wavelength division multiplexing (DWDM) in the rebuilding of the Internet are important developments and help modify the local loop strategies of many contestants. Digital subscriber line (DSL) offerings and the expected development of broadband wireless offerings, such as LMDS, will add to the changes, and VOIP, ATM, and ATM-based RF broadband will alter the development of the LAN architecture within the business community.

Today, business establishments are fortunate enough to have the ability to follow the lead of the long-distance and RF broadband industry to include ATM and RF-HFC technology in their infrastructure upgrade plans. The many benefits will show favorably on organizational balance sheets, protect organizations from renewed wiring changes, and support integrated voice, data, and broadcast quality video (even high-definition television, or HDTV) well into the twenty-first century.

2

The Evolution of RF Broadband

The advent of television marked the beginning of RF broadband technology. In order to accommodate early television broadcasters, the FCC allocated this new industry a frequency spectrum that was commensurate with the state-of-the-art technology in existence at that time.

Called the very high frequency (VHF) band, it utilizes frequencies just above the high frequency (HF) band. The VHF band starts with frequencies above 50 MHz and extends to about 220 MHz. The spectrum between 220 and 800 MHz was established as the ultrahigh frequency (UHF) band, which has also been used for television frequency assignments in the 1960s. (While radio transmission technology advanced before the Second World War, television did not establish itself until the early postwar years.)

We often hear the expression "the air waves are free," but in reality, the electromagnetic spectrum has been a solid source of income to our government. During the period between 1993 to 1996, the cellular telephony and PCS auctions launched the beginning of mobile telephony. The narrow band of frequencies between 1.8 and 2.1 GHz resulted in a multibillion dollar return. In 1998, the FCC auctioned new millimeter microwave frequencies for wireless local loop (WLL) transmissions of the future. This new frequency band, LMDS, will provide wireless integrated multichannel video and high-speed data and voice and finally establish serious competition in the local loop. LMDS is also an RF broadband technology. The bandwidth of the frequencies transmitted on a 30-GHz microwave carrier can be 1 GHz or 1,000 MHz. This is the same bandwidth that is commonly claimed by the cable industry when describing the new HFC technology.

The future of high-speed digital telecommunications is bright. In the corporate network we will find a mixture of wiring infrastructures, from IP-based legacy LANs, to ATM-based backbones, to RF broadband-oriented segments that can easily interconnect with ATM-based LECs or IP-based Internet service providers (ISPs). Soon, new LMDS wireless LECs can take the entire RF broadband transmission spectrum, translate it to the 30-GHz microwave band, and interconnect with alternative long-distance and global service providers.

2.1 The Delivery of Cable Television

Back in the early 1950s, a few television broadcast stations operated in the low VHF band of channels 2–6. The coverage area of these early television transmissions only reached an area of about 30 miles from the transmitter antenna. The population in the over-the-horizon areas received nothing except fading pictures with high white noise content.

The FCC assigned channels 2–6 in the frequency spectrum of 54 to 88 MHz. Each channel occupied a bandwidth of 6 MHz, which includes the picture carrier and the sound carrier. The television transmitter frequency assignments in the larger cities consisted of either channels 2, 4, and 5, or 3 and 6 in order to keep the coverage areas from interfering with each other. The 6-MHz spacing has remained the standard for RF broadband technology for both analog and digital services.

Since 75-ohm cable was a military surplus item after the war, the ingenuity of the technical community was applied to experiment with this cable to transmit these 6-MHz channels as a group and to reamplify the signal to extend the distance. The first five-channel low VHF band amplifiers consisted of simple tube-type tuned stages consisting of inductances, capacitors, and resistors. These so-called LRC circuits could be adjusted to different passband frequencies. One LRC circuit was tuned to one frequency, the next to an adjacent frequency, and so on. The tuned circuits that covered the 54–88-MHz band were then amplified. Thus, RF broadband technology was born.

For the first time, it was possible to install antennas for each of the received stations, combine the antenna signals, amplify the signals, and transport the multichannel product on a 75-ohm cable to the next amplifier. Since attenuation over frequencies varies along a cable, equalization circuits and tilt adjustments were added to the amplifier circuitry. The result was that up to 20 amplifiers could be installed in series or cascade. Because the white noise content or noise figure of a tube-type amplifier is quite high, and because the noise content doubles for every three amplifiers, the cable delivery industry developed an early reputation for delivering a low-quality product. However, it also succeeded in bringing television to areas of low population density not qualifying for the establishment of a television station and to areas where a frequency assignment would only have produced cochannel interference.

2.1.1 Early Systems

In the days before the application of the transistor, only the analog NTSC format existed for the transmission of video. Each amplifier required a direct connection to the power grid to be able to amplify the multichannel frequency band. Luckily, there were power lines along the streets where, at transformer locations, a connection could easily be made. To split or tap off a portion of the signal from the main cable and to feed the signal into the adjacent houses presented a major problem. One bright idea was a pin device that was screwed into the cable to connect physically with the center conductor. This patented method extracted a small portion of the signal from the distribution cable and transported this signal over a smaller diameter cable into homes. This innovation spawned the cable television delivery industry. Early systems consisted of the following components:

- A small antenna installation;
- A small equipment hut that required a power feed;
- Electronic equipment for combining the antenna signals;
- The distribution plant consisting of cable and amplifiers, passive splitters for the trunk, and service drops.

Figure 2.1 shows the technology of the early cable television system.

Antennas were simply installed on hilltops that could receive some signals. These signals were amplified and distributed over the coaxial cable. Many amplifiers were required to boost the signal level to overcome the attenuation of the cable, causing cumulative signal degradation. The cable television office was located anywhere along the system to advertise the reception, but it had no control over the system performance. The picture quality changed with temperature and was always worse at the end of the network.

With the inception of television stations operating on channel 7–13 frequencies, the amplification technology advanced rapidly to 220 MHz. This meant that the entire spectrum between 54 and 220 MHz could be transmitted over a 75-ohm cable for considerable distances and carry the low VHF band (channels 2–6) from 54 to 88 MHz, the entire FM band from 88 to 108 MHz, and the high VHF band (channels 7–13) from 174 to 216 MHz. The new

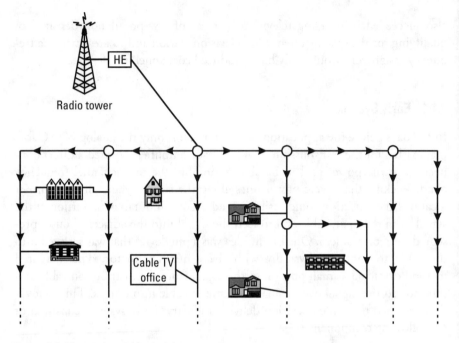

Figure 2.1 The early cable television system.

cable television industry developed fast and without much technological research or standardization. The systems grew larger and larger and rapidly encountered many performance deficiencies that have left the public with the permanent impression that the cable television industry offers a substandard service compared with that of the telephone industry.

2.2 The Many Deficiencies

The evolution of the cable industry is one of multiple rebuilding efforts. The first rebuilding efforts were caused by cable problems. Typically, a coaxial cable consists of a center conductor and an outside sheath that must maintain an exact relationship. A dielectric material between the center conductor and the tubular sheath must ensure that the center conductor is really in the center. Initially, the center conductor was manufactured from drawn solid copper, and the outside sheath was an aluminum tube. The two materials have different temperature coefficients. When installed at 70 degrees F, the cable would work well, but when the temperature dropped to 0 degrees, the center conductor would shrink, pull out of the connector, and disappear into the cable interior.

Many installers of early systems remember the desperate activities in the middle of the winter night to resplice the cables. The problem was solved by designing improved connectors that would seize the center conductor and new cables that used an aluminum center conductor with a plated copper layer.

The next rebuilding phase was a result of the new transistor technology. Sixty volt AC could now be used through the center conductor of the cable to power a number of amplifiers. This technology dramatically reduced power consumption and the number of power stations. Because the costs of the headend installation represented the largest portion of the capital budget, there was the tendency to extend the system by adding more and more amplifiers in cascade. However, the noise characteristics of the amplifiers proved that more than 20–25 amplifiers in cascade would lead to customer dissatisfaction. The white noise content (snow) in the pictures increased and impaired the picture quality. While the subscriber complained, the service was better than no television at all. Eventually, more rebuilding became necessary when the manufacturers solved the amplifier performance differences for winter and summer operation. Cable attenuation varies with temperature. In the summer, customers complained about snowy pictures because cable attenuation increased. In the winter, the pictures showed S-curves and harmonic distortions because of the lower attenuation. Maintenance personnel had to adjust the gain of the amplifiers on a seasonal basis.

In the 1970s, the franchising efforts of the cable industry extended into lucrative suburban areas and cities. To overcome its bad reputation, the cable industry had to make promises about performance. City council members will still remember that the franchising process was loaded with pledges for 300-MHz service, 35-channel service, two-way service, and separate institutional interconnect networks for town and city government. Despite these promises and many technical improvements, a number of deficiencies that still exist have damaged the public opinion of RF broadband for decades.

2.2.1 Cascades of Noise

Even newer cable systems continue to show pockets with impaired picture quality. In most cases, these deficiencies are the result of bad system design practices, such as too many trunk stations in cascade, more than two line extenders in cascade, designing for less than 100% occupancy, not enough signal level at the subscriber outlet, and multiple television sets connected to one service drop. The introduction of HFC technology and the use of single-mode fiber-optic cables limits the number of cascaded amplifiers in the coaxial distribution section, solves the noise problem, and, for the first time, assures telephone plant QoS levels. Figure 2.2 shows the complexity of a single coaxial network and the quantity of active elements acting as noise generators.

During the rebuilding phase of the 1980s, the cable television office was finally located near or even at the headend and became more responsible for assuring the quality of the outgoing signals. The newly adopted system architecture separated trunk and feeder lines. Trunks became the transport medium for long distance, and feeders became the distribution facility for the connection of the subscribers. Subscriber drops were only connected to the feeder cable so that discontinuities resulting from the insertion of multitaps could not affect the quality of the signals along the trunk lines. Figure 2.2 does not give justice to the many trunk lines and amplifiers in cascade. Systems with 30–50 amplifiers in cascade were built to satisfy the public's demand to view television.

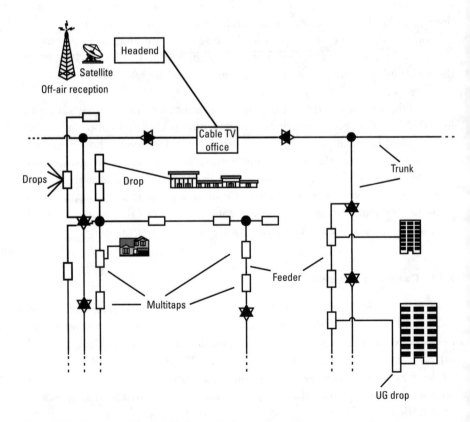

Figure 2.2 The coaxial cable television distribution network.

2.2.2 Composite Triple Beat

The greater the number of channels that are transported by a coaxial cable system, the more beat products are produced in the amplifiers. These beat products are additive in each amplification stage and reduce picture quality in long amplifier cascades. Mathematically, it can be calculated that beat products are maximized in 300-MHz channel systems. Headend frequency conversion schemes are used to assure that all carriers are exactly 6 MHz apart. This rearrangement helps somewhat but requires moving off-air carriers to offset frequencies, thereby necessitating more and more equipment. Lower picture qualities due to composite triple beat products are a direct result of the number of amplifiers in cascade and can be eliminated through the application of HFC technology.

2.2.3 Ingress and Egress

Any connector along the coaxial cable plant can cause ingress of over-the-air signals as well as egress of the signals that are transmitted through the cable into the air. In the simplest version of ingress, the channel 2 reception shows cochannel bars. This means that the off-air channel 2 has been added to the cable channel 2 and that the two signals—even though they are on the same frequency—have different velocities. To the FCC's frustration, there are also many Federal Aviation Administration (FAA) and governmental frequency assignments in the 108–174-MHz band. Emissions from cable systems not only interfered with the subscriber's reception of broadcast signals but, in some cases, were said to interfere with aviation and air traffic control. The problem led the FCC to impose fines on cable systems in violation. To appraise the egress quantity, the FCC performed fly-overs and searched for offenders. Today, strict emission rules are observed by all cable operators, but the problem can only be eliminated with good discipline in quality control, such as using the highest grade of connectors with integral shielding components and highly trained installation personnel. Even in the HFC plants of the future, the coaxial tail must be controlled to prevent and eliminate egress and ingress problems. Figure 2.3 shows a typical ingress problem caused by a defective cable connection.

A faulty connector installation at the input of an amplifier can pick up the off-air signal of a television broadcast station, transfer the signal into the cable, and cause ingress. The connector acts as an antenna, and both the cable and the broadcast transmissions are amplified. Because the velocity of the over-the-air path is different from the cable, cochannel interference is caused. A faulty connector installation at the output of an amplifier produces egress. The connector

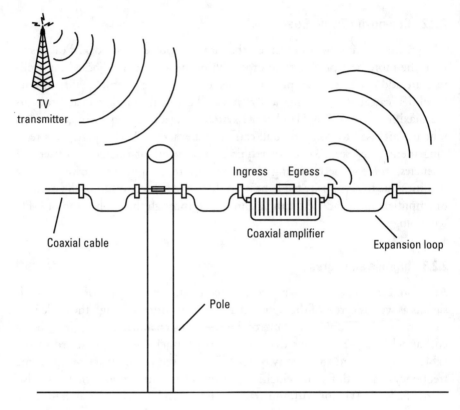

Figure 2.3 Ingress and egress.

acts as an antenna and transmits all cable channels over the air to adjacent television receivers. As a result, nonsubscribers complain that the cable system interferes with their television-set antennas and cable subscribers complain that double images and cochannel bars affect their reception.

2.2.4 Multipath Problems

Tall buildings in large cities reflect broadcast transmissions like prisms and mirrors. Because the internal isolation of television sets is very poor, the television set will receive any electromagnetic energy in its internal circuitry. The problem was first apparent in New York City. The use of cable television was curtailed until the industry designed a device commonly known as the *set-top converter*. The function of the early set-top converter was very simple: It translated all channels to channel 3 and used the television set only in the channel 3 tuner position. Even today, consumers must buy cable-ready television sets that are capable of 100-channel tuning, tune them to channel 3, and never use the tuners again. Due to a lack of standards, the tuner is still included in every television set ever sold, yet never used by a large percentage of the viewing public. Over the years, the set-top converter has become more sophisticated. Today it includes *addressability,* which enables it to set up and change program selections, to unscramble secure channel transmissions, to select pay-per-view programming, to mute commercials, to change the volume, and to execute any other operational function, all from the remote control. In the future, the new analog/digital set-top converter will translate digital and analog programming, act as the video-on-demand (VOD) control center, offer the reception of HDTV and advanced television (ATV) programming, and even interface with the computer workstation.

2.2.5 Customer Service

Many cable systems operated at monthly subscriber fees of $3.95 for many years. At that rate, customer service was not an affordable expense to the operator. In fact, $3.95 per month did not even allow operators to maintain their outside plant. When the rates increased—doubled and tripled—the demand for better customer service increased. By this time, however, the system had been sold to new owners at high prices, necessitating correspondingly steep bank loans, which were reflected in ever higher rates.

The cable industry is a typical example of the burden the consumer has to bear when buying the products or services of over-leveraged corporate structures. By now, cable television rates are, on average, $30.00 a month for basic and extended basic service, and there is still little customer service. Customer service has been the curse of the industry. The lack of customer service has given RF broadband a bad name and the cable industry a trailing position as a local loop service provider. Since cable television is oriented toward entertainment services only, cable customer service offices are often open only during normal working hours. Furthermore, they cannot help with reception problems (i.e., when the power has failed in another part of town) and cannot improve the picture quality (the bigger the size of the television screen, the more quality impairments will be visible). When cable service personnel do finally come to look at a problem, they may blame the television set, or they may determine that the problem stems from the distribution plant, which requires a service call to a different department.

Cable subscribers should realize that customer service will remain substandard until the plant has been changed to HFC technology, until cable is used for two-way transmission, and until cable companies provide data and telephone service. As long as cable companies remain entertainment delivery oriented, there will not be much stewardship of customer service needs.

2.3 The Regulatory Environment

The cable industry is subject to a three-tier regulatory regimen. The Cable Service Bureau (CSB), a branch of the FCC, implements and enforces all aspects of cable regulations at the federal level. Meanwhile, state cable commissions have been set up in Massachusetts and New York, and public utility commissions oversee the cable television business in Alaska, Connecticut, Delaware, Nevada, New Jersey, Rhode Island, and Vermont. In addition, local government bodies, such as city cable commissions, city councils, town councils, or boards of supervisors, are regulating cable franchise issuance as well as the cable industry's performance.

2.3.1 Federal Regulations

In 1966, the FCC reasonably concluded that the establishment of a regulatory authority over cable is imperative if cable is to perform a public service. Accordingly, in 1972, various rules such as the must-carry provisions for local broadcast stations, network program nonduplication, syndicated program exclusivity, cross-ownership, equal opportunity, and technical standards were enacted.

Some of these regulations changed in the following years and were revised again in the Cable Communications Policy Act of 1984, a congressional amendment to the Communications Act of 1934. In 1992—after a steep increase in subscribership and rate increases—the Cable Television Consumer Protection and Competition Act was enacted. This legislation changed the regulatory environment of the cable industry drastically by detailing procedures to be followed in areas of company registration, state and local regulatory requirements, rates, and customer service standards, such as installations, service interruptions, the handling of service calls, and changes in rates and billing practices. In addition, the federal regulations address areas such as signal carriage requirements and syndicated program exclusivities.

Separately, the engineering and technical services division of the CSB has determined minimum performance specifications. These minimum specifications reflect the state of the industry of the 1960s and have been frequently updated to include yearly reporting by all cable operators. The technical standards are a part of the cable federal regulations—47 CFR, Section 76.605—and are designed to ensure that cable subscribers receive a reasonable picture quality. Unfortunately, however, picture quality is a subjective measurement, and the minimum specifications for subscriber levels, frequency response, visual signal-to-noise ratio, and the audio level are outdated and not commensurate with new RF broadband technology. With the advent of larger television sets and more critical subscribers, a video level of 0 dBmV at the subscriber's set as specified by the FCC does not suffice. While this standard has been used by the industry as an excuse, it has damaged cable providers' image.

New HFC-RF broadband technology will force the cable industry to set its own QoS and outage standards to be able to compete with the telephone LECs and reduce its dependence on federal specifications.

Meanwhile, however, the area of ingress and egress remains an important FCC requirement. 47 CFR Sections 76.610 to 76.616 describe the operational requirements in the aeronautical bands. Aeronautical frequencies are in the frequency bands of 108–137 MHz and in the 225–400-MHz bands. They are an integral part of commercial and military air-to-ground communications and require protection from leakage from cable plants. The cumulative leakage index (CLI) is a test that must be performed by all cable systems on a recurring basis. The CLI report includes all leakage incidents that produce a 20 μv/meter reading at a distance of 3m from the cable. The test frequency must be between 108 and 137 MHz. In addition, ingress and egress will remain an important consideration even in HFC-based RF broadband systems and will lead to better installation practices in the future.

2.3.2 State and Local Regulations

The 1992 Cable Act adopted a regulatory plan allowing local and/or state authorities to select a cable franchise holder and to regulate any requirements that the FCC did not preempt. As a result, municipal franchising authorities have adopted laws and/or regulations in areas such as service requirements, technical standards, public access requirements, and franchise renewal standards and have the specific responsibility to oversee the rate structure of the cable service. Included in the grant of a franchise to a cable operator are rights relating to use of public rights-of-way, easements, and the minimum housing density per mile that must be served.

It is noted that all these regulations apply to public residential services that are provided by a cable television entity. These regulations do not apply to privately owned systems. Any enterprise, corporation, or housing complex is at liberty to build an RF broadband network on its property without a franchise and without federal or local regulation (except that technical standards and especially leakage requirements must be met).

2.4 The Telecommunications Act of 1996

The Telecommunications Act of 1996 was the first major review and total overhaul of telecommunications law since 1934. The new Telecommunications Act aimed to establish a new era of competitiveness in the communications business and to let supply and demand become the regulator of rates. The Telecommunications Act of 1996 affects local and long-distance telephone service, cable programming, video services, broadcast services, and education. The FCC has played a major role in the application of the Telecommunications Act and has developed a fair set of rules for this new area of competition. The FCC proceedings are designed to open local telephone markets, establish local loop service alternatives, and increase competition in the long-distance market. Numerous proceedings have been issued since the passing of the Telecommunications Act; they cover past regulatory standards and adopt them in the new competitive environment. Examples of the completed proceedings are rules covering pole attachments, video programming, over-the-air reception, open video systems, cross-ownership between multichannel multipoint distribution services (MMDS) and master antenna television (SMATV) systems, infrastructure sharing provisions, LEC tariffs, interconnection issues between LECs, mobile service providers and allocation of cost issues of LECs providing video services; mandatory broadcast signal regulations; and cross-ownership rules between telephone and cable television companies.

Just two years after it was enacted, the implementation of the Telecommunications Act of 1996 was in full progress. The trend toward a more competitive communications environment is accelerating. The following are noteworthy competitive developments:

- *Long distance:* Since 1995, smaller long-distance companies have increased their market share from 17 to 24%. Retail prices for interstate long-distance calls have fallen by more than 5%.

- *Local loop:* CLECs tripled their customer lines in 1997 to about 1.5 million customers. CLECs account for 2.6% of local telephone revenues. Publicly traded CLECs have raised $14 billion in capital. Their total capitalization amounts to $20 billion.

- *Cable television:* Direct broadcast satellite systems have doubled their customer base to about 5 million. Several local telephone and power companies are starting to compete with cable companies by overbuilding their plants. The market share of incumbent cable operators has fallen to 87%. The cable industry is in the process of transforming its plants to HFC technology at a fast rate. Already 30% of all networks have been rebuilt for two-way operation. It is estimated that all major cable television networks will be converted to HFC and carry voice, data, and digital video by 2002.

- *Wireless:* As a result of the spectrum auctions held by the FCC for cellular telephony and PCS, half of the country has a choice of four wireless providers. Auction 17, conducted in April 1998, resulted in the emergence of broadband wireless services in the local loop. LMDS will compete with serving LECs and the cable television systems in the local loop and become the foremost competitive service in control of commercial and residential communications expenditures in the next decade.

The Telecommunications Act of 1996 describes the competitive guidelines. It will take a few more years before the consumer can monetarily benefit from the deregulation effort. The advent of new technologies such as HFC-RF broadband and LMDS wireless will accelerate competitive activities and provide better choices to the consumer.

Residents as well as commercial enterprises will eventually benefit in the new competitive environment. Competitive offers, however, will not be uniform throughout the nation. The more densely populated areas will benefit earlier. While residents cannot do more than wait, enterprises can prepare for the day when competitive LECs are ready to offer alternative and economical services. Planning and implementing compatible WAN and CLEC interfaces is a recommended first step to leverage bargaining power.

2.5 Upgrading RF Broadband to Hybrid Fiber/Coax

While the cable industry has had difficult times from the beginning, its desire to survive in the new competitive environment has driven the development of new technological solutions to fulfill its foremost goal, the transition from its sole role as entertainment services provider to becoming a communications company. Early fiber-optic cables utilized strands of multimode fiber working below 1,000-nm optical frequencies. These cables are suitable for baseband transmissions and can be used for a single video channel. Multichannel video transmissions only became possible with the development of the single-mode fiber strand.

The spectral capacity of both coaxial cable and fiber-optic strands is in the range of 10 GHz. The single-mode fiber operates at an optical frequency of 1,310 or 1,550 nm. By connecting the combined RF spectrum of a cable headend, which may consist of one-hundred-and-fifty 6-MHz-wide channels, to the input of a fiber transmitter, the entire frequency band is translated into an optical transmission path. A single-mode fiber strand does not lose much of its light content through dispersion. Hence, the attenuation in the optical segment of the HFC system is very low. At about 0.35 dB per kilometer, distances up to 15 miles can be overcome without reamplification. Also, new single-mode fiber-optic cables can be used for wavelength division multiplexing (WDM) and DWDM, which permit multiple optical transmission paths on one fiber strand.

Coaxial cables have varying attenuations over the frequency range. The higher the frequency, the higher the attenuation. In the example of a 0.75-in diameter cable, the frequency of 750 MHz is attenuated at 1.3 dB per 100 ft. This means that a good coaxial cable has more than 10 times the attenuation of a single-mode fiber strand. At first glance, it may be concluded that the fiber should be extended to the subscriber. Why should coaxial cable be used at all? The reasons are many.

1. Fiber is unidirectional. Two strands are required for each direction. Coaxial cable is bidirectional. A single cable only is required.

2. To bring two fiber strands to every subscriber would require a massive and expensive construction effort.

3. The installation of a fiber transmitter and receiver at the subscriber premises would be wasteful and not economical.

4. To split the optical signal to feed two parties requires a fiber coupler. Such a coupler divides the optical signal in half and has a 3-dB optical loss in each direction. This corresponds to a distance reduction of about 10 km.

5. The coaxial cable can be equipped with proven and cheap multitap devices that can split exactly the right amount of signal to a subscriber. The multitap is bidirectional, can carry forward and reverse traffic, and can serve up to eight service drops.

While it took a few years to perfect the transmission characteristics of amplitude-modulated fiber transmitters and receivers, the combined application of fiber and coax has been standardized as the RF broadband platform of both the cable and telephone industries.

Figure 2.4 shows the components of the RF broadband system using HFC technology. A typical HFC system consists of the following components:

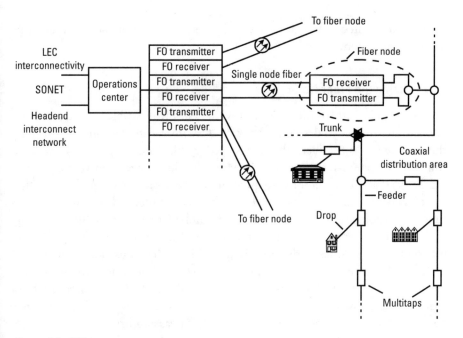

Figure 2.4 HFC system components.

- The operations center, gateway, or headend that interconnects the cable company with outside carriers;
- The single-mode fiber segment with fiber transmitters and receivers;
- The coaxial distribution segment with amplifiers and service drops.

2.5.1 Single-Mode Fiber Trunk and Coaxial Distribution

The new RF broadband technology using a combination of fiber for distance and coaxial cable for distribution is the most economical means to satisfy any bidirectional bandwidth requirement of the future. Whether the HFC system serves a community or a campus, it always consists of a single-mode fiber-optic trunk and a coaxial distribution section. The fiber trunk can be configured in a star architecture or can apply a ring architecture for service diversity.

While HFC is being used by the cable industry to serve single residencies and commercial establishments with a single coaxial service drop, it can also serve the internal ever-increasing needs for bandwidth-rich applications within a corporation, a floor, a building, or a campus area. The operations center, gateway, or headend interconnects the corporation with the outside world and provides integrated voice, data, and video services in both the LAN and the WAN environment.

2.5.2 Sizing the Fiber Node

When designing an RF broadband HFC system for the cable industry, it is of greatest importance to determine the *size of the fiber node*. This means assessing the number of subscribers that are served by the coaxial plant served by a single-fiber node. Many books and technical articles have been written about this subject, and the opinions are quite subjective. As a general guideline, a figure of 200–500 subscribers has been established by the cable industry. In the mind of a cable industry executive, these numbers refer to residences desiring cable television service and Internet access. The number 500 is inappropriate for commercial businesses that may require 20 simultaneous videoconferences, high-speed data service, and multiple voice lines but is quite appropriate for the residential market. Cable companies have established affiliate organizations that are tasked with the development of business customers. They can also apply the 500-subscriber concept to residential areas and bring fiber directly to a larger business.

Within a corporation these considerations do not have a basis. One of the most important planning tasks is to take a visionary look at the future and to determine what the traffic requirement will be 20 years from now. The sizing task also becomes a little easier when demographic considerations are included.

- A campus-wide HFC-RF broadband system may only use fiber and eliminate any coaxial distribution. Existing copper wiring can also be used for the distribution footage.

- A campus-wide system may use fiber to all building entrance locations and rely on coaxial distribution only within the buildings. This means that the fiber node is in the equipment closet and that the potential users will have sufficient bandwidth for decades to come. Another benefit of a fiber-rich installation is that an underground fiber cable installation in conduit is more economical than the installation of coaxial cables and passive devices.

- The fiber may be brought to the various floor levels of a high-rise building to avoid coaxial distribution—except for service drops. In this scenario, the incoming backbone fiber can utilize a fiber coupler in the basement to serve the individual floor levels.

While architectural flexibility is limited in the local loop and for the serving cable company, a higher degree of flexibility can be achieved within the campus environment. Figure 2.5 shows a typical HFC cable television system with 500-subscriber service nodes.

In such an HFC cable television system, the fiber network can be constructed in a star or ring architecture, with the determination of the fiber node locations critical to eliminate coaxial amplifier cascades. Route diversity of the fiber network is not shown in Figure 2.5 but is often used in aerial sections to prevent outages caused by traffic accidents on the main arteries.

Figure 2.5 indicates that large businesses can easily be served with a dedicated fiber cable. HFC architectures are flexible and forgiving. Even in cases of heavy additional traffic requirements, the coaxial service area can be split in half to accommodate the additional service load. It simply requires the use of a spare fiber, extended to the geographic midpoints of the original coaxial service area, and the development of as many new fiber nodes as required. Figure 2.6 illustrates a typical campus layout using a single fiber ring interconnecting all buildings and coaxial distribution within the buildings.

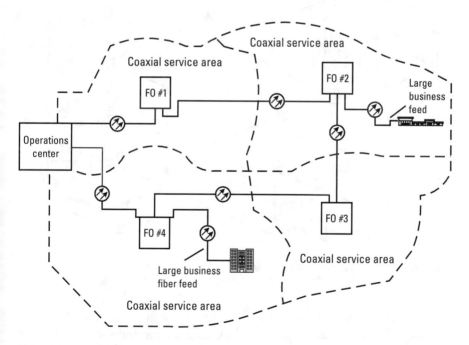

Figure 2.5 HFC cable television system with 500-subscriber service nodes.

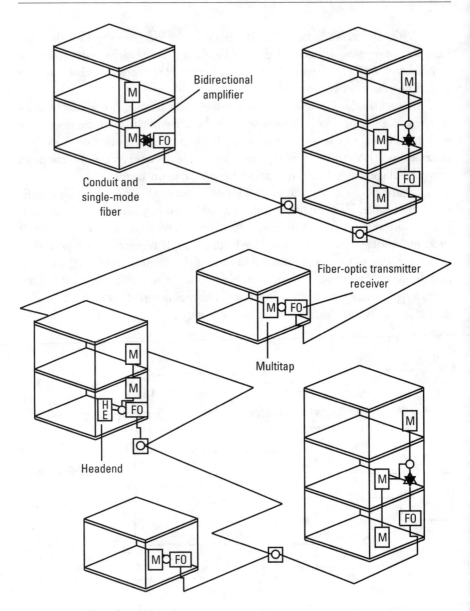

Figure 2.6 RF broadband in a college campus.

In Figure 2.6, the single-mode fiber network interconnects all campus buildings. The conversion from optical to electrical transmission is made at all building entrance locations. Coaxial tails are used for all inside plant installations. In larger buildings, a single coaxial amplifier is used to produce the levels required for the numerous outlets. In the smaller buildings, the coaxial tail consists only of passive elements such as service drop wiring or small vertical risers and service drops. A campus RF broadband network utilizing HFC technology can be built with a minimum number of active components.

2.5.3 Downstream and Upstream Spectral Capacities

The fiber and the coaxial cables have a spectral capacity in excess of 1 GHz. The present bandwidth limitations are caused by the direct feedback (DFB) laser devices needed to translate electrical signals into optical energy. Typical operational frequencies are from 20 to 860 MHz. The RF broadband spectrum of the downstream transmission follows the original FCC broadcast channel assignments. Channel 2 operates in the frequency assignment of 54 to 60 MHz and the lowest forward transmission frequency. One of the FCC's historic recommendations was to require cable television companies to transmit broadcast channels on their frequency assignment. This means that channel 2 should appear on channel 2 on the cable and be carried on the cable at its assigned frequencies of 54 to 60 MHz. Even the new HFC technology will not change this rule for the cable industry. Therefore, the downstream capacity of the cable television network in the local loop is limited to 54 to 750 MHz or, with advanced equipment, to 1,000 MHz.

Commercial establishments do not have to follow this ruling. Channel 2 within an enterprise does not have to be used for channel 2 because frequency spectrum management is in the domain of private ownership. Therefore, enterprises are more flexible in the assignment of downstream frequencies. The downstream band or forward spectrum of the fiber segment can be identical to the return spectrum or upstream band. Frequencies between 20 and 860 MHz can be transmitted in both directions. Spectrum limitations in the RF broadband system are caused by the established conventions in the coaxial segment and are based on the existing configurations of coaxial amplifiers. The standard convention used by the cable industry is the subsplit system. In this architecture, the downstream band extends from 50 to 1,000 MHz, and the return band is compressed to the frequency band of 5 to 45 MHz. In contrast, a commercial establishment has a number of different choices:

1. The subsplit system identical to the standard of the cable industry;
2. The high-split system with a forward band of 222–1,000 MHz and a return capacity of 5–186 MHz;
3. The dual system with duplicated coaxial branches and with a capacity of 20 to 1,000 MHz in both directions.

Figure 2.7 shows the three possible downstream and upstream uses of spectral capacity that can be employed in the corporate environment; they are listed as follows:

1. The subsplit asymmetric network (cable television standard);
2. The high-split modified asymmetric network;
3. The symmetric dual network.

2.6 The Operations Center—Gateway/Headend

The present headend of a cable television system consists of satellite antennas and off-air antennas that receive the programs for transmission over the cable distribution system. The need for regional dissipation of commercials was the first indication that interconnection between headends is a desirable feature. However, the desire to offer data services and possibly telephone service brought a new dimension to the term *headend*. The new network operations

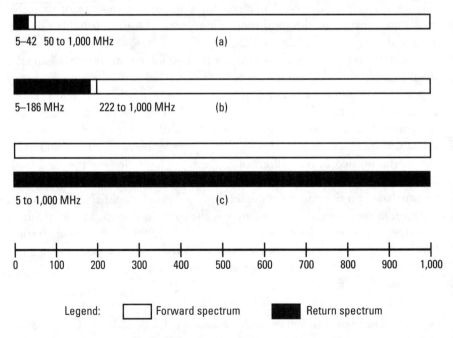

5–42   50 to 1,000 MHz   (a)

5–186 MHz   222 to 1,000 MHz   (b)

5 to 1,000 MHz   (c)

| 0 | 100 | 200 | 300 | 400 | 500 | 600 | 700 | 800 | 900 | 1,000 |

Legend: ☐ Forward spectrum   ■ Return spectrum

Figure 2.7 RF broadband spectral capacity: (a) The standard subsplit spectrum will be used by the cable industry in all new HFC upgrade projects, while (b) the high-split system and (c) the dual cable system are recommended options for corporate HFC-RF broadband networks.

center of a cable company must be able to interconnect with LECs, long-distance providers, regional and national commercial providers, and wireless service providers of the future, which include low Earth orbit (LEO) satellite systems and LMDSs. Interconnections with these global service providers will be made using the SONET platform. Already implemented coast to coast by a number of long-distance companies, the SONET's open architecture can transport data at scalable optical carrier rates of OC-3 to OC-768 (155.52 Mbps to 39.813 Gbps). The headend of the cable company is the interconnect location to the Internet and to the public network and, in the case of the AT&T/TCI merger, an intracompany interface.

Similarly, the MPOP of any commercial establishment must be able to interconnect with LECs, wireless carriers, and ISPs. Internal communications, whether presently conducted on separate data and voice networks or on the RF broadband platform, must interface seamlessly in an IP over ATM-based format with the outside world. Figure 2.8 shows a typical SONET/ATM interconnection of a cable company's operation center with the outside world. The illustration in Figure 2.8 is applicable to commercial enterprises as well.

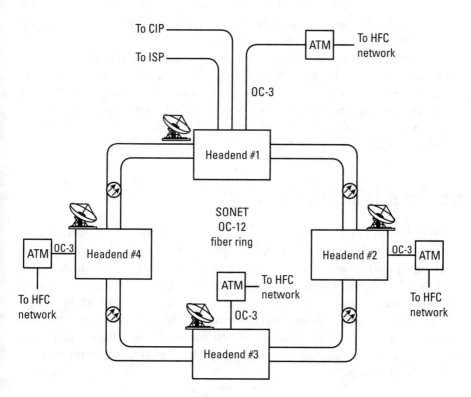

Figure 2.8 SONET/ATM headend interconnection.

Figure 2.8 shows four different cable properties interconnected with a SONET network operating at OC-12 speeds. The individual branches are served with fiber using OC-3 speeds. Internet interconnection is provided for all participants at OC-3, as are the digital video programs emanating from a VOD production center.

Of great importance to the cable industry is the regionalization of the commercial insertion (CI) business. Better returns are obtained when a commercial can be aired over a larger viewership. Another advantage is the ability to consolidate local headends into regional centers.

2.6.1 ATM Interoperability

ATM is a cell-oriented packet transmission technology that permits any type of existing transmission to be integrated and forwarded. This applies to T-1, primary rate, Ethernet, MPEG-1 and 2 video, token ring, or any other currently used digital transmission method. While SONET is a synchronized transmission platform, ATM provides the directivity and addressing of every transmission on a SONET system. The transparency of ATM and the finalization of the standards by the ATM Forum establishes, for the first time in the history of telecommunications, a global connectivity.

Commercial establishments will soon be required to interconnect in both the local loop and in the WAN in the ATM format to reduce private line and interconnection charges with the public carriers. The migration from an ATM-based enterprise network (WAN) to an ATM-based integration of voice, data, and video within the corporate LAN and the integration of RF broadband and HFC technology are inevitable.

2.7 RF Broadband Services

RF broadband HFC can provide the ultimate transmission infrastructure to every organization. When designed for quality of performance and for growth, RF broadband provides the transmission in any form of analog or digital signal, eliminating moves and changes as well as offering scalable growth to every new service that may be required in the near or far future. An RF broadband network can satisfy any current transmission application and meet any service requirement that may be conceived in the future. RF broadband applications are discussed in more detail in Chapter 7. The following are examples of RF broadband services:

- *High-speed Internet access:* Cable modems are available for various speeds and in asymmetric and symmetric configurations. IP- and ATM-compatible cable modems have passed through interoperability testing and will enter the market in early 1999. Conventional cable modems will be sold by retail outlets in the price range of $150–300.

- *Asymmetric data communications:* With some modifications, the cable modems that are available for Internet access can also be used for internal high-speed data transmissions in the corporate environment. The widely used Ethernet protocol can be used to interconnect and expand existing LANs. IP-based cable modems are capable of transmitting 40 Mbps in a 6-MHz channel in the downstream direction and 20 Mbps in the upstream direction.

- *Symmetric data communications:* Currently, 10-Mbps Ethernet cable modems are available for both forward and return transmissions. Newly developed modulation methods (256 quadrature amplitude modulation [QAM]) have increased the speed range within a 6-MHz channel to 40 Mbps. While multiple 25.6-Mbps ATM and 40-Mbps IP cable modems can satisfy most data transmission requirements in a corporate network, it is expected that the speed ranges will increase to 51.2-Mbps ATM and even to fast Ethernet.

- *Telephony:* A number of manufacturers are producing cable telephony modem equipment. Voice transmission requires the handling of a constant bit rate (CBR). ATM compatibility is required to transport voice communications in the digital format. A total of 240 voice channels can be carried in a 6-MHz channel assignment on the RF broadband system. This means that ten T-1 channels at 1.544 Mbps each can be carried in a single channel assignment. In the alternative, CTI can also be implemented using standard cable modems.

- *NTSC analog video:* Old familiar analog video/audio transmission can be accommodated in the NTSC quality broadcast format on the RF broadband network. Multiple bidirectional video transmissions can be conducted simultaneously without translation to digital formats. When digital television sets are available at economical prices, it will be possible to translate the analog transmission to the high-quality broadcast MPEG-2 or JPEG formats. By that time, the era of the jerky videoconferencing products of the 1990s will be history.

- *MPEG-1:* Currently, the common digital video transmission standard is the MPEG-1 format. MPEG-1 is an adaptation of a 1.544-Mbps

T-1 transmission for video. Compressing the signal details to one/one hundredths of the actual picture content will alter the appearance to the human eye. MPEG-1 video is fine for "talking heads," still pictures, and the capture of slow motions but cannot be used for full-motion events. However, MPEG-1 is ATM-compatible and will remain an economical video transmission scheme in long-distance applications. In the corporate setting, and utilizing ATM-based networking and RF broadband, the transmission deficiencies of MPEG-1 can be avoided by using NTSC transmissions, the JPEG format over ATM equipment, or MPEG-2 based equipment when it becomes affordable.

- *MPEG-2:* MPEG-2 has been announced as the true replacement for broadcast quality analog video. MPEG-2 is a VBR format. Bit rates can vary between 5 and 15 Mbps depending upon the motion attributes of the transmission. This means that MPEG-2 video can load down the throughput of a conventional data network. Assuming a large number of multiple simultaneous videoconferences over a single 155-Mbps ATM LAN, the remaining data throughput will be reduced. The RF broadband technology permits the separation of MPEG-2 transmissions on different 25.6-Mbps ATM loops, thereby reducing or even eliminating the effect on the data throughput. It is expected that MPEG-2 will become the ultimate multimedia format. Its immediate use in the corporate network may be the simultaneous transmission of multiple video streams in a VOD employee training program. The new DVD disk technology utilizes MPEG-2 encoding methods. Replay in the MPEG-2 format over the ATM-based or RF broadband network is a logical progression.

3

The Evolution and Future of Corporate Communications

The advent of telephony increased the need for communication between people within every business enterprise. In the beginning, and for many decades, the telephone and the written memo were the only tools to organize daily activities. It is, therefore, understandable that telephony was the dominant technology for many years and that the new data processing business was looked upon with some reservations.

Telephony, data processing, and video distribution developed in different time periods, required different wiring solutions, commanded personnel with different backgrounds, and, for decades, evolved on distinctly different paths. The current transformation of telephony and video from analog to digital transmission will change this image of telecommunications and force the unification of multimedia into a single network. Voice transmission in digital form has already changed the PBX industry, and the translation of video into affordable digital formats is in progress. With the advent of VOIP and ATM as the great equalizer of CBR and VBR transmission requirements, the combination of voice, data, and video into a single and fully integrated transmission format is both eminent and imperative. The evolutionary development of the three telecommunications technologies and their upcoming integration is the subject of this chapter.

3.1   The Development of Telephony and the PBX

The need to have more than one telephone in any business establishment is obvious. Where there are people devoted to working together, there must be telephones. Until its breakup, Bell Telephone was *the* service provider. Mother Bell came with key systems and centrex solutions and brought about fifty years of limited progress to the industry. The last post-Carterphone decades have seen an unprecedented increase in competitive products and new developments in the areas of key systems, PBXs, new shapes and sizes of telephone sets, wireless technologies, new and never-before-available features, as well as a progressive development of digital technology.

The old key systems initially served their purpose. Whenever there were more than two outgoing lines, these lines could be accessed from every telephone by pressing buttons. Even today, the key system is used in small offices with fewer than six outgoing lines. For larger businesses, the PBX is a good solution. PBXs come in all shapes and sizes, but the principle of operation is always the same. Because the PBX is the interface location between the inside and outside world, it is the location where outside lines terminate and outside calls are switched to extensions. The PBX is the location that distributes calls from extensions to other extensions. It switches calls designated for the outside world to outside lines for transmission to the central office and to any global location.

Thus, the PBX is primarily a switch. For decades, the PBX was switching circuits or telephone pairs. In the early years, there was the Strowger switch, then came different kinds of mechanical and moving contact switches. There was even a German design that used gravity to establish the dialed connection. Later came crossbar switches with fewer moving parts. After that came touch-tone dialing, and within another 20 years the digital telephone set was introduced. Figure 3.1 shows the basic principle of a PBX installation in a campus. Heavy multipair cables were installed between buildings, with the heaviest copper load nearest to the PBX location.

In Figure 3.1, outside lines are connected to the MPOP. Heavy multipair cables connect the PBX to building entrance locations or main distribution frames (MDFs), and multipair risers connect to intermediate distribution frames (IDFs) that provide the origination location for the wiring to the telephone sets.

3.1.1   Analog and Digital Technology

In the early 1990s, PBX equipment manufacturers modified their switches to handle digital telephones. While large PBX installations are currently a mixture of digital and analog technology, the connection with the LEC has become

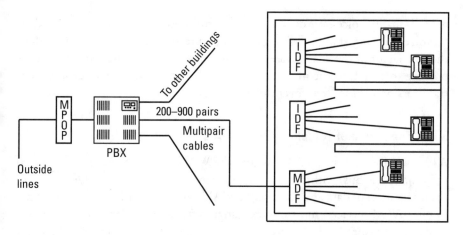

Figure 3.1 A typical PBX installation.

increasingly a 24-channel, T-1 interface. The wiring concept, however, has not changed. As late as 1994, I witnessed the installation of a 900-pair cable to serve telephones in a new 14-story building to extend the PBX. Category-3 (CAT-3) unshielded twisted-pair (UTP) cable is commonly used for the interface between the riser patch panel and the outlet. CAT-3 cable is specified up to a 16-MHz bandwidth and can be used for voice and data transmission rates up to 10 Mbps in accordance with IEEE 802.3 (10Base-T Ethernet) and 802.5 (4-Mbps UTP). However, PBX manufacturers did not integrate data into their switch designs because intermixing of voice and data does not increase the customer base, and the business community was content with the fact that telephony and data were different commodities.

3.1.2 Voice Networking in the WAN

For larger enterprises, the question of how to interconnect with the branch office in the next town was difficult to answer until the 1980s. Leased lines were the only solution and steeply increased the costs of corporate communications. While T-1 multiplexing had been used by the LECs to interconnect central offices since the 1960s, it was never applied in the local loop to protect the leased-line revenue base. This digital technology of 24 voice channels has now been adopted by many corporations to reduce the high-cost intracompany telephone service. Frame relay (FR) offerings followed. For the first time, it was possible to integrate voice and data into a common facility.

The compression of digitized voice has also proceeded at a fast pace. The International Telecommunications Union (ITU) has finalized standard G.729

for toll-quality voice over data networks. Until now 56 Kbps was used for digital voice; this new standard digitizes and compresses voice at 8 Kbps. To reduce the cost factor of the T-1 private line connection to the branch office, it is now possible to switch to the more economical FR services and compress the voice transmission to ride "free" over the data stream.

3.1.3 The Future of Telephony in the WAN

ATM, whether low speed (T-1) within the WAN environment or higher speed (25.6 or 155 Mbps) within the enterprise network, will be the unifying force in the integration of data and voice services. Although many LECs have not begun to offer ATM services, most interexchange carriers now offer T-1 ATM-based services. Low-speed ATM works with existing equipment and is scalable for WAN and global enterprise applications.

3.1.3.1 Low-Speed ATM in the WAN

The entry of low-speed ATM into the WAN is based on a number of advantages. Currently, private line T-1 connections are expensive within the same service area. Going through a number of different carrier territories, the monthly costs become prohibitive. A low-speed ATM service can be shared, has a much better bandwidth flexibility, and can, therefore, be offered at substantial discounts. In addition, the user can subdivide the 1.544 Mbps into data and voice spectrum assignments, digitize the voice to 8 Kbps, and integrate voice and data transmissions into one billing. Low-speed ATM works with existing equipment. T-1 ports as well as primary ISDN ports are available for PBX and data networking equipment. While some investment is necessary, branch offices and small extensions of a business can easily be interconnected between their Ethernet LANs, their PBXs, their FR routers, and even their H.320 video codecs. Low-speed T-1 ATM service can provide 20 voice channels, is priced substantially lower than private line T-1 service, and incorporates data. In addition FR is available with voice compression to 8 Kbps in accordance with the G.729 standard of the ITU. It is estimated that 80% of all private line T-1 connections are not fully utilized. Obviously, low-speed ATM in the WAN is only a small step toward high-speed communications, but this first step is an economical introduction of ATM to the business community. With demand, lower costs, and the influences of a competitive market, higher speed ATM connectivity will follow.

3.1.3.2 IP Tunneling in the WAN

While ATM in the WAN is a requirement for multilevel QoS in the public telephone network, IP tunneling offers an alternative approach for voice transmissions over the Internet. IP tunneling is a VPN solution that can drastically reduce the cost of data and voice traffic in the WAN. VPNs are described in more detail in Section 3.4.1. IP tunneling can achieve privacy and security over point-to-point connections over the Internet using the point-to-point tunneling protocols such as PPTP or L2TP. Since the Internet is a work-in-progress and, at present, does not have fully integrated routing control, IP tunneling can fill the void and assure high-quality voice and data transmissions in the WAN and substantially reduce the cost of long-distance communication.

3.1.4 The Future of Telephony in the Corporate Network

At first glance, there is no necessity to convert the voice network within an enterprise to a digital format. Telephony is well-established, relies on copper plant, and can migrate from analog to digital. True, there is no immediate need or advantage, but the transition toward CTI and VOIP and the development of computer branch exchanges (CBXs) are strong forces toward the establishment of integrated voice, data, and video networking. Large campus networks today may already feature fiber cables interconnecting the various buildings and supporting both voice and data. In most cases, however, one fiber is used to extend the voice system and is in the T-1 format; the data are in another fiber that delivers Ethernet loops across the campus. This architecture will remain predominant until the data network provides ATM-compatible services throughout the enterprise. At that time, voice services can be readily integrated, the PBX can be gradually replaced with CBX equipment, and the old copper plant can be dismantled.

3.1.5 The Alternative RF Broadband Solution

While it has been concluded in the above scenario that the translation to ATM-based voice transmission in a premises network will only take place after establishment of an ATM-based data network, the more attractive alternative may be RF broadband. An analog or digital voice transmission system can readily be installed over an RF broadband facility. Manufacturers of PBX equipment are already adding ATM-based modules to facilitate the growing requirement for WAN connectivity. The PBX will undergo a transformation to permit packet- and cell-based voice traffic within the corporate LAN as well. A 25.6-Mbps bit stream can transport 10 T-1 assignments or 240 voice channels over a fiber cable at baseband and, with the help of a cable modem, over a 6-MHz RF channel. If voice compression to ITU standard G.729 is applied, the total number of telephone conversations that can be held over a single 6-MHz channel can increase substantially and by a factor of at least five times. Figure 3.2

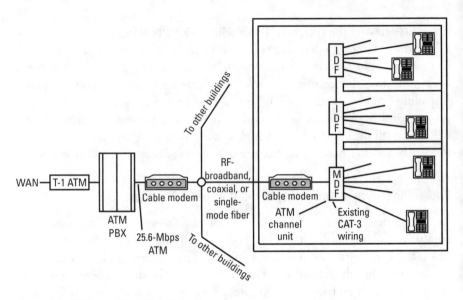

Figure 3.2 ATM-based cable telephony.

shows the gradual expansion of the existing PBX for ATM-based voice service by the addition of a large building served by RF broadband voice service.

It becomes quite evident that the ATM and RF broadband technologies avoid the stringing of heavy cables to serve the new building. The PBX serves the existing telephone system and permits the addition of ATM-based technology in a scalable manner. In Figure 3.2, the PBX has been replaced with an ATM-based digital PBX that uses a 25.6-Mbps ATM card to feed the RF broadband network through a cable modem. The existing service wiring has been used from the IDF to the digital telephone set. This solution, however, does not address the integration of voice and data or CTI and may only be attractive in cases where the costs of copper cable construction are prohibitive. More details on cable telephony are provided in Chapter 10.

3.2 The Evolution of Data Networking

The need for the transfer of data within an enterprise is fairly new and a direct result of the fast development of computer technology. In the 1970s, the Xerox Company developed the Ethernet cabling system. A 50-ohm cable was used with coupling devices to interconnect computer mainframes and terminals. Wang and IBM developed proprietary wiring schemes using coaxial cables to reduce the attenuation at baseband frequencies.

3.2.1 Data Network Wiring Systems

The twisted-pair cable manufacturers felt the impact of the coaxial cable sales and started testing twisted-pair configurations for extended baseband performance. Originally, shielded twisted-pair (STP) was the first solution until it was found that radiation and ingress into the cable does not fully interrupt a digital bit stream. Extensive testing of different wires and twist pitches showed that UTP cable can be used for a variety of transmission speeds. Category-4 (CAT-4) UTP is specified for a bandwidth of up to 20 MHz. CAT-4 can be used for voice and data transmission rates up to 16 Mbps and is standardized by IEEE 802.5, 4/16-Mbps UTP. This cable was the first twisted-pair application for token ring data networks and replaced the IBM coaxial wiring system. CAT-5 UTP wire characteristics are specified up to 100 MHz. If installed in a length of fewer than 100m between the workstation and the equipment rack, CAT-5 cable can be used for transmission rates up to 100 Mbps. CAT-5 cable has become the standard of the data networking industry for connectivity between riser rooms and outlets. The cable is standardized under IEEE 802.5 16-Mbps UTP and ANSI X3T9.5 for 100-Mbps traffic. Deducting 5m for equipment cabling requirements at each end, the remaining horizontal distance is 90m or 300 ft. Being able to span distances of 300 ft does not suffice for larger data network installations. Multimode fiber-optic cables have been used to interconnect premises-wide data networks, and a number of devices were developed to permit bridging, routing, and switching of packetized data transmissions.

The early LAN was an Ethernet backbone. It consisted of a 50-ohm Ethernet cable with couplers to interconnect the computer mainframe with workstations. Many isolated intelligent islands were formed even in larger establishments using this simple technology. The need for interconnecting these isolated data transfer areas grew, and the industry responded with bridges and gateways. The emerging client/server environment required that many workstations were able to communicate with the server at the same time. In order to assure that each client could have unlimited access to the server, the architecture of the LAN had to change.

Accordingly, the wiring in the backbone changed to multimode fiber that supported the ever-increasing speed ranges up to gigabit transmission. However, multimode fiber has bandwidth limitations and cannot cover extended distances for gigabit traffic because of attenuation and inefficient light emitting diode (LED) transmission devices. The use of more and more single-mode fiber in the corporate LAN environment is inevitable. Single-mode fiber does not have any bandwidth limitations and is being used by all communications carriers and for all long-distance applications. Laser devices have improved exponentially and DWDM has elevated the throughput to OC-192 in the 30-Gbps range. Single-mode fiber is used in all HFC rebuilding projects of the cable industry and will enter the LAN market to accommodate gigabit speed ranges and introduce RF broadband alternatives.

3.2.2 Routers—Switches and IP in the LAN

Some of the best products on the market are intended for small client/server-oriented networks or workgroups. These switches have one or two high-speed ports that can handle FDDI, ATM, or fast Ethernet and may have eight to 100 ports for the connection of workstations. The most common configuration is the 25-port unit in a single unit. These switching hubs are economically priced and competitive with high-end, non-switching hubs. When considering an upgrade to FDDI, ATM, or fast Ethernet, these switching hubs can become expensive and are somewhat tricky to implement. FDDI and Ethernet have different maximum transmission units (MTUs). FDDI's MTU is about three times that of Ethernet at 1,514 bytes. A translation to fast Ethernet may be the easiest implementation and delivers a reasonable throughput. The evolutionary development can be summarized as follows:

- *Layer 2 switches* make their forwarding decisions based on the media access control (MAC) address of an Ethernet node. By isolating collision domains in the shared Ethernet medium, these switches have become an inexpensive solution to boost the network's bandwidth. Large systems employing only layer 2 switches, however, cannot cope with high-traffic volume. Any peak broadcast traffic load can easily reduce the throughput.

- *Routers* prevent network overloads by allowing the forwarding decisions to be made with respect to protocols. Routers have software-based engines that can distinguish between IP, Apple-Talk, Systems Network Architecture (SNA), and others. While routers can save the network overload conditions, they can function in the complex environment of multiple protocols. Routers, however, cannot cope with the ever-increasing speed requirements of the LAN networks. The tradeoff is flexibility versus speed. While routers are the dominant architecture in the WAN, layer 3 switches are making fast inroads into the LAN environment.

- *Layer 3 switching* reduces the multiprotocol flexibility of a router to only the IP protocol. The layer 3 switch is basically a high-speed router that has been limited to an application-specific integrated circuit (ASIC)-based IP forwarding engine. The use of ASICs reduces costs to one/tenth of a non-ASIC product and increases forwarding speeds to 10 million packets per second. As a result, there are various layer 3 products on the market that can handle gigabit and fast Ethernet backbone traffic to cope with the ever-increasing outbound traffic requirements. In the process of rebuilding, the layer 2 switches are being moved closer to the desktop. The major data networking equipment manufacturers like Cisco, 3-Com, Bay Networks, and Cabletron market layer 3 switches as well as switch/routers that can be attached to the legacy router, learn the existing routing tables, and off-load all IP packet traffic.

Routers will not be eliminated. VOIP and QoS issues can easily be added into a router. Multiservice LANs require the flexibilities of the router technology close to the desktop and at the WAN interface. High-speed layer 3 switches, at this point in time, can be used in the backbone and will reduce congestions due to the IP broadcast mode of operation but lack the ability to arrange for service categories and multilevel QoS. New layer 3 switch designs feature backplanes with speeds up to 40 Gbps and top forwarding speeds of 24 million packets per second. They can handle 10/100 Ethernet ports, gigabit ports, 155-Mbps ATM uplinks, and 622-Mbps ATM ports. Figure 3.3 depicts a typical IP-based network hierarchy with an ATM-based WAN connectivity.

Figure 3.3 indicates a hypothetical campus data network employing gigabit switching fabrics in a core network with workgroup switches reducing the speed to fast Ethernet or 10 Mbps at the desktop. The access control unit is ATM-based for interconnection with the public telephone network and the Internet. VOIP can be used to integrate telephony into the data network. Connectivity for WAN traffic can be FR, low-speed ATM (T-1) in the public switched telephone network (PSTN), and IP-tunneling through a VPN.

3.2.3 ATM in the LAN

While ATM is more expensive, it offers a protocol-independent transmission that can easily be applied without any conversion to accommodate WAN traffic through the public network. IP exists at layer 3 of the International Organization for Standardization (ISO) model, which is the networking layer. ATM exists at layer 2, the data link and MAC layer. The ATM cell structure allows any protocol to be transmitted flexibly and at high speeds. The cell structure permits the fastest switching times and assures low latencies for voice and video transmissions. It is, therefore, not surprising that the public networks and the Internet will use ATM-based switching technology over SONET and over

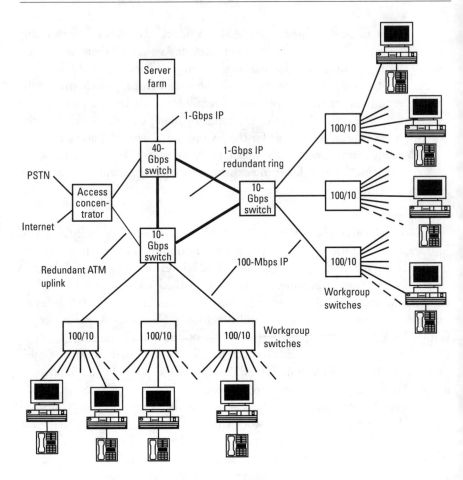

Figure 3.3 The IP-based LAN.

DWDM in the long-distance segment. While the LAN connection with the outside world will use low-speed ATM at a T-1 rate of 1.544 Mbps or as an FR connection, the SONET/ATM standards are global. The implementation of ATM, even in the environment of a small business, assures transparent interconnectivity with the outside world and provides the basic essentials to establish the multiservice network of the future. Telephony, VOD, and MPEG-2-based videoconferencing require the service categories and the multilevel QoS that only ATM can offer. See Chapter 5 for more details. ATM connectivity in the local loop and an uninterruptable ATM backbone core network may require a higher initial capital budget but assure future savings as well as optimized flexibilities in multimedia integration, the establishment of VPNs, and the selection of CLECs. The latest ATM-based switching equipment combines the technology of layer 3 switching equipment with the flexibilities of ATM. With backplane speeds of 40 Gbps, the ATM core switch can accommodate OC-48 ports at 2.5 Gbps, OC-12 ports at 622 Mbps, and OC-3 (155) down to 25.6-Mbps ports. A 2.5-Gbps backbone between a minimum of three locations in a redundant fiber ring can safeguard the LAN through cable or equipment outages. Recently measured reconvergence times after failure are on the order of 5 milliseconds. The reason for such a fast recovery is found in the 53-byte cell that can also accommodate a much faster switching interval than any packet switch. Cell switching is the only technology that can assure voice and video latencies in microseconds and accommodate any legacy protocol, IP switching, virtual circuits, FR, T-1, ISDN, and any other protocol ever conceived. Figure 3.4 shows a typical ATM-based backbone LAN with ATM-based or IP-based service areas.

Figure 3.4 illustrates the fault-proof OC-48 network core in an identical configuration as that used for the gigabit core of Figure 3.3. The connection of server farms and ATM uplink runs at more than twice the speed as the gigabit example and workgroup switches can be selected for IP or ATM traffic to the desktop. The ATM-based core permits true multimedia transmission concepts to emerge, featuring every conceivable service and QoS level requirements. The ATM-based core permits the addition of JPEG and MPEG-2-based videoconferencing and VOD, as well as the integration of computer telephony in a scalable manner. While initially more expensive than the gigabit technology, the savings are obtained in every multimedia upgrade step as well as in the ability to obtain more economical interconnect agreements with outside service providers such as CLECs and wireless LMDSs. Savings are also realized in the operation of VPNs.

3.2.4 RF Broadband in the LAN

3.2.4.1 Early Data Networks

RF broadband is not new in large data network applications. Most cars of the 1980s were assembled on an automated assembly line that used the GM manufacturing automation protocol (MAP). The Boeing Company added the technical and office protocol (TOP). The MAP/TOP specifications were incorporated into the IEEE 802.4 specifications in the early 1980s and describe the token passing algorithms for data traffic on RF broadband. Most major assembly lines in the United States have been using RF broadband LANs for multiple data transmission systems. In the early 1980s, the single-mode fiber was not a commercially available item and HFC technology did not exist. Therefore, all RF broadband LANs were constructed with large-size coaxial

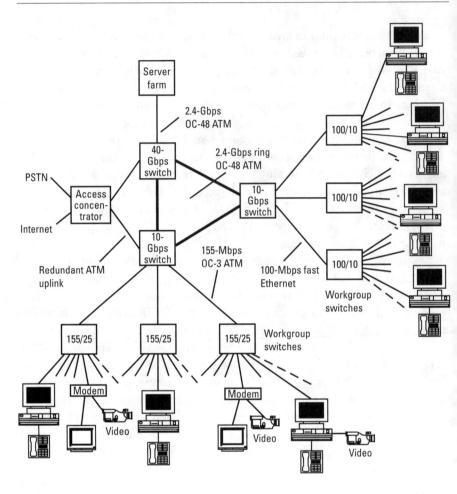

Figure 3.4 The ATM-based LAN.

cables and used numerous two-way amplifiers, subject to the buildup of noise. The coaxial architecture used in most cases was the high-split single cable system, operating in the 222–550-MHz band in the forward direction and in the 5–186-MHz band for the return. IEEE 802.4 and the MAP/TOP protocol even specify the desired carrier frequencies. At that time, the downstream transmission was identified in six 6-MHz channel assignments between 240 and 276 MHz and the upstream assignments between 53.75 and 89.75 MHz. The big advantage at that time was that the frequency assignment of the translator at the end of the cable could be a broadband device that translates the entire band.

What is today referred to as a cable modem also existed in the early 1980s. Allen Bradley and Ungermann Bass manufactured the necessary modem technology. Bradley even had a computer board that would directly connect to the coaxial cable. The number of RF broadband LANs in service increased steadily until the new data network manufacturers introduced the client/server technology that could transmit 10-Mbps Ethernet traffic, first on twisted and shielded copper pairs and later on UTP.

The largest RF broadband project was funded by tax dollars. A dual coaxial cable system was installed in 1985 in all Strategic Air Command (SAC) Air Force bases within the United States.

The system was a dual 5–550-MHz system that interconnected every work space of every building in the large expanse of these military installations. In Oklahoma City, this included the largest building in the United States, over one-half a mile long and wide. The prime contractor on this project was TRW, which also manufactured and provided the cable modem equipment. TRW's concept was that the location of forward and reverse amplifiers should be identical. It was the basis of its low bid but caused design problems due to the noise figures of the many amplifiers in cascade. The remedy was to use larger diameter cable and to direct the subcontractors under the auspices of the Federal Acquisition Regulations (FARs) to provide this more costly cable and install it. The subcontractors did provide the cabling and saved the noise budget of the RF network but were never reimbursed by TRW for their efforts. In fact, two subcontractors had to file for Chapter 11 bankruptcy protection. Coaxial-based RF broadband LANs have had their problems. The new HFC technology changes this equation.

3.2.4.2 Video Networks

There was no need to distribute video signals within an enterprise for a long time. Even in college and university campus settings, the delivery of cable television to the dormitories was considered a distraction from the learning process until the late 1980s. With increasing competition in the educational community to cater to the needs of students with respect to the use of computers came the desire to offer cable television to dormitories. Some colleges and universities opted to permit local cable companies to deliver cable television. Others built their own campus-wide cable television distribution systems in combination with data network connectivity. Three wiring architectures can be found today in most dormitories: the typical telephone cable, the CAT-5 data cable, and the F-connector outlet for television.

In the early 1990s, college and university administrators noticed that while data connectivity to the student is a service, telephony and cable television could be revenue producers. Subsequently, many educational institutions decided to make the initial investment, finance the initial construction, and develop activity fee schedules separate from tuition fees to earn a continuous revenue stream. Many of these cable television systems were installed by local contractors to minimal specifications. Lack of workmanship and basic system concepts make these installations suitable only for the delivery of a limited number of channels in the downstream direction. While these systems are sufficient to deliver cable TV, they require upgrading to accommodate two-way video or data transmissions. The campus cable television delivery system typically consists of (1) a small headend installation that may include one or two satellite receiver antennas, maybe a steerable antenna, antennas for off-air channel reception, and the electronic processing equipment required to establish the desired channel plan; (2) the trunk and distribution plant interconnecting the various campus buildings; and (3) the service drops to the dormitory outlets.

While commercial businesses did not have any need for two-way interactive and instructional television, some colleges and universities have installed these facilities between academic buildings. Typically, videotapes are transported on carts by the media department to the lecture rooms whenever a video is to be shown by the lecturer. Colleges with large tape libraries became concerned about the uneconomical tape delivery system and decided to build two-way cable distribution systems to all lecture halls.

The benefit of a two-way television network is, of course, that the tapes can be transmitted on a downstream channel and that videotapes can be made of lectures that are recorded in the media center. These recordings can now be made available on request to any student in the dormitory. This primitive form of VOD is frequently used by students with scheduling conflicts that prevent them from attending two different lectures at the same time. In cases where the headend of the cable television distribution system is not colocated with the media center, a separate interconnect cable can close the gap. Tapes that have been recorded in the media center can then be distributed on the cable television system to the dormitories. The ordering of video sequences from the media center is usually accomplished by telephone, and the scheduling is established by student volunteers. All viewers of such videos benefit educationally. With the advent of digital technology, VOD can become a major learning resource.

3.2.4.3 ATM and RF Broadband

HFC technology eliminates all the deficiencies of the old coaxial tree and branch architecture. Furthermore, the new RF broadband concept can be used as the only server to client infrastructure without any bandwidth limitations. RF broadband is an alternative architecture that can be used in the reengineering of the corporate LAN and its transformation into a multiservice facility. The benefits are plentiful. One of the more important benefits is scalability. Initially, the entire network may use one RF frequency and a 40-Mbps cable modem connected to a computer that has been equipped with a 25.6 ATM board. Later, when the traffic load increases, it is possible to assign a different (second) RF frequency to a group of workstations. Using this easy method of changing frequencies, it is possible to transform a 50-workstation network into 10 five-workstation networks without changing the wiring and adding only the required number of cable modems. Figure 3.5 shows an example of an RF broadband-based LAN using only passive components.

Figure 3.5 shows the identical campus network employing an OC-48 protected ATM core with OC-12 connectivity to workgroup switches, except that one of the geographical areas of the campus has been designed for RF broadband service. The reason was a lack of single-mode fiber to accommodate

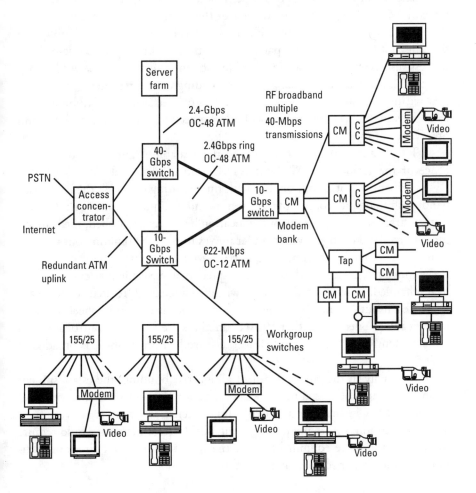

Figure 3.5 The ATM-based LAN with RF broadband.

the workgroup switches. In addition, the service requirement of the workstations was higher than 25 Mbps but lower than 155 Mbps. The compromise
was a single-fiber strand equipped with cable modems operating at various RF
frequencies. Each cable modem delivers 40 Mbps to a single workstation or to a
group of workstations. Existing UTP wiring can be used to group workstations
that can share the bit rate of one cable modem. Workstations requiring full
40-Mbps service are connected via RG-6 coaxial cable. These workstations can
select any frequency and any 40-Mbps segment by selecting different operating
frequencies of the cable modem. RF broadband and cable modems offer additional flexibilities in the classification of service levels or to assure congestion-
free access to the ATM core.

The use of RF broadband also permits seamless upgrading of workgroups
served by CAT-5 UTP wiring. Whenever the traffic load within a workgroup
that has been served by a single cable modem has increased, the workgroup can
be split at the cross-connect panel by adding a second cable modem. This
process also requires an additional cable modem and port at the ATM edge
switch. This process of reducing the data loading can be repeated without any
new wiring and until each workstation has its own data stream. RF broadband
can access the ATM core directly without the need for workgroup switches and
observance of speed hierarchies.

The beauty of RF broadband is the scalability and the nonobsolescence of the technology. Without ATM compatibility and interoperability,
RF broadband is not an option, but with the maturing of ATM, the RF broadband technology becomes the universal and ubiquitous network for the
twenty-first century. Once the outlet location has been established, there are no
more moves and changes. When a user moves to the next building but remains
in the same group, there is nothing to be moved except his or her workstation.
If the move is a change of duties, the frequency of the cable modem may have
to be changed. However, there are no wiring changes. All frequencies are available at each outlet, and the change of frequencies is a thumb wheel selection
in the cable modem. All other VLAN functions are software controllable and
remain identical as they were before the introduction of the RF broadband
addition.

Especially in the environment of a high-rise building, the establishment
of an RF broadband-based network is much more cost-effective than building
a fiber network in the riser and deploying switches and routers at every floor
level. In addition, future upgrading of the fiber/copper network would require
more fiber and more equipment, while the RF system only requires one additional frequency per additional 25.6-Mbps ATM loop.

3.3 Connectivity With the Outside World

Every enterprise and business establishment is plagued with the high charges of serving LECs. Whether the service is 56-Kbps, ISDN, FR, or T-1, the costs are just exorbitant. The reason, of course, is the lack of competition in the local loop.

While the Telecommunications Act of 1996 intended to solve this problem, the development of competitive services has been slow. In some areas of the nation CLECs are operating, but their rates cannot provide deep discounts since most of them do not own any cable plant and have to lease the facilities. The cable industry has been slowly converting its plants to HFC technology, but as a quite over-leveraged industry, it is not in the position to borrow sufficient capital to complete the transition overnight. The established local carriers appear to be the first in providing new services. Asymmetric digital subscriber line (ADSL), which can utilize the existing copper plant with some addition of fiber, will be in the position to offer at least T-1 services in an ATM-compatible format.

3.3.1 Alternative Local Loop Carriers

However, competition in the local loop will finally arrive. The communications technology is in the midst of a revolution. The bandwidth capacity of wire-based transmission systems is augmented by wireless communications. There are new satellite technologies and new millimeter-wavelength transportation systems that will affect our lives on very short notice.

3.3.1.1 RF Broadband in the Local Loop

Previously only in the domain of cable television delivery, the cable industry will now enter the local loop with competitive voice, data, and interactive video services. Cable modems for fast Internet access signifies phase one of this development. However, there are other recent developments that lead to an early phase 2 occurrence. Already there are RF broadband-based integrated transmission systems in operation. A good example is the HFC-2000 system developed by Lucent Technologies. The system is interoperable between voice and data and combines synchronous traffic requirements (telephony) with asynchronous transmission (data) concepts in a single processing system.

The HFC-2000 equipment has been deployed in the territory of Southern New England Telephone (SNET) company and provides competitive voice, data, and video services. While these islands of communications integration are still small, they are all using RF broadband networks with an HFC architecture.

3.3.1.2 Local Multipoint Distribution Services

New developments in millimeter-wave transmission equipment led to the 1998 round of FCC auctions for licenses in the 28–30-GHz band. The auctioning of frequency spectrum by the FCC has been a strong revenue producer for the U.S. government.

It also makes the entrance into a new business more expensive from the start. More initial capital is required for the winning venture to commence operations, and a small round of mergers has to be expected before service is offered to the public. In the first phase, however, the FCC auctions have been completed and have started a new industry. Implementation will undoubtedly follow in a very short time span.

The intriguing benefits of the 30-GHz wireless technology exist in the bandwidth capacity of the propagational properties at millimeter waves. The transportable bandwidth can be 750 MHz wide. As this is equal to the RF bandwidth capacity of the RF broadband cable network, LMDS will be in a good position to provide competitive services for cable television delivery, wireless telephony, and high-speed data. LMDS has been called *cellular television* to distinguish it from the rather narrowbanded cellular telephone and PCS services. This point-to-multipoint technology is made to order for competitive services in the local loop, as the size of cells is limited to about 2 miles in diameter. Obviously, LMDS will require a few years to develop and to find recognition, but it is forecasted that denser population areas and especially business establishments will be the first target of this new industry.

3.3.1.3 Near-Earth Satellite Communications Systems

There have been continuous technological innovations in the field of geosynchronous communication satellites during the past thirty years. The bandwidth capability has increased from the 50 MHz of the first Hughes satellite launched in 1965 to 1.8 GHz launched in 1996. During the last few years, however, new service needs have led to new system concepts and have served to create a new era of change. Geostationary satellites, while being able to cover about a third of the circumference of the globe, have deficiencies such as substantial latencies because of the long distance from the Earth. The new system concepts address deficiencies such as low latency echo cancellation, lower path loss, improved elevation angle, and a higher level of frequency reuse. To date, these new system developments can be grouped into several categories:

1. LEOs, operating at 800 MHz or below;

2. Big LEOs and *medium Earth orbitals* (MEOs), operating at 1.6 GHz;

3. Mega-LEOs which will operate in 20–30-GHz frequency assignments;

4. *High-altitude long-endurance* (HALE) platforms.

In many ways, these low-orbit satellite concepts are aimed to compete with undersea fiber-optic cables and address global connectivity. However, once established, they could serve regional and even local markets as well. Satellite technology must reduce latencies to be able to compete with the ever-increasing use of fiber-optic technology. The speed of deployment of these new LEO communications platforms will depend upon the availability of capital and the completion of operational field trials. The winning technology will combine the best performance characteristics with the lowest launch and operational cost factors and longest life cycle expectancy.

From these examples, it can be concluded that the local loop and WAN communications providers in the twenty-first century will be numerous. Wire-based and wireless offerings will compete for the corporate communications segment and carry integrated voice, data, and video services in interoperable ATM cell formats.

3.4 The Future of Enterprise Communications

Enterprise communications is the interconnection of the corporate headquarters with every affiliated organization, company division, and branch office wherever it is located on this shrinking planet. Enterprise communications is often referred to as the *intranet.* In order to develop a company's communication with customers, suppliers, contract manufacturers, and affiliated marketing and sales organizations, an additional layer of company-private transactions is required. This additional layer of complexity is sometimes referred to as the *extranet.* The terminology originates from the newest and fastest growing communications medium, the Internet. It is forecasted that a high percentage of daily business transactions will be conducted on the Internet and that e-commerce will become the ultimate way of life. Since the enormous growth of the Internet is met by an ever-increasing investment in fiber routes and equipment, it appears that the Internet will become the predominant communications facility of the future.

A connection on the Internet typically has a connect time of 20 minutes, while a voice call averages four minutes. As a result, the local voice network facilities are overloaded. An Internet connection is routed to an ISP between the local central office (CO) and the location of the ISP. This increases the loading factor in the intraLATA (local access transport area) and interLATA communications segment. To address this problem, signaling gateway equipment is being used to distinguish between data and voice traffic. Using the globally accepted signaling system 7 (SS7)—see Chapter 5—the data streams are routed directly to the ISPs and bypass the PSTN circuit switches. While this method does not relieve the congestion in the local loop, it recognizes the importance of IP traffic and the existence of two different worlds, the established public switched network and the growing Internet.

3.4.1 Virtual Private Networks

The private and secure intranets of multilocational corporations can exist in both of these worlds and are usually referred to as VPNs. In both the PSTN and the Internet, the availability of ATM-based access control equipment is an advantage to the user, as it provides multilevel QoS and substantial savings in connection charges. Additional long-distance savings can be obtained by routing the VPN traffic over the Internet.

3.4.1.1 VPNs in the Public Switched Telephone Network

The common characteristics of a complete VPN network are *security, speed, multilevel QoS,* and *network management.* There are at least two dozen vendors of VPN equipment that have concentrated their products on the enterprise market. Since the public network is ATM-based, it is easy to meet the requirements for security, multilevel QoS, and network management. Speed, however, remains a problem. WAN access is still based on FR and low-speed ATM at 1.544 Mbps. Higher speeds like T-3 at 45 Mbps remain unaffordable. For enterprises that require the assurance of high availability, a low-speed ATM-based VPN can be the best solution. The problem starts when the enterprises are located in geographic areas that require service arrangements with different service providers and long-distance companies. In some cases, ATM-T-1 service is not available or becomes unaffordable.

For those that can live with the offered speed ranges and the pricing, the establishment of a VPN over the public network assures availability, security, minimized packet loss, and flexibility. However, it is necessary to remember that total interoperability of all ATM devices has not been established (see Chapter 5). Furthermore, issues such as switched virtual circuits (SVCs), call routing, and addressing are still in the standardization process within the ITU and ISO's network service access point (NSAP). When negotiating with service providers, it is important that the ATM interoperability issues are settled and that all network management issues are resolved. The network manager must be able to obtain automatic reporting on performance, faults, configuration, security, and accounting. Integrating these requirements between the corporate system, the VPN equipment supplier, and the WAN service providers may not be easy and will depend largely on the network manager's savvy.

In terms of security and QoS levels, there are no problems. All vendors have adequately addressed privacy and security issues in their products. Again, the security standards for TCP/IP are still a work-in-progress at the Internet Engineering Task Force (IETF). This problem of interoperability is, however, less complex because VPN hardware will most likely be based on the simple key-exchange Internet protocol (SKIP) that is being used by all vendors. While ATM-based VPNs in the public network will be more reliable, the high costs may be prohibitive when compared with the Internet. However, it is pointed out that connection costs are estimated at 30–40% of the cost of private leased lines, a saving that may justify a closer look at ATM-based VPNs.

3.4.1.2 VPNs Over the Internet

The big advantages of using VPN hardware and routing communications over the Internet are lower costs and higher speeds. An Internet-based VPN must modify the broadcast mode used in the Internet to a point-to-point connection. This is accomplished by the point-to-point tunneling protocol (PPTP). This protocol establishes a direct connectivity with the addressed location. PPTP is a layer 3 protocol and is used for IP traffic. The more advanced layer 2 tunneling protocol (L2TP) permits various protocols to be used over the VPN. Windows NT is commonly used for network functionality. In addition, privacy and security are required to exchange data over the Internet. The Internet protocol security (IPSec) has been adopted for this purpose. With these protocols and the proper access hardware, every corporation has the ability to develop its own private Internet.

Again, there are more than two dozen companies competing in the VPN business. To summarize the differences in performance, speed, and network management is a subject that can easily fill another book. The problems are incomplete standards and interoperability. Still, implementing a VPN over the Internet has the advantages of eliminating the cost of long distance or the operation of a private leased-line network, increasing the mobility of the work force, and interacting *online* with customers and suppliers using VOIP for telephony and conducting all transactions in a secure manner. There are two different types of VPN, dial-up and LAN-to-LAN connectivity. Dial-up requires every user to establish the connection. In the LAN-to-LAN arrangement, the dial-up is automated. The basic problem with the VPN technology is the fact that the local loop restricts access speeds. The connection to the ISP, again, is T-1 or FR unless an inexpensive T-3 connectivity can be established. There are three major VPN categories:

1. In the case of a *dependent VPN,* VPN access servers are located at the premises of the serving ISPs or at the points of presence (POPs) of the long-distance provider. L2TP or the ATM protocol (ATMP) is used in a point-to-point configuration between their locations. In the dependent VPN, the user logs on to the ISP access equipment and establishes access privileges and tunneling parameters. The traffic from the user is encapsulated, optionally encrypted, sent to the participating distant ISP, and forwarded to the enterprise user. The entire VPN operation is invisible to the user, who only sees normal IP traffic.

2. In an *independent VPN,* all VPN requirements are handled with user-owned equipment. The ISPs are only handling the Internet traffic. All traffic is encapsulated and encrypted at the user's sites, and direct control can be maintained on a day-by-day basis.

3. For users that do not require a fully independent operation, there are also *hybrid VPN* solutions where some sites use dependent and others use independent VPN services. Consolidated billing is another desirable service to negotiate with the primary ISP. A single invoice for all VPN services may be an important convenience.

Major sites like headquarters and large divisions should be considered critical sites and connect with the ISP over both a primary and secondary route. In the LAN examples presented in Figures 3.3 to 3.5, the ATM uplink access controller is connected to the VPN access equipment, which should feature a primary and a secondary WAN port. Figure 3.6 illustrates a typical enterprise hybrid VPN network with independent VPN operation at major locations and ISP-dependent conditions for smaller entities.

In the planning of a VPN network it is important to (1) determine whether the participating locations are intranet-friendly or extranet, which may require firewall protection; (2) determine the best product fit relative to dial-up versus LAN-to-LAN operation and ownership and distribution of responsibilities; (3) decide on a select group of suppliers and establish a test bed; (4) determine the hardware components and the number of users at each location and estimate the capital outlay as well as the operating costs; (5) evaluate the network management aspects of the proposed system and make changes to the layout until you are satisfied with the control and billing parameters; and (6) develop an RFP with system details and general terms and conditions.

A thorough evaluation of the bids will provide details on performance, percentage of packets forwarded without loss, speed ranges, overhead due to encryption, ATM support, and scalability. A thorough planning and evaluation

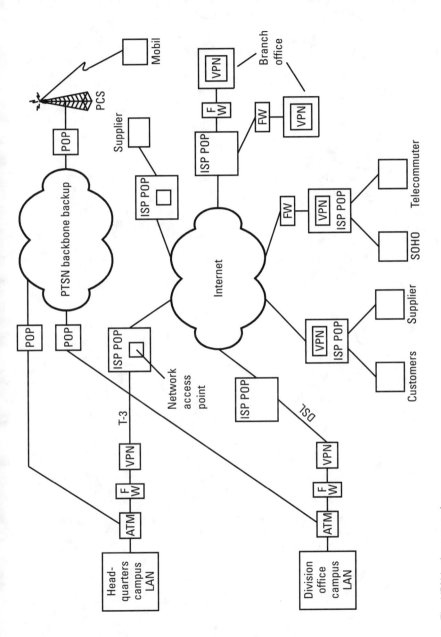

Figure 3.6 The VPN in the enterprise.

period will optimize performance and assure the most economical solution. A well-designed VPN can become an important milestone in the path toward the ultimate integrated LAN/WAN multiservice network.

4

RF Broadband—The Spectral Capacity

RF broadband utilizes fiber-optic strands and coaxial cables. *HFC,* a term derived from the composition of these two transmission media, is often referred to as a fiber/coax infrastructure and uses single-mode fiber. In contrast, most modern data networks utilize a multimode fiber/copper infrastructure. In both cases—in the fiber/copper and in the fiber/coax infrastructure—the fiber is mostly used for backbone applications, and the copper or the coax is used for the distribution or service area of the system.

The evolution of the development of fiber proceeded in two different optical regions. Multimode fiber can transport optical frequencies at 850 and 1,300 nm and features limited performance characteristics at fairly economical prices. Single-mode fiber can operate at 1,310- and 1,550-nm frequencies and has superior performance characteristics at somewhat higher price levels. The difference between these fiber strands is a direct result of their makeup.

Multimode fibers have a larger plastic core diameter and use cladding material that permits a higher optical dispersion; hence, the light content is partially absorbed in the cladding, causing substantial attenuation over distance. Single-mode fiber features a very thin glass core, operates only at higher optical wavelengths, and features low attenuation because it operates like an optical waveguide. Cladding losses are small unless the cable bending limits are exceeded. Consequently, the typical application of multimode fiber is for short distances. Multimode cables are almost exclusively used for data network applications in single buildings and in campus settings.

Single-mode fiber is used exclusively in long-distance applications. It can be found in all undersea cables that are interlinking the continents. In addition, it interconnects all major cities of North America, Europe, and other industrially advanced countries.

Copper cables, consisting of multiple twisted pairs, have been around since the building of the first telephone network. They were typically used to transmit voice at a frequency range of 300 to 3,000 Hz or cycles per second (cps), a rate that was considered sufficient for speech. In order to support ever-increasing data speeds, new copper/twist architectures evolved during the last decade. UTP wiring has found a home in the last 300 ft of almost every data network in the nation.

Coaxial cables first came into use in military applications as well as in the broadcast industry. Coaxial cables are waveguides for electrical signals. The center conductor transmits frequencies of various wavelengths within the electrical field of the cable, and the outer shield prevents energy from escaping. Every coaxial cable architecture attenuates the transmission of the electrical energy over distance. Higher frequencies are attenuated more than lower frequencies, and cable attenuation is reduced the more air is in the space between the center conductor and the outer shield.

The evolution of the cable television industry is responsible for the existence of a wide variety of different coaxial cable types and sizes. High usage of coaxial cables has produced economical pricing, which is a positive factor in the HFC rebuilding program of the cable industry. While RF broadband HFC technology uses single-mode fiber for the trunk or long-distance segment of the cable network, coaxial cables are still required to provide economical solutions in the service distribution area.

4.1 The Bandwidth of Fiber-Optic Cables

Bandwidth or spectral capacity can be expressed in cycles per second (cps), kilohertz (kHz), megahertz (MHz), or gigahertz (GHz). The transmission of data is a sequence of ones and zeros. The higher the speed of the transmission, the higher the requirement for increased bandwidth. Sections 4.1.1 and 4.1.2 compare the performance parameters of multimode and single-mode fibers. Bandwidth capability is compared with data rates and distance recommendations for various data services.

4.1.1 Multimode Fiber

There are two common types of multimode fiber cables in use today. Both types work in the 850- and 1,300-nm optical frequency ranges. The difference between the two products is the diameter of the core. The core diameter is either

50 ± 3.0 or 62.5 ± 3.0 microns. The cladding diameter of 125 ± 2.0 microns is the same in both cables. The core diameter affects the attenuation properties as well as the bandwidth capabilities. A smaller core diameter will provide lower attenuation and can increase the bandwidth performance. These parameters also change when the cable is used at 1,300 nm versus 850 nm. The higher optical frequency decreases attenuation and increases the operational bandwidth. There are also two major different cable designs to assist the user in indoor and outdoor applications of the cable. Loose tube cable makeup permits better bending properties and is used in the outside plant. The tight buffer and breakout cable is used in indoor and riser applications. Table 4.1 outlines the relationships between core diameter and optical frequency and the resulting attenuation and bandwidth values.

The standards TIA/EIA-568A, the "Commercial Building Telecommunications Cabling Standard," and IEC DIS 11801, "Generic Cabling for Customer Premises," are designed to establish uniform manufacturing and installation methods. Section 12.2.3 of TIA/EIA-568A requires that the $62.5/125$-μm fiber meet the requirements of ANSI/TIA/EIA-492AAAA, entitled "Detail Specification for 62.5-μm Core Diameter/125-μm Cladding Diameter Class 1a Graded-Index Multimode Optical Fibers," as an integral specification. It is noted that 50-μm cable has superior performance characteristics and is widely used in Europe. The cable is being promoted by the industry to accommodate gigabit transmissions but has not been standardized as yet in the United States. When forced to rebuild the fiber-optic infrastructure, however, it cannot match the performance of single-mode cables.

Table 4.1
Multimode Fiber Types

Cable Types	Attenuation (dB/km)	Bandwidth (MHz)
Tight buffer with 50-μm core at 850 nm	3.0	200
Tight buffer with 62.5-μm core at 850 nm	3.5	160
Tight buffer with 50-μm core at 1,300 nm	1.0	400
Tight buffer with 62.5-μm core at 1,300 nm	1.2	300
Loose tube with 50-μm core at 850 nm	2.5	200
Loose tube with 62.5-μm core at 1,300 nm	3.3	160
Loose tube with 50-μm core at 1,300 nm	0.9	400
Loose tube with 62.5-μm core at 1,300 nm	1.5	300

Fiber using the 850-nm mode has been used for almost all early data networks. Some of the more predominant applications are shown in Table 4.2.

It can be seen that an optical transmission on multimode fiber at 850 nm can only be used for fairly low speeds. The application of FDDI in large backbone systems is limited, and a conversion to higher ATM speeds is not considered feasible.

Optical transmission at the 1,300-nm mode appears to offer a solution for most of the formats currently used in data networking. Table 4.3 details some of the applications together with attenuation values and bandwidth performance.

Table 4.2
Limitations of 850-nm Fiber

Application	Data Rate (Mbps)	Maximum Distance (km/kft)	Attenuation (dB/km)	Bandwidth (MHz/km)
TIA-568A	N/A	2/6.6	3.75	160
Ethernet 10BaseF	10	2/6.6	3.75	160
Token ring	4	2/6/6	3.75	160
Token ring	16	2/6.6	3.75	160
FDDI-850	N/A*	N/A*/N/A*	N/A*	N/A*
ATM-850	52	2*/6.6*	N/A*	160*

*Standards are not finalized.

Table 4.3
Limitations of 1,300-nm Fiber

Application	Data Rate (Mbps)	Maximum Distance (km /kft)	Attenuation (dB/km)	Bandwidth (MHz/km)
TIA-568A	155	2/6.6	2.5	500
Ethernet 100BF	100	2/6.6	2.5	500
FDDI-1300	100	2/6.6	2.5	500
Fiber channel	133	2/6.6	1.0	500
ATM	155	2/6.6	1.5	500
ATM	622*	0.3/1/0	1.5*	500*

*Standards are not finalized.

While multimode fiber has provided good solutions for early data networks requiring interlinking buildings, its use in the new era of ever-increasing demand for speed is limited. For both in-building and especially in larger campus environments, a better performing conductor is needed. A brief evaluation of the data in Table 4.3 reveals that the limitations of multimode, fiber-optic strands are (1) that they are only good for limited distances and speed ranges and (2) that they are limited to transmissions at baseband. While 850-nm fiber extends its bandwidth capability to 160 MHz, the 1,300-nm multimode fiber can transmit a bandwidth between 0 and 500 MHz. It is noted that data streams over 622 Mbps require more bandwidth to accommodate the higher speeds. Therefore, the capability to overcome long distances is reduced.

Transmission equipment for multimode fiber uses conventional LED stages to illuminate the fiber. While the LED transmission is a quite economical method to translate electrical impulses into optical energy, the optical transfer is limited. LEDs emit light in the near-infrared region. They are primarily used with multimode fiber because they emit light in a broad cone that is captured better by the large numerical aperture of the multimode fiber. The total power emitted by the LED transmitter is not confined to exactly the center wavelength. It is distributed over a range of wavelengths spread about the center wavelength. In the case of an optical transmission at 1,300 nm, the spectral width forms a Gaussian curve between 1,250 and 1,350 nm. In contrast, lasers are more confined to the center wavelength and do not provide good illumination within the core of the multimode fiber strand.

4.1.2 Single-Mode Fiber

Single-mode optical fiber is specified by TIA/EIA-568A, Section 12.1. The standard requires that single-mode optical fiber must also meet the requirements of ANSI/TIA/EIA-492BAAA entitled "Detail Specification for Class IVa Dispersion Unshifted Single-Mode Optical Fiber." This standard defines mechanical, geometrical, and optical characteristics of the fiber.

Single-mode fiber is designed to operate at 1,310 and 1,550 nm. While the optical attenuation is lower at the higher optical frequency, the dispersion at 1,550 nm is a factor of concern. In the long distances that are experienced in the fiber routes crisscrossing the country, special dispersion-shifted single-mode fibers are frequently used. The TIA-568A specification recommends the use of nondispersion-shifted fiber for premises installation, which singles out single-mode fibers operating at 1,310 nm.

With a core diameter of only 8.3 microns and a cladding diameter of 125 microns, the single-mode fiber has a very low attenuation to the optical energy. The glass fiber core acts as an optical waveguide and reduces the loss of optical energy except in cases of overbending. To cope with the fragility of the fiber strand, a number of preventive specifications and installation guidelines assure the performance and longevity of the fiber. This is expressed by (1) coating specifications requiring a minimum of 245 microns, (2) the addition of internal strength members, and (3) different cable construction constraints. Loose-tube construction is typically used for outside plant applications and features a filling compound within the buffer tube and a flooding compound between buffer tubes. Blank fillers are added to substitute for nonexistent buffer tubes, and an aramid yarn layer increases the tensile strength. For the purpose of premises installation, fiber counts of 2–48 fibers are most frequently used. Most loose-tube cables all feature six buffer tubes, and all feature similar outside diameters of about 0.56 in. The tight buffer cable design uses direct extrusion of plastic over the basic fiber coating. This construction allows for a greater resistance to crushing and to impact forces than the loose-tube design. However, tight buffer cable cannot withstand mechanical stresses and temperature changes as well as the loose-tube design. Hence, the tight buffer cable is used primarily in indoor installations, but a design developed for military applications can also be used in the outdoors. In addition to these two predominant single-mode cables there is the breakout cable. This type is used for breakout and interconnections between equipment and is built for flexibility. Breakout cables are lightweight, do not use bundles, and are commonly used for riser and distribution applications.

The attenuation of single-mode fiber is ten times lower than that of multimode fiber. The bandwidth capability is greater than the capability of currently available optical transmission equipment. This results in extremely long distances that can be overcome without the use of optical amplifiers. In the premises environment, the use of single-mode fiber permits the placement of optical couplers that can often reduce the required number of strands in a campus setting. Table 4.4 compares attenuation and bandwidth of the different cables.

Table 4.4 does not address 1,550-nm fiber transmission performance. While this fiber has even lower attenuation, it requires dispersion shifting

Table 4.4
Single-Mode Fiber Types

Single-Mode Cable Type	Attenuation (dB/km)	Bandwidth (MHz)
Loose tube at 1,310 nm	0.3	over 1 GHz
Tight buffer at 1,310 nm	0.35	over 1 GHz
Breakout at 1,310 nm	0.4	over 1 GHz

corrections and is not used in premises and campus installations. The cable industry often uses 1,300-nm single-mode fiber in its HFC rebuilding projects. It usually carries all RF frequencies from 50 to 1,000 MHz in the forward direction and delivers analog as well as digital multichannel video, high-speed data, and digital cable telephony. In the premises environment, the single-mode fiber can satisfy the most stringent applications of speed, distance and bandwidth. These applications are summarized in Table 4.5.

Single-mode fiber has superior broadband capabilities. It has become the choice of all long-distance carriers and the cable industry. The enormous spectral capacity is already being supplemented by optical multiplexing arrangements like WDM and has been installed in all undersea cable projects interlinking our continents. Most long-distance service providers are beginning to use the new DWDM technology, which permits the transmission of speeds up to 10 Gbps over a single fiber strand.

Single-mode fiber can carry transmissions in analog and digital form. It can carry transmissions at baseband at even the highest SONET speed ranges as well as any desired RF frequency in an RF carrier-based system. The application of single-mode fiber in premises communication and corporate networking would offer unrestricted bandwidth capabilities.

4.2 The Bandwidth of Unshielded Twisted-Pair Cables

The use of UTP cables for data transmission came in direct response to the dominance of 50-ohm coaxial cable and its inherent installation difficulties. All of the initial Ethernet installations used coaxial cables in the development of

Table 4.5
Single-Mode Fiber Limitations

Application	Data Rate (Mbps)	Maximum Distance (km/mi)	Attenuation (dB/km)	Bandwidth (GHz)
TIA-568A	N/A	60/50	0.5	over 1
FDDI-SMF	100	60/50	0.5	over 1
Fiber channel	531*	10/6.5	0.5	over 1
Fiber channel	1,063*	10/6.5	0.5	over 1
ATM/SONET	155	55/36	0.5	over 1
ATM/SONET	622	50/33	0.5	over 1

*Mbaud.

the first intelligent islands. With the advent of bridges and the evolution of the client/server environment came the introduction of STPs and UTPs. Until 1991, there were no wiring standards, and the use of UTP was governed by recommendations of the newly developing data network equipment industry. Consequently, most wiring schemes were proprietary, causing confusion and frustration in the user community. In 1991, the Electronic Industry Association (EIA), the Telecommunications Industry Association (TIA), and leading communications companies joined forces to develop a comprehensive cabling standard for the type, quality, reliability, and performance of UTP cables. As usual, the development of standards lags behind the advances in technology, but since its last revision in 1995, the ANSI/TIA/EIA-568A identifies the basic ground rules of UTP wiring.

The goal of these standards was to define a *structured cabling* system to support voice, imaging, and data applications. The TIA/EIA-568A standard covers different cabling models or categories, specifically CAT-3, CAT-4, and CAT-5 compliant cable and connecting hardware specifications. CAT-3 cabling, which was defined in the original TIA/EIA-568 guidelines, has been used in some data network wiring, usually for voice installations. Its rated bandwidth of 16 MHz permits a transmission rate not exceeding 10 Mbps.

Current UTP cabling uses CAT-5 standards. The horizontal UTP cable is a four-pair cable consisting of solid 0.5-mm (24 AWG) strands. Performance marking has to be provided to show the performance category. UTP cable terminations consist of an eight-position modular jack per IEC 603-7. All four pairs must be connected. Under the standard, connecting hardware should be used with less than 0.5-in of untwisted cable (13 mm). In addition, link category markings should be clearly visible on both ends, with all labeling, markings, and color coding provided in accordance with ANSI/TIA/EIA-606.

To avoid stretching, the pulling tension shall not exceed 25 lbs. The minimum bending radius shall not exceed four times the cable diameter in horizontal installations. General caution is exercised to avoid cable twist, tension caused by suspended cable runs, and overly tightly cinched cable ties during pulling and installation. Observing these specifications and cable handling directives, the UTP horizontal cable will perform to the parameters and transmission performance specifications detailed in Table 4.6.

The attenuation values are calculated based on 90m of horizontal cabling, plus 10m of flexible cable inclusive of connectors. Every installation shall be performance-validated for length, attenuation, and crosstalk.

The desire to use UTP for high-speed ATM service has led to the development of CAT-6 and CAT-7 configurations that are suitable for a bandwidth capacity of 250 MHz. These cables reach toward the state-of-the-art limit of twisted-pair performance.

Table 4.6
UTP CAT-5 Performance Limitations

	Length (m)/(ft)	Attenuation (dB)	Bandwidth (MHz)
UTP CAT-5	100/295	21.6	100

There are many concerns about the performance of UTP cabling, and the manufacturing industry is spending effort and money to somehow make it work. The problems with the transmission of 100 Mbps speeds over UTP are plentiful. First, there is the power sum near end crosstalk (PS-NEXT). It is measured in decibels and represents the resistance of one pair to the interference of the other three pairs. This value is 3.1 dB for CAT-5, but it will have to be 37.1 dB for CAT-6. Then there is propagation delay, which indicates how long it takes for the signal to travel 100m. It is 538 ns for CAT-5 and over 600 ns for CAT-6. Return loss deals with the reflections that occur at connection points. CAT-5 indicates already marginal return loss figures of around 10 dB.

There is not one coaxial cable with less than 26 dB. The CAT-6 wish list calls for a minimum of 12-dB return loss. The delay skew is the difference of propagation between the four pairs. The CAT-5 specification calls for a maximum of 45 ns. CAT-6's proposed specifications by ISO/IEC call for a maximum of 50 ns. The power sum-attenuation-to-crosstalk ratio (PS-ACR) indicates how much stronger the data signal in one pair is than the noise on the other three pairs. For CAT-5 the minimum is 3.1 dB, but for CAT-6 it will have to be 15.6 dB.

The ISO/International Electrotechnical Committee (ISO/IEC), which has been taking the lead on standards for above 100-Mbps transmissions over UTP, is now leaning toward STP cables for CAT-7 category cables. The specification for CAT-6 is supposed to cover frequencies to 200 MHz and to 600 MHz for CAT-7. It appears that a gigabit transmission requires a bandwidth capability of at least 500 MHz. This means that only CAT-7 STP cables will be able to handle the 1,000-Mbps speed. There are even questions relative to the performance of UTP CAT-5 at 100-Mbps speeds. In several cases of installed cables, high error rates were experienced.

Whatever the new standards are, it is clear that CAT-5 performance is limited to slow-speed ATM and that gigabit transmission will require special cabling. In the alternative—and whenever rebuilding is necessary—economical coaxial RG-6 cabling can handle a bandwidth ten times higher than UTP or even STP.

Attenuation performance is absolutely critical to high bit rate applications and cannot be traded off against increasing crosstalk values. When wisely installed, in a total length of no more than 80m or 250 ft, CAT-5 cable can assure the transmission of bit rates up to 155 Mbps. Please note, however, that the use of UTP cabling is limited to transmission systems operating at baseband frequencies. UTP can provide the service drop connectivity to the workstations but should never be employed for any backbone connectivity.

4.3 The Bandwidth of Coaxial Cables

Like single-mode fiber, coaxial cables have an unlimited spectral capacity. As is the case with single-mode fiber, a coaxial cable can transmit analog and digital transmissions side by side and over the RF frequency range from 5 MHz to 1,000 MHz (1 GHz). The use of coaxial cables in the outside plant environment of the cable industry has earned the coaxial cable the reputation for being a good conduit for a large spectrum of television channels. Coaxial cables have gone through an evolutionary development cycle over the past 40 years and have proven their worth in the worst weather and climate conditions.

Coaxial cables have excellent bandwidth capabilities. Also, in the new world of cable modems, it makes sense to find a cable that can carry more than just baseband to the desktop. In every enterprise data network, the cabling medium used for the last 250 ft of distribution is of the utmost importance. The last 250 ft set the limits for present and future services and in many instances have guided the decision maker toward the need for expensive network rebuilding projects. Due to the high cost of bringing fiber to the desktop, replacing the last 250 ft with coaxial cable may be important in helping to preserve and upgrade the network for the future and moving toward the goal of nonobsolescence.

Coaxial cables are available in many different sizes and compositions. Because their attenuation varies with frequency, and because the cable industry needed to overcome long distances between amplifiers, large diameter coaxial cables were developed.

In addition, a variety of service drop cables, which act as the conduit between the outside plant and the subscriber, were developed. Since fiber will be the medium of choice for the backbone of any enterprise system, the use of coaxial cables can be limited to service drops and some use in passive coaxial horizontal and vertical risers. These two predominant coaxial cables are described in Sections 4.3.1 and 4.3.2.

4.3.1 Coaxial Service Drop Cables

UTP cables do not have any protective shielding against ingress or egress. The coaxial service drop cable, on the other hand, has gone through a rigid evolution to reduce and eliminate any interference properties, including egress or ingress. Of the various types of service drop cables in existence, RG-6 type cables provide the best compromise between attenuation, size, makeup, and shielding properties. The 6-series cable is available in both flame-retardant riser and air-return plenum configurations. The basic construction of the riser-rated 6-series cable consists of a 20-gauge [nominal 0.359 in. (0.91 mm)] copper-covered, steel-center conductor. The dielectric around this center conductor is a flame-retardant polyethylene. The dielectric is wrapped with an inner-shield aluminum laminated tape. An outer-shield aluminum braid wire is then applied; the jacket consists of flame-retardant black polyvinyl chloride.

The basic construction of the plenum-rated 6-series drop wire features an 18-gauge [nominal 0.040 in. (1.02 mm)] copper-covered, steel-center conductor, a foamed Teflon dielectric, an inner-shield aluminum laminated tape, an outer shield of bare aluminum braid, and a jacket of plenum-rated material. Due to the difference of the center conductor and the different dielectric, the attenuation values are slightly different as well. Table 4.7 highlights these differences.

Table 4.7 shows the attenuation over the frequencies that can be transmitted over the cable. The column dB/250 ft represents the recommended maximum length of a coaxial service drop. This length compares closely with

Table 4.7
6-Series Cable Attenuation Over Frequencies

Frequency (MHz)	6-Series Riser Attenuation		6-Series Plenum Attenuation	
	dB/100 ft	dB/250 ft	dB/100 ft	dB/250 ft
5	0.67	1.67	0.53	1.32
55	1.86	4.65	1.58	3.95
83	2.22	5.55	1.9	4.75
300	3.89	9.72	3.77	9.43
550	5.00	12.5	5.45	13.6
750	5.8	14.5	6.6	16.5
1,000	6.75	16.8	7.9	19.7

the 80- to 90m length requirement of a UTP service drop. While a CAT-5 cable can only transport about 100 MHz and has an attenuation of about 22 dB, the 6-series cable only has an attenuation of 17 or 20 dB at a frequency of 1,000 MHz (see Figure 8.3). This is a tenfold increase in the bandwidth capability. More information on the attenuation and performance differences of UTP and coaxial service drop cables is provided in Chapter 8.

With the advent of cable modems, ATM compatibility, and cable telephony, the use of 6-series RG-6 coaxial cables for service drops can assure a high degree of nonobsolescence in the distribution segment of the corporate LAN and could, if ever required, bring multigigabit services to the desktop.

4.3.2 Coaxial Backbone Cables

Coaxial cables can be utilized to extend the fiber network termination point. This application can apply in the riser of a large building or in a recent building addition. Without the need for coaxial amplifiers, the passive coaxial backbone can serve to lower the cost of rebuilding.

The single-mode, fiber-optic receiver translates the optical spectrum into electric RF broadband frequencies. This translation depends upon the light level at the receiver's photo diode. A standard light level will translate into a comparable electric level with an output of between 30 and 40 dBmV. The backbone coaxial cable can deliver this level to other floor levels for distribution and alleviate the need for adding new fiber strands. There are many other applications where the use of a coaxial cable can eliminate the need for additional fibers and additional fiber receivers, thereby reducing the costs to the price of economical coaxial cable, passive devices, and connectors. Examples of these architectures are provided in Chapters 6 and 7.

A wide range of one-half to one-inch coaxial cables exists. Larger diameter cables have lower attenuation and use various types of dielectric between the center conductor and the outer shield. All campus cabling requires riser-rated performance as a minimum. In special return air spaces the use of plenum-rated cables is required. This narrows the choice to the riser-rated P3-500JCAR or the P3-500JCAP cables. Larger diameter cables are available, but they sharply increase the installation difficulties and, therefore, are not recommended. The makeup of these coaxial backbone cables is similar for the center conductor and the outer sheath, but different for dielectric material and jacket. Both cables feature a 0.109-in center conductor and a solid aluminum sheath with a 0.5-in diameter. The riser-rated cable uses an expanded polyethylene dielectric and a flame-retardant polyethylene jacket. The plenum cable uses a foamed Teflon fluorinated ethylene propylene dielectric and a solid Kynar jacket. These cables have a minimum bending radius of six inches, which is not worse than the bending requirements for fiber cables.

Due to the differences of the dielectric, the attenuation over frequencies varies between the riser and the plenum cable. Table 4.8 identifies these attenuations.

While the attenuation of the plenum cable is somewhat higher, its use is limited to return air spaces mostly not found in vertical riser spaces. The above example of extending a single-mode fiber with coaxial cable is a quite economic solution. Assuming that the service for a multistory building needs to be divided into multiple ATM rings, the fiber entrance location does not have to be changed. The coaxial backbone cable is run vertically in the riser and equipped with passive splitters at each floor level. Separate cable modems can now be installed at each floor to serve the workstations in separate ATM loops. With an attenuation of about 2.5 dB per 100 ft, the coaxial cable can distribute separate frequency bands to each floor level, in most cases without the need for any amplification equipment.

4.4 Bandwidth Limitations

Overcoming bandwidth limitations is an important consideration to assure future nonobsolescence of the network.

Table 4.8
Riser- and Plenum-Cable Attenuation Over Frequencies

Frequency (MHz)	Riser-Rated (0.5 in$\oslash$) Attenuation (dB/100 ft)	Plenum-Rated (0.5 in$\oslash$) Attenuation (dB/100 ft)
5	0.16	0.17
50	0.52	0.58
220	1.11	1.38
550	1.82	2.46
750	2.16	3.43
1,000	2.52	4.31

4.4.1 Infrastructure—The Bandwidth Conduit

Limitations of bandwidth in the infrastructure are the main reasons for frequent rebuilding. The telecommunications manager has to listen to the claims of the vendors and is focused on equipment choices and equipment upgrade considerations. The infrastructure is not considered an important issue because it is the vendor's responsibility to make everything work. In these discussions, the infrastructure is seldom viewed as the key to future connectivity at higher speeds. In reality, however, the physical layer is the foundation of any communication transmission.

The advent of gigabit speed equipment indicates that the infrastructure has to grow with technology and that it must be possible to handle new and presently unknown speeds without constant reconfigurations. The examples of bandwidth capability, described earlier in this chapter, identify the spectral and distance limitations of the media that has been commonly applied in the data networking industry. In a comparison of multimode fiber, single-mode fiber, UTP, and coaxial cable, it becomes evident that only multimode fiber and UTP have bandwidth limitations. These limitations, however, can easily curtail the upgradability of a network.

Figure 4.1 shows four infrastructure models, two of which are currently in use. Most common is the combination of multimode fiber in the backbone and CAT-5 UTP in the distribution.

A commonly accepted standard is the relationship between speed and bandwidth (bits and hertz). This means that two bits can be transmitted within the bandwidth of one hertz. Analyzing the Figure 4.1 presentation in this manner, it becomes clear that 155-Mbps ATM can barely be supported by multimode and CAT-5 and that gigabit speeds only live in the single-mode fiber environment and become a problem in the UTP distribution area. CAT-5, or even STP cables, are only marginally suitable for speeds above 200 Mbps. Attenuation, crosstalk, and electromagnetic interference (EMI) problems require the UTP cabling to be shortened to achieve even 155-Mbps ATM speeds.

A critical comparison of the infrastructure architectures indicates that the fiber/copper hybrid systems of today have reached their state-of-the-art limitations and that, in order to cope with future higher speeds, the use of single-mode fiber is a sound solution.

The HFC or RF broadband architecture offers a simple solution. In the example of Figure 4.1, the single-mode fiber is interconnected with a short coaxial backbone and extended with 6-series coaxial cable to the user. This infrastructure architecture does not have any limitations regarding distances or bandwidth. It can provide the multigigabit services of tomorrow, or it can be used for any RF-based transmission technology. Tomorrow's integration of

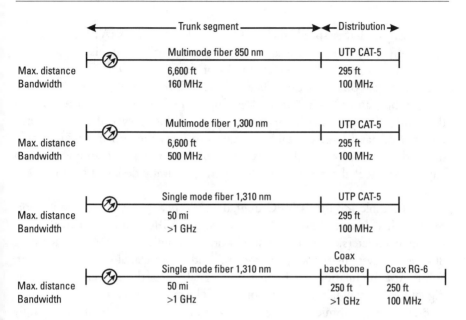

Figure 4.1 The limitations of the infrastructure.

data and voice, computer telephony, and the convergence of ATM-based information of any kind requires a detailed evaluation and reassessment of presently used infrastructure components.

4.4.2 Fiber-Optic Transmission Equipment

Data equipment manufacturers have been concentrating on LED transmission methods for the multimode fiber. Because the use of single-mode fiber is increasing, there is a large number of companies that concentrate on the conversion technology between multimode and single-mode architectures. There is even compatibility between SONET/synchronous digital hierarchy (SDH)/ATM and FDDI laser transmission equipment. The most economical single-mode fiber transmitter employs the Fabry-Perot laser. More powerful devices for wider bandwidth applications employ the DFB laser.

The long-distance carriers have the need for powerful lasers to illuminate the fiber for long-distance performance, and the bandwidth of these optical transmitters is increasing with the requirement for high-speed traffic in global SONET speed standards. While lower speed ranges differ between U.S. and international formats, higher rates are uniform and oriented toward global interoperability. For example, the North American T-1 standard, also called DS-1, of 1.544 Mbps compares with the international standard E-1 at

2.048 Mbps. Higher speeds such as OC-3 at 155.52 Mbps (ATM) or OC-12 at 622.08 Mbps are identical to the international standards of the ITU 1/SDH-1 and 4/SDH-4. The evolution of the global SONET speed ranges does not stop at 622.08 Mbps. They include OC-48 at 2.488 Gbps, OC-192 at 9.953 Gbps, and OC-768 at 39.813 Gbps. The bandwidth requirements of these long-distance systems are enormous and require powerful laser transmitters that demand large capital investments to meet the ever-increasing traffic demand. Note that currently available gigabit hardware is not compatible with any of these global standards and has been specifically designed for local data traffic.

The cable industry's HFC upgrade projects apply AM modulation formats in their laser technology. Since video is a VBR transmission with exact timing parameters, it is today still more economical to leave the existing 80-channel systems working in the old and proven analog NTSC format. While AM modulation can be used for digital video, the conversion to solely digital formats will take a number of years. The main reasons for this slow transition are the economics of television sets, cameras, and VCRs that are readily available to the public. DFB laser transmitters for RF broadband HFC have a bandwidth of 750 MHz and illuminate the fiber between 20 and 750 MHz. For the return, there are lower powered fiber transmitters available that operate in the same 20–750-MHz band.

Even at that reduced bandwidth, the conventional DFB laser used in the HFC system can provide a bandwidth equivalency for OC-12 SONET systems at a far more reasonable capital layout.

4.4.3 Coaxial Transmission Equipment

Coaxial transmission equipment is used in the service areas of any cable television system. HFC engineering standards recommend that no more than two amplifiers be used in cascade. The coaxial amplifier is bidirectional and has a forward bandwidth of 50–750 MHz and a return bandwidth of 5–42 MHz. Again, the active equipment element is limiting the bandwidth capability of the coaxial network.

However, the use of coaxial amplifiers in the RF broadband LAN can be limited, if not eliminated. The bandwidth restriction of coaxial amplifiers is only of academic value. RF broadband corporate LANs can be established with solely passive coaxial components and retain a 1,000-MHz spectral capacity. Soon, laser transmitters will be available for a spectrum of over 1,000 MHz, and coaxial cable will be able to handle the added bandwidth.

4.5 RF Broadband—Architecture Limitations

The cable industry developed as an entertainment delivery industry. The earliest cable television systems were capable of the delivery of channels 2 through 13. The broadcast frequency of channel 2 is at 54 MHz, which has been the lowest forward frequency throughout the evolution of RF broadband technology. Over the years, the upper spectrum limit advanced from 216 MHz for channel 13 to 300 MHz for channel 36, to 330 MHz for channel 42, to 550 MHz for channel 78, and currently to 750 MHz for channel 112.

The convention of the cable industry has been, and still is, to transport channel 2 on its broadcast frequency over the cable television network. The FCC reinforced this ruling to permit subscribers to view cable channels on their assigned frequencies. Even though available addressable and channel mapping set-top converters can make any channel appear as channel 2, the industry includes many systems that do not use set-top converters. For this reason, the lowest forward frequency of 50 MHz has become the de facto standard.

4.5.1 The Subsplit Architecture

The term *subsplit* originated in the 1970s when municipal authorities had to be informed about the plans of a new cable franchisee. The term means that return transmissions are below 54 MHz and that delivery channels are above. Figure 4.2 illustrates the spectral capacity of the subsplit system that has been adopted in the architecture of the new HFC-based networks.

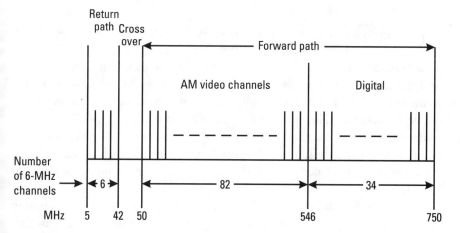

Figure 4.2 The subsplit architecture.

The subsplit architecture, which is a direct result of the spectral selection of the cable industry, refers to the amplifiers that will be available. In the RF broadband corporate LAN, any amplification in the coaxial segment is only required in rare circumstances. In a passive system the separation between the forward and return portion of the spectrum does not require any equipment. The selection of the RF carrier frequencies decides the direction of the propagation.

The frequency spectrum in the forward direction covers 50 to 750 MHz. This spectrum can accommodate 112 channels, each 6 MHz wide. It is planned by most cable companies to utilize the band between 50 and 550 MHz for standard analog video transmissions and to reserve the band between 550 and 750 MHz for digital services. Digital services can be MPEG-2 video channels, HDTV, high-speed data, and telephony.

The present state of the development of cable modem technology permits the transmission of either five MPEG-2 channels or 240 voice channels, or 25.6-Mbps ATM in a single 6-MHz frequency assignment. Assuming the spectral architecture indicated in Figure 4.2, the cable industry's forward system could carry the following services:

- In the 50–550-MHz spectrum: 78 analog video channels;

- In the 550–750-MHz spectrum: A combination of video, telephony, and data in thirty-four 6-MHz channel assignments, such as 240 voice channels in one channel, plus 10 loops of 25.6-Mbps ATM, plus 115 MPEG-2 digital video, or any other desired combination.

In contrast, the return capacity is limited to the spectrum of 5–42 MHz. To alleviate this deficiency, the cable industry proposes to "downsize" the fiber node service area and to provide four 5–42-MHz bands in a 5–186-MHz fiber return transmission. Because each of the 5–42-MHz bands can only accommodate six 6-MHz channels, the total return bandwidth capability can be 24 channels. This spectrum then can accommodate 15 MPEG-2 video-conferences in three channels, ten 25.6-Mbps symmetric ATM high-speed data channels in 10 channels, and 240 voice channels in the 24th channel. Other combinations are possible and subject to spectrum management considerations.

It can be concluded that the RF broadband spectrum is a powerful tool to change the competitive nature in the local loop. It can also be concluded that the subsplit architecture is only one of many choices for a premises wiring system, especially when only passive coaxial elements are used in the distribution segment of the network.

4.5.2 The High-Split Architecture

As a result of the franchising process, the high-split architecture was often publicized as the municipal interconnect facility. It was offered by the franchise aspirants as a giveaway to interconnect all schools and municipal buildings in a community. Some of these systems were built and remained unused because the modem technology was not advanced enough to provide data and voice services. Other institutional networks were used for NTSC-based videoconferencing, but municipal funding was not sufficient to sustain and grow the applications. Figure 4.3 shows the spectral capacity of the high-split architecture.

All RF broadband frequency architectures have crossover areas that divide the operational frequencies into forward and return transmission segments. Only coaxial cables can be used for bidirectional transmissions. In the case of the high-split cable, the crossover region is in the spectrum of 186–222 MHz. High-split amplifiers typically provide for the forward transmission in the spectrum of 222–550 MHz (55 channels) and for the return transmission from 5 to 186 MHz (30 channels).

Again, this architecture is based on the design parameters of coaxial amplifiers. If no amplifiers are used, the crossover area can be selected as desired. In a passive RF broadband LAN, the selection of RF carrier frequencies can determine the directivity in the coaxial cable.

While the high-split architecture is more symmetrical, it will not be used by the cable industry in its HFC rebuilding programs. The reason is that the channel 2 transmission is in the return spectrum. Also, the upper frequency of 550-MHz equipment limits the spectral capacity. However, when employed in a corporate network that only uses passive coaxial components, the high-split architecture has no limitations. A passive crossover filter at the transition

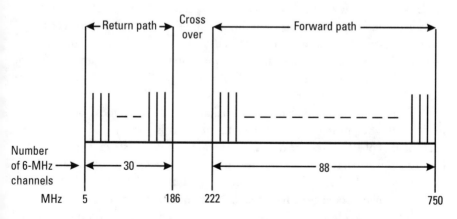

Figure 4.3 The high-split architecture.

between the fiber and the coaxial segment or just the selection of RF carrier frequencies can separate the forward and reverse transmission directions. Figure 4.3 illustrates such a passive coaxial network for a premises installation. The forward spectral capacity extends from 222 to 750 MHz, which indicates that 88 (6-MHz) channels can be obtained. In the return direction, the spectrum from 5 to 186 MHz is available for a total of 30 (6-MHz) channel assignments. Considering that 240 voice channels, 25.6-Mbps ATM data, or five MPEG-2 video channels can be transmitted in a single 6-MHz channel assignment makes the high-split architecture a desirable spectral choice in the LAN environment.

4.5.3 The Dual System

A dual system architecture permits operation of the entire spectrum in both directions with symmetrical bandwidth. All existing fiber backbone systems are dual systems. Since optical transmissions can only be made in one direction, a second fiber is required for the return. The same applies to UTP cabling. Copper pairs are required for each transmission direction. Coaxial cables, however, can be bidirectional. One RF frequency can be sent in one direction, and another can travel in the opposite direction. The subsplit and high-split architectures use only a single coaxial cable for both transmission directions. In the dual system architecture the coaxial service segment is simply duplicated. Such a system can interact seamlessly with the fiber backbone and offer the entire bandwidth capability of 1,000 MHz in both directions. Figure 4.4 depicts the RF broadband dual system architecture and its enormous spectral capacity.

The dual system permits the entire 20–1,000 MHz spectrum to be used in both directions. Even considering the bandwidth limitations of currently available fiber-optic transmitters, the entire spectrum between 20 to 750 MHz is available in both directions. The implications of this bandwidth capability exceeds present day requirements. A total of 121 channels, each containing a 6-MHz spectrum, is more than any baseband system could ever provide. Again, each 6-MHz channel can carry a 25.6-Mbps ATM data stream, 240 voice channels, or five full-motion video channels in the MPEG-2 format.

The decision to duplicate the coaxial service segment does not have to be made in the beginning. Because the coaxial plant is passive and for the most part consists of RG-6 service drops, it is good practice to assure that the modular outlet plates have an empty position for the coaxial F-connector. Both F-connector and wiring can be installed at any time in the future. Scalability is one of the many benefits of RF broadband technology. The dual system is not an architecture of choice of the cable industry. It would be too capital-intensive to install dual coaxial plants everywhere without prior assurance of the

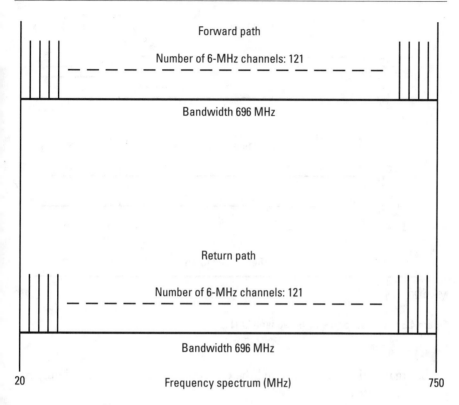

Figure 4.4 The dual system.

monetary return. In addition, the cable industry firmly believes in asymmetric traffic requirements. The most cited example is the Internet model that requires fast download speeds but keyboard return speeds. RF broadband offers a rich menu of choices within the confines of a campus network. Figure 4.5 shows the 750-MHz spectrum that can be provided.

Whether the dual fiber/single coaxial or the dual fiber/dual coaxial architecture is chosen does not matter much. Either will overcome the short distance and limited bandwidth capabilities of the baseband LAN technology.

4.6 The Future Corporate LAN

It is obvious that currently existing data, voice, and video systems have a purpose and should not be replaced for the sake of wiring for the future. Every commercial establishment has its own set of circumstances that has brought about the evolutionary development of its communications networks. In some

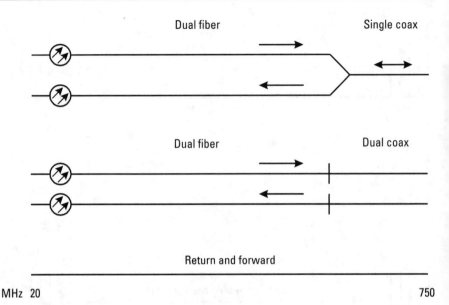

Figure 4.5 Choices for the RF broadband LAN.

cases, organizational lines dictated the development. In other cases, budgetary constraints froze the progress.

It is not the purpose of this book to single out RF broadband LANs as the only option for the future. Every enterprise should first evaluate the existing network infrastructures and the equipment that is being used. Second, it is important to set long-term goals for both intra- and intercompany communications solutions. The savings realized in the WAN by digitizing voice on FR may be more important than the migration to a multiservice LAN within the campus. However, it is important to set milestone goals to keep up with the rapid changes in telecommunications. Somewhere on this list of milestones is the RF broadband LAN. It may be an extension to the existing FDDI network or a breakup of service nodes into smaller groups that may be more economically handled with RF broadband. It may be the decision to "un-PBX," the need to reduce moves and changes, or the price drop of digital television sets that establishes RF broadband as a firm requirement. The implementation of company-wide full-motion videoconferencing, the integration of telephony and data, the establishment of campus-wide ATM, or the need for single-mode fiber-optic cabling will be the motivating forces. Whatever and whenever the decision is made, RF broadband components will be available on short notice and at economical price levels.

5

The Importance of SONET and ATM

ATM is the application of a cell transport structure that permits any existing protocol or transmission format to use a physical layer technology like SONET to transport digital information of any kind around the globe. While ATM works on copper wiring, it is designed to take advantage of the high-speed formats that are achievable with the optical transmission standards of the SONET transportation system. Since SONET has been widely applied for global interconnectivity, it forms the basis of modern telecommunications networking. If SONET is the highway for high-speed digital transmission, ATM is the truck carrying payload between single or multiple destinations.

The evolution of the corporate LAN has led to higher and higher data speeds, but this evolution progressed without concern for CBR traffic like telephony and video. The IP protocol and the Internet lack the ability to transport voice or video within strict latency requirements. The long-distance carriers, in their networks, have taken a different approach, concentrating on combining VBR and CBR traffic through the application of the SONET technology. ATM over SONET combines the best features of both of these different approaches. The result is a homogeneous global communications network that uses globally accepted transmission standards in the local loop, in the WAN, and in long-distance transmissions. While IP is the protocol of choice in the data community, it is desirable to apply ATM to the corporate LAN backbone in order to take advantage of end-to-end multilevel QoS. To realize the full benefits of consolidating voice, video, and data on the next generation of IP and ATM networks, the ability to use multilevel QoS will be of great importance. Now is the time to adopt the rules and economies of the new global communications systems and apply them to premises interconnectivity. While the integration of voice, video, and data has taken place in the public network, corporations still rely on separate networks for voice, data, and full-motion video in their corporate networks.

The utilization of ATM as the unifying technology in the United States lags behind a number of European countries. This is not surprising as the SONET standard is an outgrowth of European transmission speed hierarchies and had to be adapted to permit internetworking with commonly used speed standards in the United States. This chapter briefly reviews the SONET technology and the multiprotocol transport application of ATM and its applicability to the RF broadband LAN.

5.1 The Synchronous Optical Network

When it became known in the late 1960s that plastic and glass fiber strands could be used to transmit a wide bandwidth of frequencies over long distances, the ITU developed the transmission standards to internetwork different countries' communications systems. Known as SDH, the new optical transmission standards were developed by the International Telegraph and Telephone Consultative Committee (CCITT). The primary SDH standards are covered by CCITT G. 707–709 specifications.

At the same time the American National Standards Institute (ANSI) developed and released the SONET standards: T1.105 covering transmission rates and formats; T1.106, which covers the optical interfaces; and T1.119, T1.204, and T1.231, dealing with communications, administrative, and monitoring standards. The U.S. SONET standards required some modifications to permit internetworking with the networks of other countries but are, nevertheless, an example of global cooperation to achieve a uniform interoperability standard for transmissions in digital form.

5.1.1 The SONET/SDH Hierarchies

The lowest rate of the global SONET/SDH level is called the synchronous transport signal level 1 (STS-1) with a signal rate of 51.84 Mbps. The optical equivalent of the electrical STS-1 is the optical carrier level 1 (OC-1) signal. OC-1 is obtained by a direct conversion of the electrical signal to optical. All higher level signals are obtained by byte-interleaved multiplexing of the lower level signals. The rates of higher signal levels are expressed by the multiplier n. SONET only permits certain values of n to assure compatibility between equipment. These values are 1, 3, 9, 12, 24, 48, 96, and 192. Table 5.1 presents the optical carrier (OC) and synchronous transport signal (STS) rates.

Table 5.1
Synchronous Digital Hierarchy

Synchronous Transport Signal	Optical Carrier	Line Rate (Mbps)
STS-1	OC-1	51.84
STS-3	OC-3	155.52
STS-9	OC-9	466.56
STS-12	OC-12	622.08
STS-24	OC-24	1,244.16
STS-48	OC-48	2,488.32
STS-96	OC-96	4,976.64
STS-192	OC-192	9,953.28

All the standard rates listed in Table 5.1 are commonly used in the long-distance segment and in global interconnections, except for OC-9. In the area of corporate networking, equipment is available for 25.6 Mbps and for OC-1, OC-3, and OC-12; that is representative of the typical speed requirements of the presently emerging data networking equipment technology.

The SONET network has the same functionalities as the existing DS3-based network. DS3 at 45 Mbps, commonly referred to as T-3, is a multiple of the familiar T-1 U.S. standard at 1.544 Mbps. T-1 has been used for decades in digital voice connectivity between COs and is based on the simultaneous transport for 24 telephone-quality voice channels. It is interesting to note that the digitization of a voice channel requires 56 Kbps. Twenty-four voice channels then require a combined rate of 1.344 Mbps. The difference between 1.544 and 1.344 Mbps is reserved for overhead functions.

In Europe, the common standard equivalent to T-1 is the E-1 transmission system. In this standard, a voice channel is digitized using the slightly better algorithm of 64 Kbps, and 30 combined voice channels require a minimum rate of 2.048 Mbps. While the speed requirements are different between ANSI (U.S.) and CCITT at the lower speeds, the European E3 rate at 34.368 Mbps and the U.S. DS3 rate at 44.736 Mbps are directly mappable into the SONET/SDH hierarchy. SONET OC-1 optical and STS-1 at 51.64 Mbps represent the native speed and the global standard. Lower speeds are mapped into the native speed and are therefore scalable into SONET. Figure 5.1 illustrates the global interoperability above the OC-1 rate of 51.84 Mbps and the freedom of using proprietary speed ranges below OC-1.

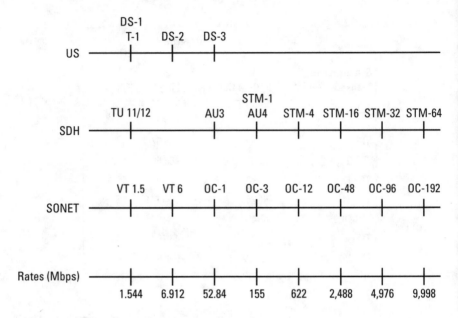

Figure 5.1 The SONET and SDH comparison.

Figure 5.1 shows the ITU-defined speed ranges from 0 to 10 Gbps, the SONET, and the SDH synchronous transport structure. While the European SDH designations are called synchronous transport methods (STMs)—that is, STM-1 for OC-3, STM-4 for OC-12, STM-16 for OC-48, STM-32 for OC-96, and STM-64 for OC-192—the signal rates are identical.

5.1.2 The SONET Advantages

The previously existing copper technology requires frequent multiplexing and demultiplexing functions to interconnect two users on a point-to-point transmission path. For instance, a 64-Kbps signal from a phone call has to be multiplexed to a T-1 channel using channel banks and again multiplexed to a DS3 circuit using an M13 multiplexer. When the signal arrives at the switch location, it must be demultiplexed, switched, and multiplexed again before it is finally demultiplexed to its original format. This back-to-back multiplexing requirement is expensive and wasteful.

With a SONET network, all bandwidth allocations and routings can be controlled remotely by the insertion of addresses and commands within the SONET overhead. The SONET transport system removes the proprietary nature of equipment; it provides a flexible network architecture that can assign avoidance routes in case of network failures; and it has built-in network and maintenance support capabilities.

The first fiber-optic transmission systems utilized DS3 at 45 Mbps as the highest level of connectivity. SONET has expanded this rate to an upper limit of OC-192 at 9.99 Gbps. Circuits used to have manual patch panels. Now, a digital cross-connect system (DCS) routes the circuit through the network. This routing can be changed in the address header of the packet overhead and without any technician assistance. In the event a customer cancels a service, the reprovisioning is a matter of seconds. Former DS3 networks required minutes and even hours to repatch the connection or to relocate the equipment. Figure 5.2 shows the flexible interconnectivity of the SONET network.

The example shows a simple network that transmits information at high-level SONET rates between three locations. A break in one portion of the network routes the traffic in the opposite direction through the ring, without the occurrence of an outage. SONET offers enormous savings in equipment and operational requirements. Using a concept called *direct asynchronous multiplexing*, there is no need to demultiplex tributary channels before switching.

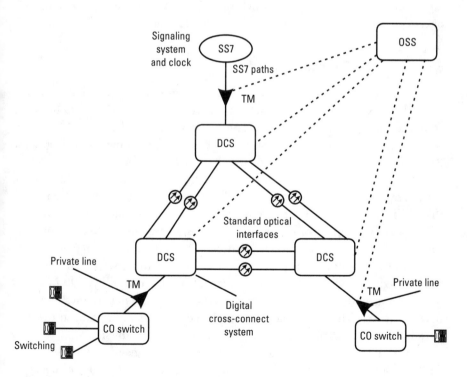

Figure 5.2 The SONET-based network.

Because the entire network is clocked and operates synchronously, the tributary traffic is simply inserted into a time slot and forwarded to the next segment.

The top element in Figure 5.2 shows SS7, a clear channel-signaling specification standard of ITU-T and the prevailing standard for signaling in telephone networks. SS7 has been implemented by all carriers throughout the world and forms the control-signaling layers of the SONET transport as well. SS7 is the network management system that decides on link selection, message routing, signaling traffic management, flow control, rerouting, changeover to a less faulty link, and recovery from link failure. In addition, SS7 provides signaling and route management and transfers status information about signaling routes to remote signaling network locations.

SONET, a multipoint-to-multipoint network, permits control at the network management level. Operation support systems (OSSs) pass the network information from one network operator to another, wherever desired, for the purposes of control, supervision, and performance. While standardization efforts normally fail to produce full interoperability between products, SONET standardization assures that the equipment of different vendors will provide the same information to the OSS, which then disseminates the information to all operators in the same manner and with identical detail.

5.1.3 The SONET Frame Structure

The basic building block of the SONET transport is the STS. This standard frame has been chosen in a manner that will assure lower rate traffic to be assembled and transported to the next location at the speed of light.

The SONET frame consists of 90 columns and nine rows. There are 810 bytes within the frame, 27 of which are reserved for transport overhead; the remaining 783 bytes are reserved for the transmission of information. This area of the SONET frame is called the synchronous payload envelope (SPE). Figure 5.3 shows the frame structure from a two-dimensional view.

Each byte consist of 8 bits that are transmitted in sequence through the rows, starting at the top left-hand corner (A1), followed by the bits in A2 and so on, until byte 90 is transmitted. Then, the bits of the first byte in row 2 are transmitted through byte 90. This sequence continues until byte 90 of row 9 has been transmitted and the sequence is repeated for the next frame. The 8-bit byte structure within each frame identifies each byte as to location in the data stream and preserves this location for an end-to-end transmission.

SONET transmission has been designed to transmit STS-1/OC-1 at 51.82 Mbps in one frame and in a time period of 125 μs. This arrangement also permits the transmission of the old former DS3 standard, which has been used quite commonly for full-motion video in a digital format. Figure 5.4

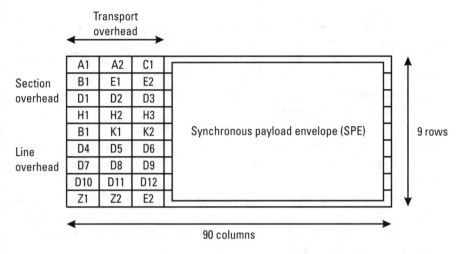

Figure 5.3 The STS.

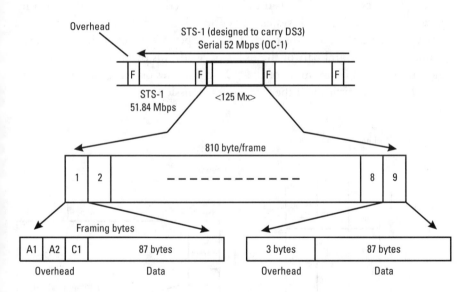

Figure 5.4 The STS-1 frame structure.

depicts the frame, the overhead in each row, the sequence of the rows, and the interdependent relationship with overhead and payload.

Note that the 27-byte system overhead is an integral part of the frame structure. The first 3 bytes of the first row are sent first, followed by row 2 through 9. While the payload in each row consists of 87 bytes, the overhead consists of 3 bytes. Called the transport overhead, certain bytes are associated with the *path*, the *section*, and the *line* of the circuit.

The *path* span extends between the transmit and the receive ends of the transmission path and supports the transportation of the SPE between the customer locations. These are the locations where the data stream is assembled to SONET frames and disassembled at the other end.

The *line* span represents the network performance between transport nodes or the digital cross-connect locations. The line overhead maintains the performance of the transport and of the SPE between adjacent switching centers.

The *section* span monitors the network performance between the line span, which normally consists of fiber-optic regenerators, amplifiers, optical loss monitoring, and other fault-locating and reporting equipment. Figure 5.5 shows the arrangement of these three overhead functions.

The first 3 bytes in each of the first three rows are reserved for section overhead. The first 3 bytes in rows 4 to 9 are used for line overhead. This makes up the 27 bytes of SONET overhead that is directly dedicated to the performance and monitoring of all system functions. The path overhead consists of a single byte in each row. This total of 9 bytes per row is set aside for the user. While it is an integral part of the user's payload, this overhead is dedicated to the assembly and disassembly of the SONET frames over the end-to-end path as well as the monitoring of the payload transmission.

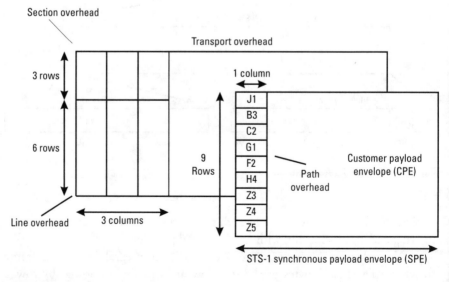

Figure 5.5 SONET overhead functions.

The payload capacity area of the client/customer payroll envelope (CPE) consists then of 86 columns by 9 rows, or a total of 774 bytes, or 6,192 bits. This provides for a 49.536-Mbps transport capacity with a frame repetition rate of 8,000 Hz. It is specifically designed to accommodate the transportation of a DS3 tributary signal in 8 SONET ST-1 frames. Figure 5.3 illustrates the transport overhead area with different designators. The following is a list of overhead bytes, their uses, and purposes:

- Section overhead:
  - A1, A2: Frame alignment pattern;
  - C1: STS-1 Identification;
  - B1: Parity check;
  - D1, D2, D3: Data communication channel—192 Kbps;
  - E1: Orderwire—voice communication;
  - F1: User channel—maintenance.
- Line overhead:
  - H1, H3: Payload pointers;
  - K1, K2: Automatic protection switching;
  - B2: Parity check;
  - K2: Alarm;
  - D4 D12: Data communication channel—576 Kbps;
  - E2: Orderwire—voice channel;
  - Z2: Line FEBE (far end block error) and growth;
  - Z1: Synchronization messages.
- Path overhead in CPE:
  - J1: Path trace;
  - B3: BIP-8 (bit per byte counter);
  - C2: Signal label;
  - G1: Path status;
  - F2: User channel (downloading for user);
  - H4: Multiframe (phase indication of CPE);
  - Z3 Z5: Growth (future use).

SONET is designed to accommodate the highest level of performance control ever devised. Even globally interconnected systems can report system failures instantly and switch to redundant equipment and routes to avoid any outage times. SONET also provides the user with the performance status of any payload information, thus assuring the highest level of QoS.

5.1.4 The Virtual Tributary Frame Structure

The virtual tributary (VT) frame structure is specifically intended to support the transport and switching of lesser payload. VT frame structures fit into the CPE in order to simplify multiplexing capabilities. To provide uniformity across all SONET transport capabilities, the payload capacity for each of the tributary signals is always slightly greater than the required capacity. Called mapping, the process provides synchronization of the tributary signal with payload capacity. Payload mapping is achieved by stuffing extra bits into the signal stream. VT1.5 accommodates the mapping of a T-1 transmission at 1.544 Mbps into the CPE.

The VT1.5 frame consists of 27 bytes, structured as three columns of 9 bytes. At the frame rate of 8,000 Hz, these 27 bytes provide a transport capacity of 1.728 Mbps, more than required for the T-1 transmission. A total of 28 VT1.5 can be multiplexed into a single STS-1 CPE area. The VT1.5 is the most important size of a virtual tributary. This is because the three-column by nine-row structure fits directly into the 86 columns of the CPE.

It has the highest density of all VT mapping processes. The 28 VT1.5 that can be packed into the CPE leave two spare columns that can be filled with stuff bytes. The total density of these 28 T-1 transmissions result in a 48.384-Mbps bit stream that is almost identical to the global OC-1 structure of SONET.

E-1 transmission systems in Europe require the mapping of 2.048 Mbps into the CPE. Here, the frame consists of 36 bytes, structured as four columns of 9 bytes. At the same frame rate, the VT2 virtual tributary mapping produces a transport capacity of 2.304 Mbps. Note that 21 VT2s can be multiplexed into a standard CPE. VT3 has been devised to accommodate the DS1C bit stream. The VT3 bitstream consists of 54 bytes, structured as six columns of 9 bytes. Fourteen VT3s can be multiplexed into the CPE. There is also a VT6, which can be used to the mapping of a DS2 signal. The VT6 frame consists of 108 bytes, structured as 12 columns of 9 bytes. These bytes provide a transport capacity of 6.912 Mbps each. Seven VT6s can be multiplexed into the SPE.

5.1.5 The SONET Transport System

The SONET network is the superhighway for synchronous communications. Fiber-optic cabling is the conduit for different speed ranges, which can extend from as low as 1.544 Mbps in a T-1 connection to 9.98 Gbps. The payload or the CPE is carried by the STS frame, which can be adapted to lower speed transmissions by using VT frames. A SONET network may be described as an interconnected mesh of signal-processing nodes. Each node is connected to another node by an individual transport system. Each of these transport systems may transport its payload in a different speed range. The CPE is used to transport tributary signals across the network. In all cases, the signal is assembled at the sending location and disassembled at the destination. Within the network, however, the SPE is transported without any alteration through the various branches of the SONET mesh configuration. Figure 5.6 shows a small section of the SONET network, a major node in Denver with incoming traffic from Dallas, San Francisco, and Chicago and outgoing traffic destined to Washington, D.C.

In Figure 5.6, OC-12 traffic arrives from the three incoming locations and is forwarded as OC-48 to Washington; each of the incoming signals is clocked at the origination location, and the outbound traffic is clocked in Denver. Ideally, a synchronous network should have a master clock that is used to synchronize the entire traffic of the network. This is, however, not achievable due to the propagation delays over spans of different lengths. Therefore, SONET provides for individual path clocking and adjusts the line rate at the DCS. In our example, the BITS clock of 1.544 Mbps is compared with

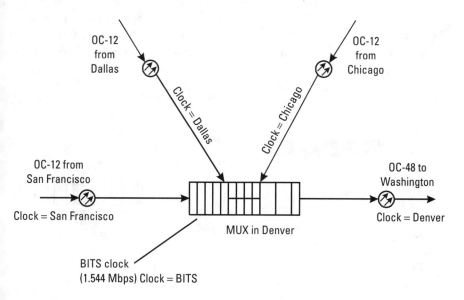

Figure 5.6 SONET networking.

the clocking rate of the incoming signals and relies on *payload pointers* H1, H2, and H3 in each line overhead to make the adjustments (see Figure 5.3).

The clock adjustment process allows the CPE of any STS-1 frame to be partially reassigned to the next frame. This can be compared with loading a portion of the shipment on a second truck, because the first truck has to leave on time. The rest of the CPE is shipped in the following STS-1 frame. The result is that the SONET network can operate asynchronously within limited clock offsets. The big advantage of this clock adjustment is the fact that the various long-distance operating companies can all use their own proprietary clocks and still interconnect via SONET in the global environment. The loading of the SPEs can be moved positively, or negatively, one byte at a time, by automatic updating of the payload pointers in each SONET node. In addition, the complexity of the multiplexing equipment is greatly reduced within the SONET hierarchy. Taking signals from STS-1 at 51.84 Mbps to STS-3 at 155.52 Mbps can be accomplished simply by *synchronous byte-interleaving multiplexing*. Figure 5.7 shows a simple example combining three basic OC-1 speeds into a single OC-3.

Byte-interleaving multiplexing is accomplished by taking one byte from signal A for the output, followed by one byte from signal B, followed by one byte from signal C. At the destination location, the byte-by-byte composition is restored to the original CPE. The STS-3 signal of 155.52 Mbps is assembled by byte-interleaving three parallel frame synchronized STS-1 signals. The result is a composite of 270 columns, or three times the number of columns of the STS-1 signal. The total signal capacity of the new STS-3 signal is, therefore, 2,430 eight-bit bytes or 19,440 bits per frame. In the same manner, it is possible to combine four STS-3 signal streams into an STS-12 stream at 622.4 Mbps. It can easily be seen that global communications can be conducted at ever-higher and scalable speed ranges.

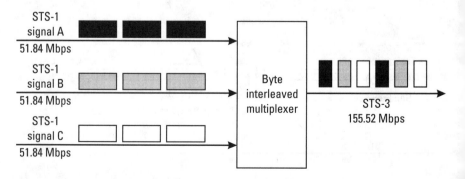

Figure 5.7 Synchronous byte-interleaving.

Newer high-speed digital services, such as FDDI, DQDB, and ATM, operate at frequencies higher than the basic DS3 SONET rate. To accommodate these rates, it is necessary to map the speed into the STS-3 framing architecture. Instead of combining three STS-1 signals, the 155.52-Mbps speed from an ATM switch is directly mapped into an STS-3 frame structure. The process is frequently called *concatenation.* As with the VT addition of lower speed traffic, any speed over 50 Mbps is directly loaded as an STS-3 transport frame. The payload portion of the 125-ms frame has a capacity of 260 columns and 149.76 Mbps. The payload overhead consists of one column and nine rows and the section and line overheads are nine rows by nine columns.

The result is that ATM at 25.6 Mbps is forwarded in an STS-1 frame, while 155.52 ATM is directly mapped into an STS-3 frame without any delineation. ATM is a cell format and can transport asynchronous traffic in every conceivable protocol. ATM is then mapped into the SONET. The combination of the two technologies becomes the ideal high-speed transport platform for voice, data, and video and for the transportation of digital information in the global environment of the twenty-first century.

5.2 Asynchronous Transfer Mode

SONET has been developed to provide a scalable high-speed routing platform for connection-oriented transmissions of information. ATM allows all existing protocols and data transfer methods that have been devised since the inception of Ethernet to travel globally at high speeds. The ATM technology can support LANs and WANs. It can integrate constant bit transmissions with variable bit traffic. It unifies data networking by being adaptable to proprietary equipment designs and by supporting telephony and video in any format or standard.

While most data transmission protocols rely on packet routing to transmit data in a broadcast mode to all parties, ATM can provide connection-oriented data transfer to only the destination location or in a point-to-multipoint environment. While IP and the Internet lack the ability to conduct any kind of traffic management, ATM-based IP transmissions can share the many advantages of ATM, such as high-speed switching, logically multiplexed connections, connection-based routing, and multilevel QoS.

The ATM architecture was originally devised by the ITU to enable a global broadband integrated services digital network (B-ISDN) connectivity. In cooperation with ANSI, the ATM Forum formed by all major global manufacturers has expanded the functionality of ATM into every area of communications and information transfer. ATM resides on top of the physical layer of a conventional layer model but does not require the use of a specific physical layer protocol. In many ways this has been quite intentional since SONET is just such a physical layer protocol. However, other physical layer protocols, such as DS3, FDDI, or the European CEPT4, can also be used.

The enormous versatility of ATM is directly related to its architecture. ATM segments and multiplexes any user traffic into small, fixed-length unit cells. The basic cell is 53 octets with five octets reserved for the cell header. Each cell is identified with virtual circuit identifiers that are a part of the cell header. An ATM-based network uses these cell identifiers to route the traffic through ATM high-speed switches to the SONET transport for delivery through route redundancy to the far-end ATM switch and to the destination point. Even in the case of an IP packet, intended for a particular server, ATM will select the shortest route to the server and not load the network with a broadcast message that is intended for a single location. ATM has been in standardization for a long time because of the many existing proprietary vendor schemes that need to be accommodated. In the period from 1984 to 1993, the majority of the interoperability standards were agreed upon and issued. However, refinements and specific operational standards are still in development. One of these incomplete areas is the IEEE 802.14 standard for RF broadband cable modem operation. While most ATM technology concerns itself with connection-oriented traffic, the use of cable modems in the local loop adds the complexity of point-to-multipoint connectivity.

This development of interoperable standards is a healthy and unifying step toward a seamless global communications environment. In the beginning, the long-distance segment was the only interest. Now, however, WAN and enterprise applications have been added. The incorporation of RF-based local loop standards can also be used for wireless point-to-multipoint applications as they will be required for broadband LMDS equipment working in the 28–31-GHz band. ATM technology will be all around us, from your desktop, your television, and your telephone at home to the corporate LAN, the enterprise, your global business partners, and your distant relatives on other continents. ATM is the only technology that can assure total multimedia integration and QoS. It is the payload layer on the SONET transport that unifies and integrates all proprietary and commonly used protocols. Accordingly, a brief description of the ATM technology is in order.

5.2.1 The ATM Cell Structure

The ATM cell consists of 53 bytes, including a 5 byte header. This cell size was only agreed upon after extensive deliberations among many working groups. Japan and the United States favored a payload cell consisting of 64 bytes, and

Europe favored a size of 32 bytes. The selection of a cell size had to be between 32 and 64 bytes because such an arrangement works with any existing equipment without requiring echo cancellation.

The cell sizing also must provide a good transmission efficiency and reduce complexities in implementation. A compromise for 48 payload bytes and the 5-byte header was reached, and the ATM cell became the global standard for multimedia transmissions. Figure 5.8 shows the cell format in the user network interface (UNI) structure.

The first 4 bits of the first byte are the virtual path identifier (VPI). VPI is the highest order of the address. VPIs can be grouped for routing and identify a bundled handling of transmission going to the same destination. The generic flow control (GFC) has not yet been defined but can be a user-controlled flow requirement. The virtual channel identifier (VCI) contains a total of 16 bits for addressing purposes. It identifies a virtual channel within a virtual path. The combination of VPI and VCI is used for mapping in the switching node and to identify the outgoing route. This switching, which may be called VP or VC switching, is used to transfer the ATM cells onto the SONET transport. PT identifies the payload type. The field's use is limited to 3 bits but suffices for the identification of eight different user data service options that are a part of the ATM adaptation layer (AAL). The cell loss priority (CLP) and header error check (HEC) are error control bits. The CLP, when set to 1, is a lower priority than 0 and will discard the cells first. CLP can be set by either the ATM access device or the switching node. HEC detects and corrects 1-bit errors. From statistical testing, it was found that 99.4% of all errors are single-bit errors, so forward error correction for single bits is included in the header. The payload field of 48 bytes is reserved for user data and can contain control fields for the AAL.

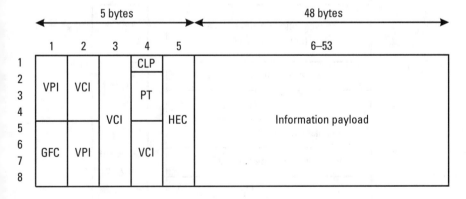

Figure 5.8 The ATM cell.

5.2.2 The ATM Adaptation Layers

The ATM reference model concentrates the AALs in the link layer section of the Open Systems Interconnection (OSI) layer standard. OSI distinguishes between physical layers, link layers, network layers, session layers, and upper or applications layers. ATM's layer structure does not cover the upper or higher layers but concentrates on the link and physical layer sections. Figure 5.9 shows the ATM reference model and its relationship to OSI.

ATM brings a different dimension to networking. In conventional networking protocols, a 256-byte packet can be sent in one link layer frame and a 2,048-byte packet can be sent in multiple link layer frames. Because ATM uses fixed-length cells, it demands a higher degree of complexity. The 256-byte packet is first broken down into multiple ATM cells; then it is assured that the cells are delivered to their destination and that they are reassembled to the original format. This adaptation requirement is applicable to all existing networking protocols and packet transmissions. This AAL process is commonly used for the support of all user data.

The packaging into small cells assures the reduction of latencies to microseconds and makes it possible to transmit voice without the requirement for echo cancellation. The adaptation layer can be subdivided into the following functions: the convergence sublayer (CS), the segmentation and reassembly sublayer (SAR), and the adaptation layers AAL1, AAL2, AAL3/4, and AAL5.

The CS contains two parts, the service-specific convergence sublayer (SSCS) and the common part convergence sublayer (CPCS). Both are a part of the AAL3/4 and AAL5, have their own headers and trailers within the 53-byte cell, and reduce the payload area of the cell to 42 to 44 bytes. The SAR also has

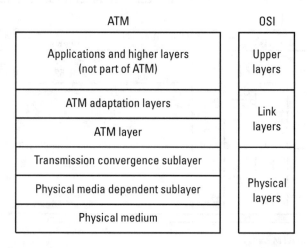

Figure 5.9 The ATM reference model.

its own 2-byte headers and trailers within the payload section. SAR is an integral part of the AAL3/4 adaptation layer. The adaptation layers are:

- *AAL1:* This adaptation layer is used for connection-oriented CBR traffic. It is suitable for voice and video as well as for traditional DS3 or T-1 transmissions. The CPCS for AAL1 uses time-stamped information for DS3 or higher speeds. Lower speeds are non-time-stamped and will be integrated to the ST3 speed for time stamping. AAL1 can accommodate any constant-rate traffic such as SNA synchronous data link control (SDLC) links.

- *AAL2:* This adaptation layer is reserved for VBR traffic that requires timing accuracy. This can be MPEG-2 bursty video traffic that requires accurate timing of all frames. The video information of a single frame may have a totally different digital content, but the beginning and end of each frame must be timed accurately. The low latency of the ATM cell structure is the only technology that can master this task.

- *AAL3/4:* This adaptation layer combines connection-oriented and connectionless traffic. It consists of CS, which is responsible for error checking, and of SAR, responsible for assembly and reassembly functions. AAL3/4 can accept and process a user payload of up to 65,535 bytes in one operation and segment this traffic into 53-byte cells for transmission. AAL3/4 is superior to AAL5 when the transfer of information is large, critical, and time-sensitive. It has the ability to correct transmission sequences by requesting retransmission of the SAR user payload.

- *AAL5:* This adaptation layer is most commonly used for all connection-oriented and connectionless transmissions. AAL5 does not have any extra headers or trailers in the user payload field. It uses the PT field in the cell header and identifies the last cell of the data unit. The 3-bit PT field indicates with a 0 that all cells are a part of a data unit. Any number above 0 indicates the last cell of that data unit. This arrangement permits that a simple disassembly procedure, known as simple efficient adaptation layer (SEAL) can be used. AAL5 is also more economical because of the higher payload utilization. Figure 5.10 shows the CPCS payload area.

Figure 5.10 indicates the 48-byte payload area of AAL5. The trailer is determined as a maximum of 8 bytes, reducing the information area to 40 bytes. Four bytes are reserved for cyclical redundancy checks (CRCs),

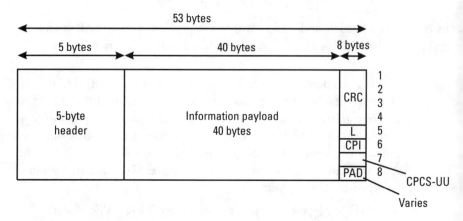

Figure 5.10 The AAL5 payload trailer.

2 bytes to express the length L of the information, and 1 byte for the common part indicator (CPI) and for CPCS user-to-user indication. The PAD area is only used when the trailer format does not add to the 8-byte format. PAD is used to fill the payload area to the required 48 bytes in cases where the trailer uses less than 8 bytes of information.

AAL5 was conceived originally for FR traffic. End-to-end QoS can be maintained with minimum overhead. It was also conceived because AAL3/4 contains unnecessary overhead. Today, AAL5 is the most commonly used adaptation concept. There are many products that use the AAL5 for CBR traffic like voice and video in conjunction with traffic synchronization operating above AAL5.

The ATM Forum is also working on additional AAL standards that may include time stamping and digital speech interpolation, but the present ATM standards are sufficiently detailed to produce interoperable equipment for high-speed transportation of any protocol that has ever been conceived. ATM is complex in nature, and it would take a more comprehensive treatment to describe the technology in detail. Some of the features and advantages of the ATM cell transport are briefly described in the next paragraphs.

5.2.3 ATM-UNI Signaling

The UNI is the standard signaling protocol for an ATM user to access the ATM network. UNI provides support for dynamic ATM setup of connectivity. The signaling procedure is based on connection sequences such as requests, responses, and confirmations. The ATM Forum's UNI 4.0 signaling features are divided between the user side and the network side. Signaling to establish a connection is accomplished on a point-to-point basis. Point-to-multipoint connectivity can also be provided by adding additional parties on an individual basis.

UNI signaling consists of a sequence of messages, starting with *setup*, followed by the *call proceeding* confirmation and the *connect call* acceptance. It then notifies the user by the *connect acknowledge* that the call setup is completed. The called user then sends the *alerting* message to the network and to the calling user, the *release* message is sent from the calling user to the network and the called user, and a confirming *release complete* message is sent in the opposite direction. If additional parties are to be included for a point-to-multipoint connection, the calling user adds the *add party* message. The added party message signaling again goes through *acknowledge, alert,* and *release* message routines, and the second connection is established. In the same manner, additional parties can be added or dropped, as required.

ATM signaling can also handle *group addressing* in the same manner. Since the group address does not include the individual user address, the successfully connected parties will return their addresses to the calling user. Signaling messages often deal with *root*, the calling party, and *leaf*, the called party. Signaling even includes an available bit rate (ABR) service. The ATM traffic descriptor can support the minimum ABR cell rate parameter. ABR setup may contain traffic parameters such as initial cell rate and rate increase and decrease factors. UNI 4.0 also supports multiple signaling channels, enabling multiple logical interfaces to be used on a single UNI interface. This virtual UNI feature can be used to segment high-speed OC-12 ports and provide shared lower speed support to many users. The resulting rate structure can offer reasonably priced WAN connectivity at lower costs. UNI 4.0 is still in evolution and will grow as the nature of information exchange becomes more oriented toward dynamic any-to-any connectivity. However, even in the present configuration, UNI 4.0 has far superior features than any IP connection will ever have, unless it is forwarded through ATM. The UNI signaling protocol of the ATM network follows the ITU standard Q.2931, which is the established global signaling standard.

5.2.4 ATM Switching

Since the ATM switch functions are standards of ATM and SONET functionalities, switching is not defined in the ATM standard. Vendors are free to implement their own switching architectures, yet are bound by the ATM transmission requirements. The ATM switch is a most important component, as it must perform cell switching without cell loss, accept asynchronous and synchronous traffic, minimize queuing and switching delays, and perform satisfactorily in its support of input and output lines operating at STS-3 speeds.

Because the switching matrices of most vendors today support gigabit backplanes, the implementation of ATM switching is directly related to the ingenuity of the product designer to incorporate flexible matrices that comply with all ATM functional requirements.

The ATM cell switch must be able to subdivide the user end into sub-STS-3 speed ranges so that 1.544-Mbps, 25.6-Mbps, and 51.2-Mbps ranges can be supported. At the WAN side, the same ranges will become important to interconnect with new offerings of LECs and CLECs. Low-speed ATM using T-1 at 1.544-Mbps rates as well as voice and data over FR are already being offered as a replacement to private line T-1 service and as economical alternatives to the user.

There are two types of switches used in the ATM/SONET network. The VPI only switches on the header information in the VPI part only; the VCI, however, operates on the entire routing field. In SONET, these operations are well defined for multiplexing and demultiplexing events within the link and the section. The difference becomes important in the long-distance segment where different operating companies may utilize different VCI formats. Seamless transfer and QoS value transfer as well as traffic usage information and ABR need to be transparent. The switching matrix may be based on proprietary design concepts, but the functionality must provide for the timing between sender and receiver, the bit rate, whether the connection is connection-oriented or connectionless, the sequencing of the payload, flow control operations, accounting for user traffic, and segmentation-reassembly of payload components. These controls are different for the various classes of service and must be maintained throughout the switching process.

5.2.5 Network Operation and Routing—Private Network-to-Network Interface (PNNI) and Network Management

Routing selects a path or paths between the source and the destination. The ATM Forum has developed the private network-to-network interface (PNNI) to set standards for interoperability among different ATM equipment vendors. PNNI was also designed to support large and global internetworking. As a result, PNNI can scale from simple private networks to the largest possible interconnectivity. PNNI defines an interface specification that supports the routing and call setup between ATM network nodes. Therefore, PNNI is going to become one of the key requirements for ATM vendors and service providers. In most of existing ATM networks, the equipment is provided by the same vendor. If different manufacturers are involved, they are interconnected at the edge of each network and do not provide interoperable routing and topology information.

PNNI specifies two key aspects for the network interface: the routing aspect and the signaling aspect. Both protocols are part of the UNI and the management protocol and above the AAL5 layer. The routing protocol is based on the traditional link-state routing protocol such as open shortest path first (OSPF), which is used in the IP routing standard. The PNNI routing protocol distributes network topology and routes information to all participating network nodes. While IP is broadcast-oriented and ATM is connection-oriented, the ATM routing protocol is more detailed and requires the network to be set up for transmission before any user information can be sent. The connection setup requires connection-related information to be transmitted to all participating nodes and to be relayed between nodes to determine the connection route. This information is transferred in the PNNI signaling protocol. The ITU-T Q2031 and the ATM Forum UNI protocols are the basis of the ATM PNNI protocol.

Recent tests of ATM switching equipment show that call setup and tear-down can be completed at a rate of more than 400 calls per second with fully meshed traffic on four OC-3 ports. Tests also show that the reconvergence time, the time to route around a failure, is in milliseconds. In contrast, layer 3 switching with OSPF is about 200 times slower. Typical reconvergence times in frame-based switching equipment are between 1.5 and 10 seconds. Call setup and tear-down as well as reconvergence times are of greatest importance to service providers and are the basis for QoS commitments and throughput, whether in the long-distance segment or in the LAN. The superiority of ATM-based cell switching over frame-based IP switching will advance the use of the technology rapidly. The reason for this 200% better performance is directly related to the 53 byte cell. Every 125 ms the connection can be monitored and the decision made to connect, tear down, or reroute. Figure 5.11 shows the PNNI routing and signaling aspects.

The PNNI addressing hierarchy follows the network topology. Peer groups are formed and *peer group leaders* are elected. The optimum number of nodes within a peer group depends on network design and the use of the application. However, PNNI can support many levels of hierarchy when using 30 to 50 nodes per peer group and can scale to tens of thousands of nodes by using three or four levels of groups. In a case where a network consists of 10,000 nodes, it can be divided by 50 to reduce to 200 groups. These 200 logical group nodes can be divided by 50 again and result into 2 groups. The peer groups are constructed recursively, the peer group leaders (PGLs) are elected in sequence, ending at the highest hierarchy level peer group at the destination end of the connection.

While UNI signaling is concerned with the establishment of a connection between the users, PNNI signaling is network- and node-related. The

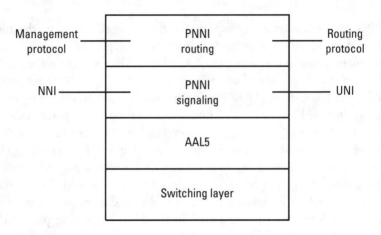

Figure 5.11 PNNI routing and signaling.

connection path computation must choose the route for the connection using the topology of the network. In addition, the connection path computation sets up the connection path support of each node along the route to accommodate the service category, the traffic type, and the QoS requirement. PNNI signaling then is responsible for the desired type of traffic (i.e., the five classes of service categories as well as the five classes of QoS and the guarantee that the connection will support the desired traffic). The user may set up the maximum end-to-end transit delay requirement in the setup message. During the transit of the setup message, each node adds the actual transit time of each hop and the cumulative transit delay value is computed. If the actual cumulative value exceeds the end-to-end transit delay requirement, the call setup is terminated and an error response provided to the caller that the QoS requirement cannot be met.

While the commonly used IP routing protocol enjoys popularity, the ATM PNNI interface specification is a much more powerful routing and signaling protocol. As the deployment of ATM networks becomes more widespread, the big advantages regarding classes of service, QoS, user protocol interoperability, connection assembly time, tear-down time, and reconvergence time will assure a steady migration toward ATM networking.

5.2.6 Traffic Management and Congestion Control

Traffic management deals with the transportation of the cells through the end-to-end network and the ability to monitor that the user is getting what he or she has been subscribed to based on the connection requirements. Traffic management concerns itself with the high speed of any typical ATM network. It reports on the transportation of the ATM cells at OC-48 or even higher speed ranges. Compared with a 28.8-Kbps connection, the ATM traffic management aspects of a gigabit connection are quite different. The network requires continuous information on bandwidth, quality of service, delay tolerance, flow control, and priority control. The end user system requires in addition route-preference information, flow-control capability, usage information, and a description of the type of traffic.

ATM service provides a traffic contract that is specific to a particular user network and the network service provider. It includes commitments regarding traffic descriptors, QoS, service categories, cell delay variation tolerance, and priority of transmission. For instance, part of the traffic parameters are the service categories such as constant bit traffic, VBR traffic, real-time VBR, and unspecified bit rate traffic. The traffic descriptors contain information on the assured peak cell rate (PCR), the sustainable cell rate (SCR), and the maximum burst size (MBS). PCR is the maximum cell rate that an ATM source can send. It is typically measured in cells per second and reflects the capability of the public network. The SCR is an average maximum rate that is measured over a longer time frame than PCR. SCR has another parameter called burst tolerance (BT), which is a function of PCR, SCR, and the maximum burst size. MBS is the maximum burst limit that can be transmitted at the peak rate and still conform to the traffic contract. In addition, there are other parameters, including the initial cell rate (ICR) and the minimum cell rate (MCR), that are a part of the ABR service class and are discussed in Section 5.2.7.

ATM traffic management includes congestion control measures that can be implemented by the user and the network. Assuming that the user has contracted for a PCR of 120 cells per second, it is required that any higher speed traffic will be buffered to meet the contract PCR value. Should a user try to send 200 cells per second traffic over the network, then the connection admission control (CAC) will refuse the call setup and inform the user that the call cannot be completed because of the PCR contract limitation.

5.2.7 Service Classes and QoS

ATM provides QoS as an integral part of the traffic contract. The QoS service parameters are different for the various classes of service. It is, therefore, possible to identify the quality of service within each of these service classes. Table 5.2 lists the classes of service defined by ITU-T and the ATM Forum.

QoS is part of the connection traffic contract. The QoS service parameters are defined by the ATM Forum in the traffic management 4.0 specifications. The QoS parameters are listed as follows:

Table 5.2
QoS and Service Classes

QoS Class	Service Class	Traffic Type
0	Unidentified	Best effort
1	A	CBR, circuit emulation, voice
2	B	VBR, audio, and video
3	C	Connection-oriented data
4	D	Connectionless data

- The maximum cell transfer delay (max CTD);
- The peak-to-peak cell delay variation (CDV);
- The cell loss ratio (CLR);
- The cell misinsertion rate (CMR);
- The severely errored cell block rate (SECBR).

CTD and CDV are measured between the source and the network and between the network and the destination. Transfer delay is contributed by cell-propagation time, cell-switching time, and cell-transmission time. Delay variations are caused by cell scheduling, cell processing, and the cell-buffering time. The delays are measured as the sum of all delays through the network. The CLR, which is the number of lost cells divided by the number of transmitted cells, is computed by the comparison of cells sent and cells received. Cells can be lost in cell errors at the physical layer or in the node-to-node cell transfer paths. The CMR is the result of the number of misinserted cells divided by the time interval. Finally, the SECBR is the result of the total of misinserted cell blocks divided by the total number of cell blocks transmitted.

It is obvious that these QoS measurements could never be guaranteed by any existing network component. None of the old analog voice transmission systems have any QoS assurances. IP broadcasting and Internet connectivity has only limited QoS parameters unless it is transmitted in the ATM cell format.

The service categories listed in Table 5.2 are offered to the user so that the best selection can be made to satisfy traffic needs. Currently, the following types of traffic are most commonly used:

- *Data:* Usually a variable load (bursty) with no real-time requirement;

- *Voice:* A consistent load with real-time requirement;
- *Video:* A consistent but bursty variable load with real-time requirement;
- *Graphics:* A variable bursty load of maximum speed with no real-time requirement;
- *Multimedia:* A variable load with a real-time requirement.

To cover these different types of traffic, the ATM Forum has determined the following traffic definitions: CBR, real-time VBR (rt-VBR), non-real-time VBR (nrt-VBR), unspecified bit rate (UBR), and ABR. Since these designations will become more important even in the enterprise LAN and when integrating voice, data, and video into an ATM facility, a few additional remarks are appropriate. While these explanations refer to the WAN and long-distance segment, they also can apply to every private network.

- *CBR service:* A CBR service with a specified bandwidth requirement. The QoS commitment is to transfer cells at a specific rate for any time and duration. CBR is used for voice and video applications that are time-sensitive. Priority together with the QoS objectives are used to establish the traffic contract. CBR is class A or QoS class 1 traffic. This category is intended for real-time applications that require a fixed amount of bandwidth, characterized by a PCR that is continuously available during the duration of the connection. This category is intended for applications that require tightly constrained CTD and tight CDV. These factors are of utmost importance in video and voice transmissions where latencies in excess of 5 ms are objectionable to the user but are also used for circuit emulation services (CES). It is the basic commitment of the service to provide a guaranteed QoS, once the connection has been established, through the life span of the connection. Typical applications for CBR service are any data/text/image transfer applications such as videoconferencing, interactive audio, telephony, audio/video distributions like TV, distance learning, pay-per-view, and video-on-demand.
- *rt-VBR service:* The rt-VBR service with a specified bandwidth requirement. It is commonly used for time-dependent applications such as VBR video (MPEG-2) that can be bursty at times. VBR traffic must support delay variations and buffering capacity to handle the bursty nature of the traffic. The QoS objective, therefore, includes a BT parameter and defines the MBS. rt-VBR is class B or QoS class 2

traffic. rt-VBR service is also appropriate for voice and video applications where the sources are expected to transmit at rates that vary with time and are considered bursty. The traffic parameters of this service are the PCR, the SCR, and the MBS. The QoS class 2 traffic permits a higher cell transfer delay, as such a delay is of a lesser importance. One of the applications of rt-VBR can be native ATM voice with bandwidth and silence compression, but its use also applies to some classes of multimedia communications.

- *nrt-VBR service:* A nrt-VBR service with a bandwidth requirement. It is commonly used for non-time-dependent applications like bursty data traffic. The QoS objective of the traffic contract includes CLR and buffering capacity but in a less stringent format because the CDVs and the delay tolerances can be widened for non-real-time operation. The QoS for nrt-VBR can be class C or D and specified as QoS class 3 or 4 service and defines BT and the MBS similar to the real-time traffic parameters. Other traffic parameters are the PCR and SCR. nrt-VBR is commonly used for any data transfer that requires critical response times. For instance, services such as airline reservations, process monitoring, or banking transactions require controlled response times, which is also an important factor in FR internetworking.

- *UBR service:* Represents an unspecified bit rate service with no QoS guarantees. It is useful for applications that do not require time dependency, that can tolerate delays, and that may retransmit lost data parts. Present Internet IP fits this definition as it does not provide any QoS parameters. UBR is classified as QoS class 0 traffic and does not have class category definition. This service is a "best effort" service that does not worry about latency and is not suited for voice or video transmissions. However, UBR service is ideally suited for noncontinuous bursts of cells that only need a maximum cell rate limitation. The PCR is the only traffic parameter of this service. Applications for UBR service include any service that is tolerant to delay and cell loss. Typically, these include any text/data or image transfer, messaging, and store-and-forward networking. The usual Internet access as well as the needs of telecommuters fall in this category. Without any QoS requirement, these services can profit from reduced tariffs.

- *ABR service:* Provides a service at an ABR with dynamic bandwidth requirements. The user takes advantage of the available network resources. Since ABR is not time-dependent, it cannot be used for CBR applications, but it assures a level of performance. Therefore, it is used mostly for data applications requiring a consistent performance.

Instead of a guaranteed cell rate, there is the MCR that guarantees the lowest transfer speed. The actual cell rate is expected considerably higher since the network is utilized dynamically, and the cell transfer will adjust to the highest value that is available at the time. The traffic descriptors are the PCR, MCR, and actual cell rate (ACR). Figure 5.12 shows the dynamic bandwidth provisions of ABR.

The ABR flow control is based on a closed loop congestion control between the source and the destination. The information is returned to the sending end and reports on all nodes along the way, relaying congestion and rate of flow information. The ABR endpoint is both the source and the destination because ATM connections are bidirectional. The ATM traffic management 4.0 allows for the segmentation of the flow reporting data. Each segment can be an ABR flow control loop with its own set of parameters. It can be seen that ABR permits the setup of many complex traffic parameters between the user and the network. If the network cannot support the contracted MCR, it may reject the connection request, and the user may try later or opt for an alternative connection route. ABR, a class C service in the QoS class 2 category, offers transfer commitments superior to the UBR service. ABR service supports any non-time-critical applications that require minimal requirements for throughput and delay, but cannot lose cells. The low CLR requirement is typical for all LAN interconnection and internetworking services between businesses.

LAN interconnections are typically run over router-based protocol stacks like TCP/IP that can easily vary their transmission rates. ABR will result in an improved end-to-end performance and forms the basis for LANE and MPOA.

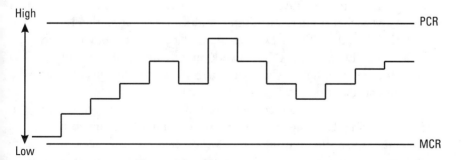

Figure 5.12 ABR cell rate traffic adjustment.

5.2.8   LAN Emulation

ATM networks within a corporate network do not have to be connection-oriented. All existing LANs operate in the broadcast mode. For instance, the Ethernet broadcast support is part of the network's physical and link layer characteristic. Existing LAN equipment supports broadcast addressing and the ability to receive packets simultaneously at other stations. ATM does not have any broadcast capability because it is a connection-oriented technology. While it is more advantageous to converge to ATM even in the LAN environment and especially in the backbone, it is becoming almost mandatory that LANs must be able to interface with ATM in the WAN and with the public network.

The need for the coexistence of existing LAN platforms and ATM is covered by ATM Forum's LANE standard, LANE 1.0. LANE is based on the concept of emulating a LAN in a seamless manner. The deficiencies of broadcasting, lack of QoS, and best effort delivery, of course, cannot be rectified by LANE. LANE, however, establishes interfaces that permit the transport of LAN traffic over high-speed ATM networks. LANE support consists of three main functions: (1) the emulation of LAN media access service to support ATM devices, (2) the connectivity support between LAN-based and ATM-based systems, and (3) the use of LAN-emulated services and ATM signaling to support LAN operations.

There are four LANE components, the LANE client (LEC), the LANE server (LES), the LANE configuration server (LECS), and the broadcast and unknown server (BUS). LANE is typically incorporated into an ATM enterprise switch/router at the WAN interface, or into backbone switches, where LES, LECS, and BUS functions provide the emulation service to the LEC for LAN operations. The LECs provide for address resolution, data transfer, address caching, interfaces to other LANE components, and an interface driver support for upper-layer functionality. LES is also a part of an ATM switch/router and provides LEC support, address registration, and resolution as well as interfaces to other LANE components. The BUS component, also a part of an ATM switch/router, distributes multicast data into the LAN, sends unicast data into the ATM network, and interfaces with the other LANE components. The LEC of one LANE network can be connected to any other ATM component and to other LECs. Figure 5.13 illustrates the operation between LANE-equipped entities.

The LEC can find the counterpart LEC's address and connect to this address by using UNI signaling in point-to-point and point-to-multipoint configurations. In the existing LAN/WAN environment SNA is used to interconnect broadcast-based LANs through remote bridges and routers. With LANE, the broadcast environment ends at the ATM switch/router, and

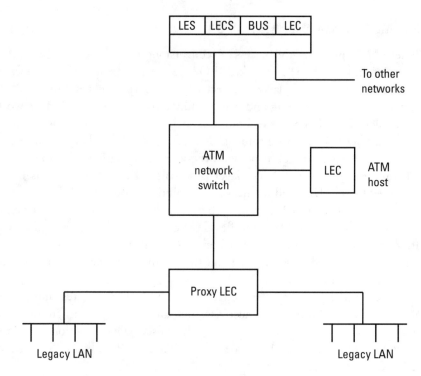

Figure 5.13 LANE operation.

connection-oriented ATM cell traffic connects to the remote ATM component for translation back to the broadcast mode LAN. Since the ATM traffic is transparent, neither LAN has to know that the connection has been made via ATM and the SNA will operate as before.

There are also VLANs in existence, using layer 2- and layer 3-based switching equipment. VLANs can also use ATM end systems together with the LANE protocol to interface with the outside world. The switching and routing tables, ports, and address caches do not change for any internal data transfer. Outbound data, however, are emulated to the connection-oriented requirements of the ATM network.

While there is pressure to convert every enterprise LAN to ATM in order to conserve bandwidth, reduce unnecessary loading from the broadcast mode of operation, and advance to QoS benefits, many LANs will continue to operate in the Ethernet, token ring, and fast Ethernet domain. LANE provides an effective tool to stop the wasteful use of bandwidth at least at the WAN interface location and use ATM connection-oriented high-speed global interconnectivity in the outside world.

5.2.9 Multiprotocol Over ATM (MPOA)

While LANE provides the necessary conversions between established LAN topologies and the ATM network, MPOA deals with the uniform support of end-to-end internetwork layer connectivity in the public network. MPOA effectively supports bridging and routing functions within the ATM network and enables all LAN network technologies such as LANE and VLANs as well as layer 2 and layer 3 network protocols. MPOA is WAN and long-distance oriented. MPOA takes the ATM network topology into consideration to support both the network-layer and link-layer protocols. The same concept is being considered by the IETF and the next hop resolution protocol (NHRP).

The key point of MPOA is to provide the best and fastest route for end-to-end connectivity. MPOA identifies short cuts to establish a direct one-hop path to the designated location. The congestions on the Internet are good reasons to eliminate broadcast loading and to determine any possible shortcut to relieve the network.

The ATM Forum specifies concepts, operational procedures, and protocols to achieve shortcut connectivity for any existing configuration of LANE, VLAN, ATM hosts, and legacy hosts. The functional groups are the MPOA server (MPS) and the MPOA client (MPC). Figure 5.14 shows the MPOA support services provided by MPS and MPC.

The MPOA server supports the operational requirements of the MPC. The MPS contains the key functions of configuration setup, address registration, and resolution for end-to-end connectivity, forwarding service support, internetworking layer support, support for QoS, and multicasting support. The MPC sets up the shortcut connectivity between the corresponding partners and provides access for the end systems that do not have direct ATM connectivity

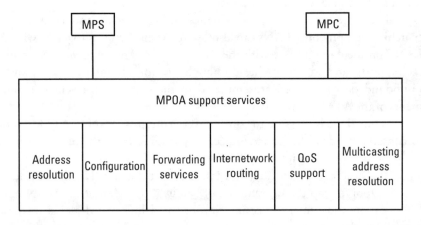

Figure 5.14 MPOA support services.

support. MPC is responsible for the connectivity between different end systems, whether they are ATM, legacy LAN, or VLAN.

The main advantage of MPOA is the ability to establish shortcuts for layer 2 and layer 3 traffic. The network traffic is analyzed, and shortcuts are set up based on the internetworking protocols, the network topology, and the different configurations. Using attached host functions and default forwarding functions, shortcuts can be established to other edge devices, to ATM attached hosts, and to any MPOA-supported router.

In the legacy LAN environment, the main characteristics are broadcast and multi-access capabilities. These capabilities are wasting bandwidth and need to be adopted to the unique characteristics of the connection-oriented ATM/SONET technology. MPOA's framework indicates that bridging and routing functions can be synthesized in the ATM environment. The key objective of MPOA is to reduce the bandwidth of IP-oriented traffic and to determine shortcuts between client servers and destinations over the ATM network. Both NHRP and LANE are used for this process. MPOA provides a broader scope than NHRP and LANE. It is more efficient and streamlines the legacy LAN connectivity in the long-distance segment while saving bandwidth.

5.3 ATM—Interoperability and Multiservice Technology

The fact that ATM/SONET can handle legacy LAN traffic over the long-distance segment assures that this technology will become the predominant technology for global communications. Even IP-oriented Internet service providers will sooner or later adapt ATM because of the higher speeds, bandwidth conservation, low latency for voice and videoconferencing, and multiple QoS choices. ATM network enterprise switches will also become the choice of many enterprise networks requiring WAN connectivity. The cost savings attainable from switching from private line T-1 service to low-speed ATM will pave the way.

Within a campus or a single enterprise, there are many reasons to change to ATM, but the progress may take time because of the capital investment already made. However, the advantages of voice/video/data integration, QoS, interoperability, and conservation of bandwidth will sooner or later pave the way toward total ATM connectivity. This does not mean that the Ethernet card in the workstation has to be changed. LANs can be upgraded in stages and use Ethernet to the equipment closet, switch to fast Ethernet in the riser, and use ATM in the backbone. The interconnectivity possibilities are endless and directly related to existing topologies and future demand. To better assess the flexibilities of converting the existing LAN to ATM, it is appropriate to review

ATM interoperability with legacy protocols and services in the long-distance segment.

5.3.1 ATM and Frame Relay

FR is a commonly used service between corporate offices. For WAN interconnectivity between branch offices, FR service offers a reasonably priced solution in most cases. It can be used for data traffic between LANs and for digital voice traffic. FR uses a frame-based, bit-oriented protocol that can be mapped easily into the ATM cell-based protocol.

FR is a modified *link access procedure-D* that is similar to high-level data link control (HDLC) except that it does not contain any control characters. Address length can vary between 2, 3, and 4 bytes to accommodate more addresses. The adaptation of frame-based FR into the cell structure of ATM is handled by the FR convergence sublayer and by the common part convergence sublayer for AAL5. FR is well established for data transfer and linking LANs in enterprise networks. For time-sensitive traffic and for QoS applications, ATM is better suited and provides a transparent handling to the user. Corporations can still use FR and, at the same time, take advantage of the ATM features. Internetworking is also provided over the ATM network and permits one FR network to connect with another. The biggest advantage to the enterprise, however, is the ability to use FR for both voice and data traffic. An ATM-based enterprise network switch can interconnect with the PBX and deliver digital voice into the FR network. In addition, the new voice compression techniques can be utilized to provide five voice channels in the 56-Kbps bit rate of a single channel. The high-speed ATM over FR network can also accommodate internetworking with high-speed LANs and provide for real-time imaging transfers such as medical imaging.

5.3.2 ATM and TCP/IP and the Internet

The Internet, which was started around 1980 by the Defense Advanced Research Program Agency (DARPA), uses the TCP/IP protocol throughout. The Internet is a diagram-based communications network consisting of interconnected networks. Internet systems consist of host and router combinations that are interconnected by a suite of protocols called request for comments (RFCs) specifications. Proper operation among the Internet systems requires that the systems implement protocols and operational rules as defined by the related RFCs. These RFCs are a part of the Internet official protocol standards and include recommendations for user datagram protocol (UDP), file transfer, domain names, address resolution and reverse address resolution, simple mail transfer, Internet control message, and transmission control. Other RFCs include simple network management protocol (SNMP), the serial line Internet protocol, the routing information protocol (RIP), the point-to-point protocol (PPP), and requirements for Internet host communications. RFCs are grouped in three different categories: (1) the required category that must be implemented for all systems, (2) the recommended category that should be implemented, and (3) the elective category that can be implemented when needed.

TCP/IP consists of two protocols: the transmission control protocol (TCP) and the Internet protocol (IP). TCP/IP began as a requirement to interconnect different types of computers. TCP/IP has three major components: the internetwork, the transport, and the applications. These components are arranged similarly to the OSI seven-layer model where the network layer is responsible for the routing of the data to the destination network. In IP, the network layer is commonly referred to as the subnetwork layer and is dependent upon the link protocol and the physical transmission media. The subnetwork layer depends upon the system types, like an Ethernet-related subnetwork, but supports Ethernet interfaces. The internetworking layer is commonly called the IP layer. Because the interface to the layer under IP is implementation-specific, TCP/IP can be implemented in different kinds of link layer protocols and network service protocols, such as X.25, wireless, FR, logical link controls, and ATM. IP is simple, connectionless, and best-effort. There are no QoS, flow control, reliability, or error recovery mechanisms. The address resolution protocol (ARP) is used to find the hardware component for an IP address. The request is broadcast to everybody and the destination party replies with its hardware address. The IP datagram format consists of 24 octets and 31 bits all used for addressing and service functions. The TCP service uses the IP datagrams to set up logical connections for the exchange of data. TCP supports an end-to-end logical connection setup and a reliable data transfer. The TCP header format, again, consists of 24 octets and 31 bits.

The ATM network uses ATM network protocols such as PNNI to support its network operations. This means that both the Internet and ATM know how to route their individual connections. It is therefore possible to use ATM as one of the subnet protocols under IP. However, using two separate network protocols causes excessive overhead especially when a more integrated approach can be defined. Such an integrated approach has been proposed by the ATM Forum PNNI working group. Called I-PNNI, this integrated approach provides a routing protocol that can be used for the simultaneous support of IP packets in an IP-over-ATM network environment. I-PNNI extends the ATM network protocol to accommodate IP routing information. IP routers that support I-PNNI also run the PNNI protocol and interface to all network ATM switches. Speed is the big advantage of I-PNNI. The ATM network can speed the transportation of IP within the ATM cell structure and, using the SONET OC-12 to OC-192 speed ranges, improve the TCP/IP connectivity by several magnitudes. This speed increase and the seamless integration of voice and video services are very much needed by existing and future ISPs, by the exponential growth of Internet communications, and by the ever-increasing requirements for quality voice and video services.

5.3.3 IPv6 and ATM

IPv4 was the classic Internet IP version. IPv6 represents the next generation IP version. The many deficiencies of the present Internet protocol include (1) a restrictive 32-bit address range, (2) a limited topological hierarchy, (3) a restricted IP header structure, and (4) the lack of quality and priority support. The key technical criteria for the next generation IP protocol is the RFC 1726, which accommodates the following changes:

- An increase of the address size to 124 bits;
- Unicast, anycast, and multicast addressing;
- Robust network services and associated routing and control protocols;
- Scaling to 1,012 end systems and 109 networks;
- Application of different routing architectures;
- Support for any kind of physical media and link speeds;
- Support for wireless hosts and networks;
- Autoconfiguration of addresses.

Without going into the details of the proposed improvements, IPv6 does not interfere with the ability to integrate with ATM. Similar to the I-PNNI protocol for present Internet technology, the ATM network support can be scaled to operate the next generation of Internet service. While the overhead increases due to the additional header and addressing requirements, the speed advantages and scalability of the ATM/SONET technology will be able to satisfy any new requirement for the Internet traffic of the future.

5.3.4 System Network Architecture in ATM

SNA is widely established in enterprise networks. SNA, a network architecture originally developed by IBM, has been widely used in token ring networks. Especially in WANs, SNA is still the backbone transport protocol for most large corporations. SNA combines three key areas of standardization: the format definition, the defined protocol, and the layering approach. It is interesting to note that the OSI seven-layer concept is based on the SNA layering approach.

Figure 5.15 shows a comparison of the seven layers between OSI, SNA, and TCP/IP.

The ATM layers do not map directly with the OSI layers. The ATM layer operates in layer 2 and 3 of the OSI model. The AAL combines features of layers 2, 4, and 5 of the OSI model. ATM is not a clean fit, nor does it have to be. The OSI model is over 10 years old and needs to be adjusted to reflect emerging technologies.

The SNA layers consist of layer 1, the physical layer, which defines the physical and electrical interfaces; layer 2, the data link layer, which defines the reliability of the data bits on the physical layer; layer 3, the network layer, which defines how to route the data bits to the destination location; layer 4, the transport layer, which provides reliable transfer of data between two endpoints; layer 5, the session layer, which enables applications to establish conversations

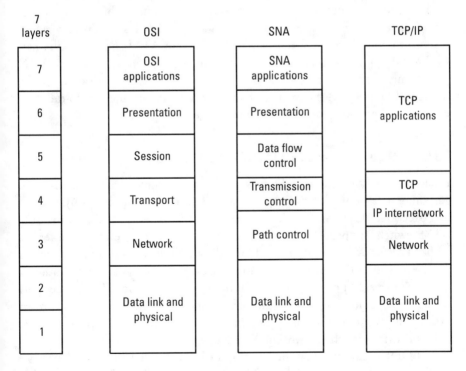

Figure 5.15 The seven-layer comparison—OSI, SNA, and TCP/IP.

on top of layer 4; layer 6, the presentation layer, which provides applications with mapping of data representation; and layer 7, the application layer, which provides user applications and network management.

Because an SNA network of one company does not have to interface with the SNA network of another company, SNA does not feature global addressing modes. Every SNA system has its own addressing that is dependent upon hierarchy networking and central control by the master of all hosts. For the interconnection of two SNA systems there is the SNA interconnection feature that can provide a gateway function. SNA includes dynamic directory updates, a class-of-service based route selection, flexible flow control, high performance routing, connection-oriented protocols, and various other features that make SNA a reliable and flexible network architecture.

When applying ATM to the SNA network, it is necessary to encapsulate the multiprotocol over the adaptation layer 5. The ATM Forum's document is entitled "Multiprotocol over AAL5 Implementation for SNA." SNA over ATM follows RFC 1483 with additional parameters for layer 2 and 3 protocol identifiers. ATM provides end-to-end support through a controller and the host front-end processor and can communicate at higher speeds. ATM has a much higher bandwidth and different traffic control mechanisms, but every form of traffic such as VBR, PCR, sustainable cell size, and MBS parameters can be taken into consideration when mapping SNA into ATM. Because SNA networks and applications are a major factor in the corporate LAN and WAN environment, an early migration of SNA into ATM is quite advantageous to the user. Using ATM, the many advantages can be incorporated into the SNA network and provide global connectivity to the user.

5.3.5 ATM and Ethernet: Fast Ethernet, FDDI, and Gigabit Ethernet

Ethernet, fast Ethernet, and the new gigabit technologies are the mainstay of corporate data networks. Both gigabit and FDDI are technologies well suited for backbone LAN traffic. The data network vendors are all providing ATM switch support for these technologies at the gateway to interface with ATM for WAN transport and long-distance traffic. Most of the activities associated with ATM focus on how to make existing protocols run without any changes in an ATM environment. This transparent access to ATM is a key part of the ATM technology. ATM can offer a smooth migration path to convert existing infrastructures to connection-based connectivity and to integrate voice, data and video into the networking capabilities.

FDDI backbones may be the first to be integrated into ATM because FDDI can be mapped into ST-3 ATM quite easily. ATM is designed to end at the desktop, but that is not a requirement. ATM could form the backbone traffic in the equipment closet, and every desktop can interconnect using the inexpensive Ethernet cards over CAT-5 wiring. Whether the interface with ATM is at the wiring closet, the building entrance location, or at the network center is a decision that the network manager has to make.

RF broadband and the use of cable modems can reduce the complexity of the LAN and concentrate all core, edge, and enterprise switching equipment at a single location. If this location is the PBX location, voice and data integration can use the fiber backbone and replace all copper wiring. Every decision is scalable and can be made on a step-by-step basis. The driving forces will be the desire to integrate the network into a multimedia facility, the budget constraints, and the demand for speed, QoS, and traffic control. It is certain, however, that WAN and long-distance connectivity will be the first elements applying ATM. ATM is developed to run from the desktop through the WAN in a transparent manner, a big advantage compared to legacy LAN networks. Already, there are WAN service offerings for low-speed T-1 ATM available that can drastically reduce the T-1 private line charges. The future will bring 25.6-Mbps ATM service to the WAN and greatly improve the connection speed between branch offices. New VPN services, using ATM in economical interconnections between branch offices of an enterprise, are fast becoming an economical replacement to private line leases. The migration of ATM into the corporate LAN is just a matter of time. Bringing native ATM to the desktop and using the wire for telephony and for MPEG-2 video may easily become the more common technology within the next decade.

5.3.6 ATM: Video and Telephony

The application of digital video and telephony formats over ATM/SONET is a logical progression in the long-distance segment for the next decade. There is also no reason why this development has to stop at the entrance to the corporate network. ATM offers the ability to map telephony as well as digital video in all formats—that is, MPEG-1, JPEG, MPEG-2, ATV, and HDTV—without any difficulties and at speed ranges never before attainable. Every video format requires VBRs and time sensitivity for 30 frames per second. Telephony can now be carried as an adjunct to data at 16 Kbps or less and is no longer limited to the 24 channels per T-1. It is a foregone conclusion that ATM/SONET in the WAN and in the long-distance segment will become a true multimedia high-speed connectivity for all communication services inclusive of IP.

Within the corporate network, the migration to ATM-based multimedia solutions may take a slower path, but it is certain that most of the larger corporate networks will adopt the ATM technology. At present, the PBX industry is going through the adaptation of digital technology, but it is expected that, within the next three years, there will be ATM-based PBXs on the market. With the outside world changing to ATM, it is logical to extend the technology to the extension lines. The switch will be based on ATM technology and provide digital telephony to the set or even to the workstation. It is easy for PBX manufacturers to offer data services as well and the time may come when PBX vendors will be in competition with the data network vendors. Data network vendors have already formed alliances with ATM-based telephone providers. The un-PBX concept that promises CTI shows the direction of the data networking vendors. It is already possible to purchase a network interface card (NIC) card, install it in the computer, and connect an analog telephone set to the computer. These systems run at 25.6-Mbps ATM, which is a good speed for combined voice and data transmission and also routable through a cable modem.

On the video side, it has been proven that 56 Kbps and even ISDN video does not measure up to the picture quality that we are used to seeing on our television sets. JPEG encoding is available for transmission over ATM networks and can deliver quality motion video at different resolutions and at constant bit rates changing the Q-factor for optimized picture quality. MPEG-2 will be available in a very short time frame and provide quality full-motion video that finally can be used for videoconferencing. MPEG-2 is a VBR transmission that requires between 5 to 8 Mbps. It can become a small loading factor on a 25.6-Mbps ATM circuit, even when integrated with data and voice on a CAT-5 wire. This 25.6-Mbps ATM circuit could serve a single workstation or a group of workstations depending upon the bit rate requirement of each of the workstations. As it is unnecessary to bring gigabit technology or fiber to the desktop, it is not wise to assume that the corporate integrated ATM network will only use a single wiring system. In most organizations, separate copper and fiber plants are available for separate voice and data networks. ATM integration does not mean a mandatory migration to a single wiring system. It is more likely that the existing wiring will be used for separate voice and data networks, but there will be interfaces between these two networks. There will be the call center that requires CTI, the heavy Internet user that needs multimedia connectivity at the desktop, the corporation that incorporates videoconferencing into the daily conduct of business, and the SOHO LAN that requires the combination of all media.

ATM at OC-3, 12, and 48 will be the choice for the corporate LAN backbone, while low-speed ATM and OC-1 will serve the WAN. This topology may further be altered by the use of IP- and ATM-compatible cable modems. In this scenario, the ATM backbone switch can become the central switching center for an integrated multimedia traffic. The centralized ATM switch combines PBX and data network functions and interconnects with workstations and phone hubs through 25.6-Mbps ATM cable modems. This centrally located star architecture looks like the old PBX, using single-mode fiber in the backbone and existing CAT-3 and CAT-5 wiring in the horizontal distribution. ATM, while considered expensive today, offers economy of scale savings in the future. The network becomes simpler, more flexible, more scalable, and easier to manage.

5.4 ATM Forum Specifications

The ATM Forum was formed in 1991 to determine the implementation of ITU-T and ANSI specifications relative to B-ISDN convergence, multiplexing, and switching operations. The ATM Forum's membership currently includes 900 companies representing the computer and communications industries, government agencies, research organizations, and end users. It is the charter of the ATM Forum to speed the development and deployment of ATM products and to develop interoperability specifications for equipment and operation.

The ATM Forum is an international nonprofit organization, headquartered in Mountain View, California, that consists of a worldwide technical committee; three marketing committees for North America, Europe, and Asia-Pacific; as well as an enterprise network roundtable, which concerns itself with end-user participation. There are 10 working groups and several ad hoc groups that are diligently at work to achieve the ATM Forum's objective to accelerate the deployment of ATM products and services throughout the world and to offer a rapid convergence of the global communications and product interoperability. The number of approved specifications is increasing every year, proving that ATM has become the universal communications technology. Without listing the details of each of the specifications, the following is a description of the activities of the more important working groups.

- *The physical layer working group:* Has evolved to reflect the need to develop a family of optical and electrical physical layer interface specifications. The list of these physical layer specifications has grown to 17 as of 1997 and is now concentrating on high-PHY interfaces required for OC-48 to OC-192 rates. The approved PHY interface specifications include any service on CAT-3 to 51 Mbps, on CAT-5 for 155.2 Mbps, and 1.544 Mbps (DS1) service to SONET interfaces up to OC-48. There are a total of 18 approved specifications issued by the physical layer working group that are summarized below with their identification numbers:

- Physical Layer, covering 25.6–622.08 Mbps interfaces for SONET, FDDI, UTP 3/5, and fiber: af-phy-0015/16/17/18/29/34/39/40/43/46/47/53/54/62/63/64.000;
- Inverse Multiplexing -T-1/E-1: af-phy-0086.000.

- *Signaling working group:* Responsible for the development of communication protocols to support switched services across the UNI and the PNNI. These protocols establish signaling standards that allow the user to communicate to any public or private network. The UNI specifications are aligned with ITU-T recommendations and under version 4.0 include group addresses, point-to-multipoint connections, multiconnection calls, narrowband ISDN internetworking, alerting for multimedia calls, and PNNI signaling protocols. There are seven UNI and two signaling specifications listed below with their identification numbers:

- UNI: af-uni-0010.000/1/2 and af-uni-0011.000/1;
- Signaling UNI: af-sig-0061/76.000.

- *PNNI working group:* Responsible for private switch-to-switch communications, private network-to-network on demand connections, and multiprotocol routing for overlay, peer, integrated PNNI, layer 3 address resolution, single edge-device layer 3 address resolution, and coordination of separate PNNI actions on PNNI route servers. A total of four specifications have been approved. They are listed as follows.

- PNNI: af-pnni-0026/55/66/75.000.

- *The network management working group:* Develops interface specifications for network management information flow between management systems and the network. In addition, there are interfaces across administrative boundaries of private-to-public and public-to-public networks. The M5 specification will define the interfaces between network management systems of different network operators. The M3 and M4 specifications are covered by nine different specifications:

- Network Management-M4: af-nm-0019/20/27/58/72/73/74.000.

- *The traffic management working group:* Responsible for the service categories and their continuous improvements. A VBR+ category has been added that specifies a minimum bit rate and is based on flow control. VBR+ corresponds better to the need of the LAN to access the available bandwidth in a highly predictable manner. Data exchange interface (DXI) and the integrated local management interface (ILMI) are included in the following listing of completed specifications:

- Traffic management 4.0 and ABR: af-tm-0056.000/1 and af-tm-0077.001;
- DXI- 1.0: af-dxi-0014.000;
- ILMI 4.0: 5.61, af-ilmi-0065.000.

- *The broadband intercarrier interface (BICI) working group:* Develops BICI specifications for private virtual and switched virtual connection services between multicarrier ATM networks and operators. The work consisted of enhancements and changes made by ITU-T in the network signaling area and by the addition of usage metering requirements. The completed specifications are listed as follows:

  - B-ISDN intercarrier interface (BICI): af-bici-0013.000/1/2/3 and af-bici-0068.000.

- *The service aspects and applications working group:* Develops specifications that enable existing applications to run over ATM-based networks to support new applications and allow the interworking of existing LAN/WAN services. Application program interface (API) and audio/visual multimedia services (AMSs) involve the service requirements for broadcast video, videoconferencing, and multimedia to the desktop and include specifications for MPEG-2 support. Former efforts included the interworking of FR and DS1/E1 as well as DS3. The completed specifications are listed below:

  - Service aspects and applications: A/V-multimedia, names, native services, circuit emulation and FUNI: af-saa-0031/32/48/49/69/88.000.

- *The LANE working group:* Responsible for the development of seamless interconnectivity of legacy LANs over ATM. Emulation architecture, format, backward compatibility, initialization actions, and how an emulated LAN is recognized are a part of the activities as well as network management aspects. The approved specifications are listed as follows:

  - LANE and LUNI: af-lane-0021/38/50/57/84.000.

- *The voice and telephony working group:* Has recently completed all required specifications to enable efficient communications between existing telephone instruments across the ATM network. It is expected that long-distance carriers will utilize ATM/SONET capabilities in their networks and that ATM voice-based telephony will enter the local loop on short notice. Dynamic bandwidth utilization in 64-Kbps time slots and CESs have recently been added and provide support for legacy circuit services. The specifications are listed as follows:

  - Voice and telephony over ATM: af-vtoa-0078.000 and af-vtoa-0089.000.

- *The MPOA working group:* Responsible for the internetworking of bridges and routers supporting shortcut connectivity over ATM networks. The goal is to improve and optimize the performance of large, router-based networks. The work includes the streamlining of the Internet and its conversion to an ATM/SONET foundation that promises gigabit speeds in the global trunking segment.
  - MPOA: af-mpoa-0087.000.

- *The testing working group:* Has developed numerous testing specifications that cover the range of compliancy and interoperability testing sequences. Every aspect of the ATM transport mechanism has been covered to assure conformance to specifications and performance. The present scope of 18 specifications is listed as follows with the respective identification numbers:
  - Testing specifications: af-test-0022/23/24/25/28/30/35/36/41/42/44/51/59/60/67/70.000.

- *The residential and small business working group:* Created within the North America market and education committee of the ATM Forum to identify and evaluate user applications that could benefit from the use of ATM, to provide QoS services to residencies, and to improve the interoperability of future local loop services such as HFC cable providers and wireless LMDS services. A number of services are being addressed by this working group, including information-on-demand, home shopping, distance learning, data delivery, telecommuter services, home-based business communications, video-on-demand, video telephony, and multimedia-on-the-desktop. The work, of course, addresses the utilization of ATM-compatible cable modem equipment to offer ATM-based integrated voice, video, and data services to the residential level. One of the most important cable modem specifications is the IEEE 802.14 standard, which will be discussed in Section 5.5.

The above listing only contains approved specifications through August of 1997. The ATM Forum continues to add new specifications, but the existing specifications prove that ATM can be implemented across every aspect of communications from the desktop to the global carrier. Additional specifications will help to assure interoperability between vendor products and make it easier for the user community to migrate towards ATM.

ATM technology over the SONET physical layer is truly the new world communications standard. The European Market Awareness Committee

(EMAC) coordinates all ATM Forum activities in Europe. Planning and implementation of ATM/SONET interconnections between all common market countries is proceeding on schedule.

Separately, the North America Market Awareness Committee (NAMAC) has announced that it is expanding its focus to include Latin America. For the first time in the short history of the development of digital communications it is possible to rely on standardized performance specifications. For the first time in the history of telecommunications, the use of proprietary technologies has been eliminated through a set of globally accepted specifications that promise interoperability between vendors and that assures the implementation of ATM across multiple service providers from the desktop to any desktop at any location on the globe.

5.5 Cable Modem Standards and ATM

5.5.1 The Residential Broadband Working Group

The Residential Broadband (RBB) working group within the ATM Forum has developed guidelines for the transportation of ATM over RF broadband HFC networks. The main charter of the RBB working group is to define and complete a set of specifications for an end-to-end ATM transfer over HFC. These specifications include the access network interface (ANI), the UNI, and the technology-independent interface (TII). The UNI interface will be able to support not just HFC, but all other local loop technologies such as FTTC, FTTH, VDSL, ADSL, and even the new wireless LMDS technology. The goal of TII is to use existing forum-approved specifications and to work closely with IEEE 802.14 to define the MAC and the physical layers for the HFC network. Figure 5.16 shows the ATM Forum's reference model for the RF broadband HFC network.

5.5.2 Institute of Electrical and Electronic Engineers—IEEE 802.14

The IEEE is an international standards body affiliated with ANSI. The IEEE 802.14 working group is responsible for digital communications services over HFC as well as the use of cable modem technology in RF broadband cable television branching bus topologies using fiber and coaxial cables. IEEE 802.14 concerns itself with bidirectional traffic using CBR, VBR, as well as connection-oriented and connectionless data streams. This means that any type of protocol must be transferable over the HFC system using

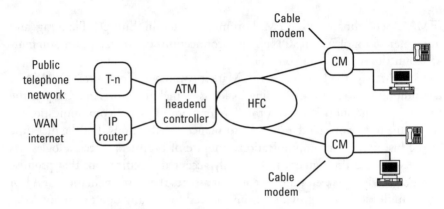

Figure 5.16 The ATM Forum reference model for HFC.

specification-compliant cable modems. IEEE 802.14 covers the standards of a cable modem to support every conceivable bidirectional transmission, a category that may include telephony, VOD, regional video servers, information databases, telecommuting services, Internet access, and videoconferencing. The IEEE 802.14 standard is devoted to the interconnection with global ATM-based communications networks through the headend of the serving cable provider. Figure 5.17 illustrates the IEEE 802.14 reference model.

It is interesting to note that a headend can also be the gateway of a campus or an enterprise system. In both cases, the HFC or RF broadband network stops at the headend/gateway and is translated to ATM/SONET regional, WAN, and long-distance service providers. IEEE 802.14 is the responsible standard for the development of the MAC and physical layer that will ensure stringent performance requirements to meet the QoS service requirements of SBR, CBR, VBR, and UBR on an end-to-end basis between the local loop HFC provider and any other global destination location over the cable modem.

5.5.3 The Digital Audiovisual Council

The Digital Audiovisual Council (DAVIC) is a nonprofit international organization comprised of conglomerate international companies to standardize the network interface unit and the set-top converter. DAVIC is organized in working groups that concern themselves with the connectivity between video servers, the network, the delivery systems, and the end-service system at the consumer premises. All major access technologies, such as HFC, FTTC, FTTH, and LMDS wireless, are addressed and contain specifications relative to the fiber and coaxial segments of the network. Figure 5.18 depicts the network

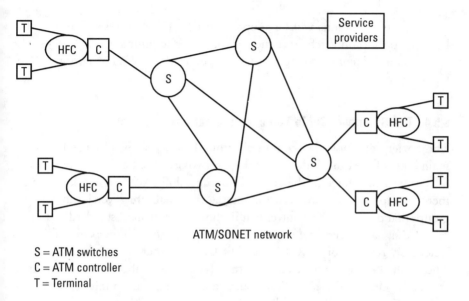

S = ATM switches
C = ATM controller
T = Terminal

Figure 5.17 The IEEE 802.14 reference model.

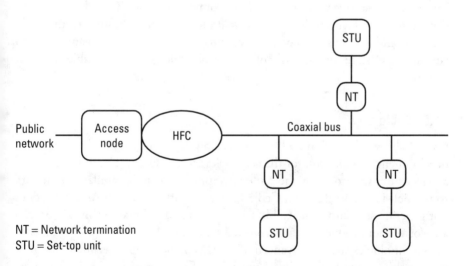

NT = Network termination
STU = Set-top unit

Figure 5.18 DAVIC-HF access.

terminations (NTs), the connection to the consumer (A1), and the set-top unit (STU).

The fiber extends from the access node to the neighborhood node and changes to coaxial cable for service to the subscriber. IEEE 802.14 MAC and PHY specifications will be adopted by DAVIC. At the set-top unit, the DAVIC specifications reference the network interface unit at the NT. This reference point is identical to the TII interface defined in the ATM Forum's RBB reference model.

5.5.4 The Society of Cable Telecommunications Engineers

The Society of Cable Telecommunications Engineers (SCTE) has been the main body of the cable industry for the last two decades. During the past years, SCTE has been active in the areas of interoperability of cable modem equipment, set-top units, and network interfaces. There are several subworking groups that deal with the involvement in the development of standards for digital video signal delivery and the coordination of efforts with the National Cable Television Association (NCTA), the FCC, and other related organizations. These subworking groups consist of the digital video subcommittee, the data standards subcommittee, the design and construction subcommittee, the interface practices and in-home cabling subcommittee, the emergency alert systems subcommittee, the maintenance practices and procedures subcommittee, and the material management/inventory subcommittee. The SCTE has been the driving force to bring technical coordination and standardization to the cable industry and was instrumental in forming the Cable Labs organization to advance the state of the art of the cable television and cable modem development.

5.5.5 Cable Labs

Cable Labs was created in 1988 to responsibly act as the technical research and development organization for the cable television industry. The organization is a consortium sponsored by the cable operating companies. Similar in structure to the Bell Labs and Bellcore, Cable Labs is not affiliated with ANSI and does not generate standards, but Cable Labs checks and enforces technical compliance of vendor products with established standards. The organization addresses new technologies such as advanced TV, HDTV, operations technologies, network architectures, and the interoperability of cable modems. In the *new services* initiative, Cable Labs is responsible for the development of high-speed Internet access, VOD, PCS, computer program exchanges like distance learning, videoconferencing, and interactive educational services. The *Network Architecture, Design and Development Group* focuses on the regional hub concept. Based on SONET technology, the hubs will employ a dual rotating interconnection and serve individual headends that can serve as the platform for integrated voice, data, and video services.

Working in conjunction with Multimedia Cable Network Systems (MCNS) Holdings, L.P., a group of major cable television operators such as Comcast, Cox Communications, Tele-Communications, Inc., and Time-Warner Cable, Cable Labs is administering the development and deployment of the cable modem technology supporting high-speed and high-bandwidth properties. The project called data over cable service interface specifications (DOCSIS) consists of an expanded MCNS group of companies determined to complete the standardization effort of cable modems in the shortest possible time frame. DOCSIS defines the interoperability specifications for cable modems between the subscriber location and the headend. The CMRFI specification describes the single-port RF interface between the coaxial drop of the HFC system and includes all physical, link (MAC and LLC), and network level aspects of the communications interface. The specifications address RF levels, frequency modulation, coding, multiplexing, contention control, and frequency agility. At the headend, the cable modem downstream RF interface (CMTS-DRFI) and the upstream RF interface (CMTS-URFI) are two separate ports on the cable modem termination system equipment. One port sends data signals to the downstream RF path, the other receives data signals arriving in the upstream direction. Security over HFC is defined in the specification called data over cable security system (DOCSS). The specifications will benefit high-speed connectionless and ATM-based data services over RF-broadband HFC networks. The approved interoperable MCNS-compliant cable modem is discussed in more detail in Chapter 9.

5.6 ATM Applications for RF Broadband LANs

The migration toward ATM in the WAN and long-distance segment is proceeding at a rapid pace. It will produce the next generation Internet at higher speeds and with a multilevel QoS component. The entrance of ATM into the enterprise network and into the campus LAN domain will follow because of bandwidth limitations, the desire to increase "best effort" transmissions to measurable QoS standards, and the many benefits that can be obtained from a step-by-step integration of voice, data, and video. The existence of single-mode fiber in the backbone of the corporate LAN opens new avenues for the integration of ATM and HFC-RF broadband technology. The step-by-step scalability of RF broadband offers a new dimension for the networking in larger campus LANs and in smaller corporate LANs. The network manager can simplify the hierarchical architectures of established LANs; integrate voice, data, and video; and offer economical services and measured bandwidth to each workstation, while enhancing LAN operations with QoS and flexibility. Corporate applications include cable television distribution, high-quality videoconferencing, VOD, CTI, ATM-based PBXs, and high-speed QoS-based data connectivity within the corporate LAN as well as ATM-based interfaces in the WAN and in the global long-distance segment.

6

ATM-Based Network Architectures and RF Broadband

Whether we are looking at the public sector, the local loop, or the private sector, the networking architectures are changing. Based on the technological advances that have been accomplished by using SONET/ATM solutions, the trend for the next decade will be to build global multiservice networks. However, the term *multiservice* can be deceiving. Multiservice has often been used to describe a network that combines data, voice, and video in one network, but multiservice networking also involves the integration of legacy protocols and every interfacing issue with other networks.

A multiservice network architecture based on an ATM core with multiservice edge switching topology is the optimal approach for every service provider. The term *service provider* is used here in the broader sense. Whether the service provider is a long-distance carrier or a network manager in a private corporation, the duty is the same—to provide communication services to its end users. Employing ATM in the backbone of any network takes advantage of the best known methods for handling data, voice, and video in an integrated network architecture. QoS is not only important for cell or packet switching; it is also important for both real-time and non-real-time traffic. Only ATM has been designed from its inception to support any traffic format at various speeds and with full assurance of quality standards.

Before discussing the place of the RF broadband technology in the multiservice network, it is important to summarize the utilization of ATM in the global multiservice network and the implications of this development to enterprise networking and campus networks.

6.1 The Global Multiservice Network

The emerging global multiservice network architecture allows the integration of multiple LAN and WAN protocols into a common networking environment. ATM traffic can be accommodated over many types of interfaces and at speeds from a few kilobits to almost 10 Gbps. Protocols such as X.25, SNA, IP, internetwork packet exchange (IPX), Ethernet, fast Ethernet, gigabit Ethernet, token ring, FR, and packetized voice can be accommodated using the ATM cell structure and the global SONET transport system. Figure 6.1 shows the global multiservice network architecture in a simplistic manner.

The center circle in Figure 6.1 represents the DWDM SONET/ATM global network core that will extend throughout the world and that will be shared by a multitude of terrestrial and submarine cable service providers. The SONET/ATM fabric can easily interface with satellite-based service providers, such as the new LEO, MEO, and GEO service providers; mobile service providers, such as cellular and PCS services; LMDS wireless providers; legacy telephone network providers; Internet service providers; as well as xDSL service providers and local cable television companies. The SONET/ATM network core, of course, supports any corporate, enterprise, or campus network in any format or protocol that was ever conceived during the short evolutionary period of data networking.

SONET/ATM multiservice networking will finally provide a rich competitive element in the local loop; offer significant operational savings to local, regional, and global corporate enterprises; and cause a migration toward an ATM integration of all services within even the smallest campus or corporate entity. The ability to use ATM multiservice within the WAN of an enterprise and the integration of voice and data already reduces the corporate communications expenses incurred by the high costs of private line and T-1 leases. The rates for low-speed ATM service, offered by LECs in some areas of the country, are already substantially lower than T-1 private line lease rates. The advent of these savings will speed ATM-based networks into the private domain of every enterprise.

6.2 The Enterprise Network

Because most medium- and large-size corporations have more than one site and require voice and data communications to interconnect their facilities, WANs have created one of the largest markets in the world for a wide diversity of equipment and services. The expansion of these corporate communications

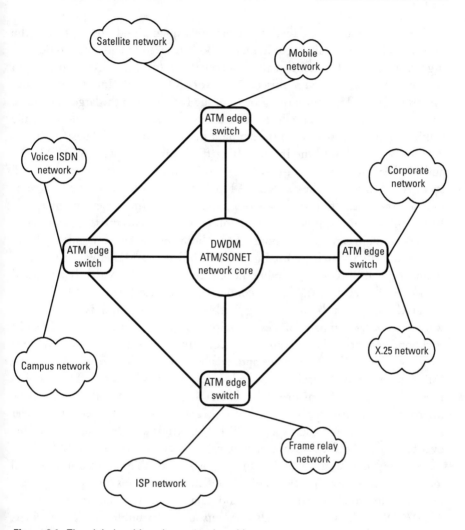

Figure 6.1 The global multiservice network architecture.

needs into global markets can only be satisfied by employing a global network-ing strategy. The inception of the multiservice SONET/ATM network fills this need and promises reasonably priced integrated services. Low-cost communica-tions has also been promised by ISPs. There are many solutions being offered that would establish voice communications over IP. However, the latencies experienced in two-way voice transmissions cannot assure a telephone conver-sation that compares to the quality of the public telephone network. The prob-lem is similar to the stalled growth of videoconferencing. The picture quality at

56-Kbps is no more than a jerky still-picture sequence and does not meet the viewers' subjective quality requirements. Both telephony and videoconferencing will mature only whenever the Internet IP protocol has been relocated on the global SONET/ATM multiservice network. QoS on the Internet is another important issue. The enormous growth of the Internet and the large number of users competing for "best-effort" service typically result in long delays while waiting for a response to even the simplest network functions. These problems will not be solved until the Internet IP traffic is integrated with ATM/MPOA and until ISPs can offer end-to-end QoS solutions for Internet services. Until then, ATM access to the SONET/ATM backbone core offers the best solution for the integration of data, voice, and video traffic for every enterprise. Figure 6.2 shows a simple representation of a typical enterprise network consisting of a main campus, a remote office, a home office, and a mobile facility.

The network architecture in the home office campus network consists of a conventional PBX system serving the telephones and a LAN that is concerned with the internal data traffic. By connecting both the LAN and the PBX to an ATM access unit, both outgoing WAN-oriented voice and data can be forwarded to other company offices in digitized formats over the SONET/ATM backbone network. The campus ATM access unit connects with the ATM edge switch via FR or ISDN for voice and data as well as with IP over ATM to the ISP. The interconnect facility can be low speed T-1 ATM or ADSL as well as native ATM at 51 Mbps or OS3 ATM at 155.2 Mbps. Small offices can be interconnected with an FR access device (FRAD) or with low-speed ATM over ADSL. In the future, high-speed ATM 25.6-Mbps services will be available over HFC-based cable distribution networks as well as over wireless LMDS.

Enterprises have the ability to opt for shared ATM service in the local loop on a competitive basis and reduce present expenditures for separate private line leases for voice, Internet access, and data circuits by a factor of 30–60%. The WAN of the future may also be outsourced to public service providers resulting in additional savings to the corporation. It is expected that WAN services offer a tremendous opportunity to public service providers who can offer lease-line replacement strategies while offering QoS protection. ATM and FR access services will be in significant demand when offered with Internet and intranet (VPN) support. WAN connectivity is primarily driven by the cost of bandwidth. FR over ATM has the ability to transport IP, legacy LANs, voice, video, circuit emulation through the frame user-network interface (FUNI), and low-speed ATM. The network manager is well advised to analyze the competitive offers of the service providers and to assure that network management services are offered that include legacy services such as SNMP, that inform about

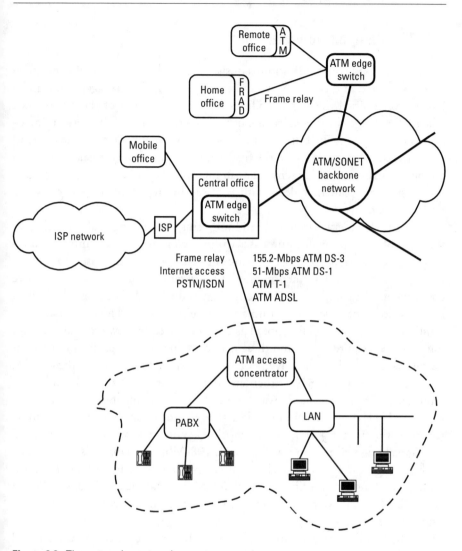

Figure 6.2 The enterprise network.

QoS parameters of multimedia transmissions, and that provide an accurate record of the performance of every network component.

Cost-effective ATM-based services will increase the number of communications options in the WAN for every corporate enterprise. In order to prepare for this transition it is wise to take a closer look inside the campus to determine if voice and data integration may be a beneficial migration.

6.3 The Campus Network

Existing campus communications consist of two distinct networks, the telephone network, and the LAN. In most corporations the managerial hierarchy has not yet unified the responsibilities for the supervision of these networks. The MIS manager worries solely about the data transfer, and the network manager is charged with the duties of making the voice network meet the demand for moves and changes. One of the first requirements of every organization is to establish a managerial hierarchy that can incorporate the technological changes into a strategic plan that deals with the step-by-step development toward a multiservice network.

Many data networks have been built using the existing understanding, at the time of the equipment purchase, without consideration of future requirements, excess capacity, and performance limitations. In many cases, the equipment manufacturer has incorporated proprietary methods that may prove costly when an upgrade to ATM technology is discussed. The present-day data network does not have any performance margin to spare. The combination of multimode fiber and UTP copper has been outpaced by new technologies and higher speed requirements, and these networks operate at capacity and without performance margin. The advent of fast Ethernet, FDDI, 155-Mbps ATM, and, as of late, the gigabit Ethernet offerings has pointed to the need for rebuilding one more time. It was hoped that the combination of multimode fiber and UTP copper would provide a good platform for these services. Unfortunately, this is not a realistic assumption. While a properly installed UTP CAT-5 cabling will perform well in basic 100-Mbps systems, it is not suitable for higher speeds. The new gigabit technology can only span 220m over multimode fiber. Single-mode fiber is required for future campus backbones. Gigabit and 622-Mbps ATM transmission can be extended to over 10,000m using single-mode fiber.

The first duty of determining the needs for an integrated voice, data, and video network is to define the cabling hierarchy. In the area of workstation connectivity, CAT-5 wiring will suffice for native ATM (51.2-Mbps), low-speed ATM at 25.6-Mbps, and even fast Ethernet at 100 Mbps, especially when ATM-based connectivity lowers the data transfer load. Backbone traffic and connections to gigabit servers should use the single-mode fiber to meet future speed increases.

The telephone system usually employs massive copper cables between facilities that have been constructed in a star architecture. Within a few years, a new generation of PBX architectures will appear. CBXs will be ATM-based and fully digital and will no longer rely on circuit switching. In addition, they will apply packetized voice routing. It is time for the telephone organization of every corporation to learn about voice digitization and compression options. This is not to say that the telephone instruments could not be connected to the ATM-PBX using existing wiring, but the possibility of integrating telephony into the data network exists today and will expand. There are many CTI vendors that will offer their wares forcefully and fuel the desire to establish the multiservice network.

The existing wiring topologies in campus and in the corporate networks require analyses and transformation to the new multiservice requirements for using single-mode fiber in all backbone configurations and either UTP or RG-6 in the service drop segment of the network. A comparison between the fiber/copper and the fiber/coaxial architectures shows interesting compatibilities and suggests that once the single-mode backbone exists, both copper and coaxial distribution areas can supplement each other and offer a flexible migration path.

6.3.1 Migration Toward the Multiservice Network

The existing dual- or even triple-network architecture does not have to be changed all at once. The planning document (Chapter 7) should survey present installation details in all campus buildings, determine present capacity and usage rates, and provide a good estimate of future needs over a five-year period. With this data at hand, it is possible to plan for a step-by-step integration of the multiservice requirements.

The migration from Ethernet, token ring, FDDI, and fast Ethernet to ATM and gigabit Ethernet technology begins in the backbone of the campus network. A simple first step may be the provision of single-mode fiber strands between buildings and to the location of any proposed gigabit servers. Keep in mind that QoS issues are important. While both IP and ATM technologies will be able to meet the need for high performance, only ATM offers multiple QoS levels. While VOIP is a workable solution in the LAN environment, IP over ATM assures low latencies in the WAN.

Figure 6.3 shows a mix of present and future LAN architectures and represents a milestone point in time in the transition from separate voice and data networks to the single multiservice architecture.

The data center has been converted to an ATM LAN switch/router that can handle telephony. A portion of the PBX has been converted to digital voice and is interconnected with the LAN switch. The server farm is connected to the data center. A new large LANE/MPOA server has been connected to the center via single-mode fiber and in the gigabit Ethernet format. Both the PBX and the

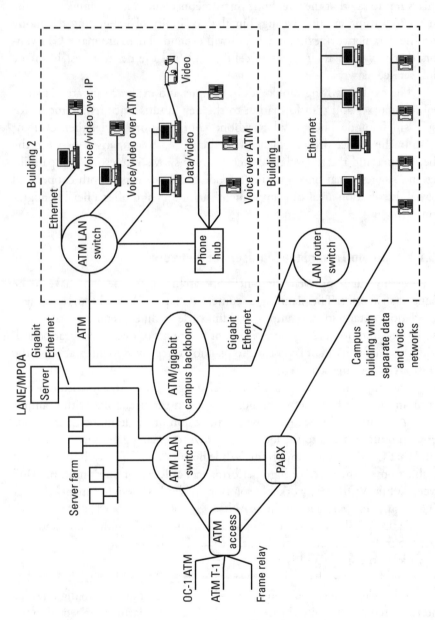

Figure 6.3 The corporate campus network—a work-in-progress.

LAN switch/router are connected with an ATM access concentrator that provides FR, low-speed ATM, and native ATM (51.84 Mbps) connectivity in the WAN and with the outside world.

The campus backbone delivers OC-12 (622.08-Mbps) ATM and gigabit Ethernet to every campus building over single-mode fiber strands. This architecture has been chosen to optimize flexibility and to enable any further network upgrading without having to rebuild the backbone.

Building 1 is still served in the conventional manner. Telephones and workstations are supported by separate infrastructures. The building uses conventional CAT-3 wiring for telephony and CAT-5 wiring for data. The LAN router/switch at the building entrance location is connected over single-mode fiber and delivers gigabit Ethernet to the building. There are numerous Ethernet and fast Ethernet connections to individual workstations and to selected groups of workstations.

Building 2 is fed by OC-12 ATM at 622.08 Mbps. The ATM switch/router at the building entrance location features 25.6-Mbps and Ethernet modules. The in-building wiring is mostly CAT-5 for integrated traffic but also uses CAT-3 wiring from a 25.6-Mbps ATM phone hub to serve regular telephone sets. CAT-5 wiring serves individual workstations with connected telephones using voice/video over IP and over ATM. Some workstations are equipped with ATM cards; others are equipped with Ethernet cards. Flexibility has been optimized to permit videoconferencing and telephony over the integrated multiservice network. It can be seen from Figure 6.3 that upgrading to ATM-based multiservice is mainly a function of rearrangement of the backbone to single-mode fiber and the integration of the data center with new CBX technology. An additional step can be taken by incorporating RF broadband technology and achieving further economical and operational advantages.

6.3.2 The RF-ATM Corporate Campus Network

Many of today's campus networks feature the main data center, a backbone network, and additional switch/router equipment located at building entrance locations. This hierarchy is necessary for all baseband-oriented networking architectures. The high-speed backbone needs to be translated down to speeds that are required by the workstation population. An ATM-based multiservice network is not different. It may feature OC-12 speeds in the backbone that need to be translated to native ATM or 25.6-Mbps ATM as well as provide connectivity for legacy protocols.

The application of RF broadband technology changes this equation. No longer is it necessary to convert from high-speed ranges to lower speeds at building entrance points. RF broadband has the ability to prepare the required speed range, the authentications, and all other VLAN functions directly at the data/voice center. Legacy PBX technology always used the principle of centralizing the switching equipment—and for good reasons. Circuit assignments can be changed at a single location, and all troubleshooting and maintenance events are centralized and hence more cost-effective.

The data network companies, however, have good reasons to avoid equipment centralization: The conversion of speed ranges requires equipment; the conversion of fiber to copper requires equipment; the amount of equipment sold increases revenues; and the closer the equipment is located to the workstation locations, the easier it will be to increase speed ranges without changing the CAT-5 wiring system. Indeed, equipment centralization of a data LAN appears out of the question because it would require a multitude of fiber in the backbone. Nobody would be interested in a 900-fiber strand cable to feed a high-rise building. This kind of old fashioned wiring was used in the telephone era. RF broadband technology can accommodate equipment centralization using only a small number of fiber strands. A multitude of speed ranges can be assembled on a single fiber strand and brought directly, in the right format and protocol, to the desktop. Figure 6.4 provides an example of a centralized ATM-based multiservice LAN.

The data center and the PBX have been combined. All voice communication is digital and in the ATM format for transportation to any telephone across the campus. Analog telephone sets can be used through an ATM phone hub, where they are powered. CAT-3 wiring is used in the building to connect the telephone sets. The data center interfaces with the ATM-PBX seamlessly and connects with servers throughout the campus. Gigabit Ethernet is used exclusively for server access. The campus backbone is connected to the ATM-based integrated communications center via cable modems. If there are 100 different 25.6-Mbps workstation and telephone groups, each group requires a cable modem at each end. Legacy Ethernet can also be supported by modules in the fast backplane ATM switch/router. It uses cable modems working with the IP protocol and can support VOIP as well. Based on the above example of 100 cable modems operating at 25.6-Mbps ATM, the overall network capacity is 2.560 Gbps or equivalent to almost three gigabit Ethernet systems.

At the building end, the expensive router/switcher has been replaced by a simple array of cable modems. Each cable modem operates on a discreet RF frequency and serves a single or a group of workstations and telephones. The wiring can be CAT-5 for all workstations except in areas where workstations require access to more than a single RF frequency path. In these cases, the use of coaxial cable in the distribution wiring is indicated. Multiple RF signals can be accommodated on this coaxial distribution, and the cable modems would

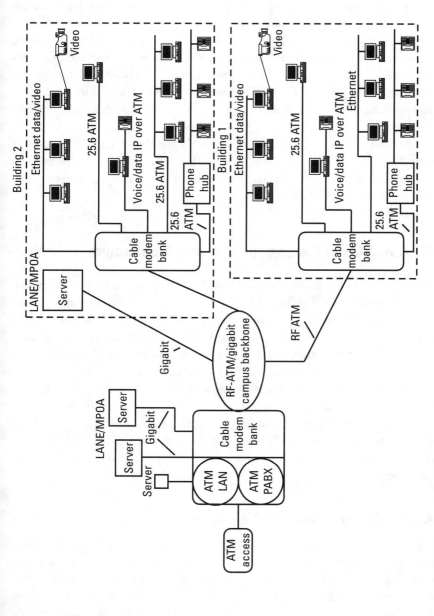

Figure 6.4 The RF-ATM corporate campus network.

move to the workstation. The cable modem bank also may include frequency assignments for "telephone only" service. These 25.6-Mbps bandwidth portions are connected with ATM phone hubs that translate the ATM to analog voice, provide powering for standard telephone sets, and connect to multiple sets using existing CAT-3 wiring.

RF broadband technology provides a new dimension to integrated multiservice networking. The architecture of the integrated multiservice LAN can resemble that of a legacy PBX and combine all the advantages of integrated voice, data, and video, CTI, VOIP, and IP over ATM and offer centralized maintenance at higher capacities than any other LAN architecture and multiple QoS levels.

Because the migration from existing LAN topologies and PBX networks to the RF-ATM-multiservice LAN mostly affects the backbone architecture, it is appropriate to address the typical backbone architecture options in the campus environment in more detail.

6.4   Versatility of the Campus Network Architecture

The selected network architecture must be flexible in regard to routing, expansion, and capacity issues. A useful network architecture must meet all performance parameters with a large safety margin and provide alternative routing options in the backbone to be able to cope with equipment outages and cable damage. In the distribution segment, it is desirable to avoid any active elements and to include alternate routing redundancy whenever affordable.

These elements of good networking apply to RF broadband as well as to any baseband networking architecture. For the most part, the infrastructure has been handled as the necessary evil to interconnect existing equipment. The following sections (Sections 6.4.1–6.4.4) are solely devoted to clarifying some of the dos and don'ts of the physical layer, especially when considering single-mode fiber.

6.4.1   The Fiber Backbone Star Architecture

Single-mode fiber can be used to protect the investment in the ever-increasing demand for higher speeds in all backbone applications. It can be used in a campus setting as well as in the vertical riser of a high-rise building. The integrated voice, data, and digital video facility of the future interconnects all buildings within an enterprise with a fiber backbone. This interconnectivity can be accomplished in the star topology shown in Figure 6.5.

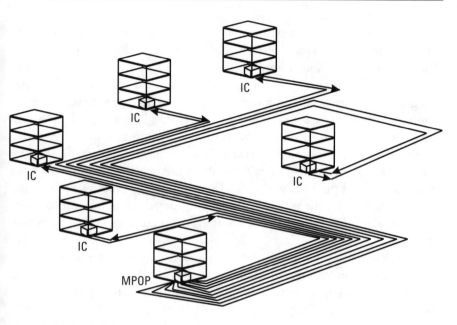

Figure 6.5 Fiber backbone star architecture.

The presentation of Figure 6.5 indicates the connectivity between six buildings in a campus environment. The building with the most fiber concentration is the MPOP of the installation. In the real world, the MPOP is not often collocated with the data center or even the PBX location. The identifier MPOP has been chosen to illustrate the begin location of the single-mode fiber backbone network.

Many fiber vendors recommend the use of the star architecture as the only way to avoid problems, and there are certainly some advantages. Most of these advantages are cited with multimode cable in mind—that is, the physical star provides the lowest connector loss because all fibers can be used in a point-to-point arrangement. While the connector and patchcord attenuations are high for multimode cables, the use of single-mode fiber eliminates this concern. Single-mode patchcord and connector losses typically do not exceed 0.5 dB. It is also often expressed that the physical star supports all applications and topologies. Indeed, having a point-to-point connectivity to each of the buildings can support logical ring protocols and enable token ring and FDDI concepts. It is also often said that a star architecture permits the main cross-connect (MCC) to be at one location, thus making rearrangements and administrative changes easier. These are good arguments in favor of the physical star, and the RF broadband can utilize this architecture without any problems.

There are, however, some negative factors. First, there is a high concentration of cable required leaving the MPOP or the data center. The example of Figure 6.5 innocently shows only the fiber strands that are required for transmission. Each building only requires two fibers. To avoid splicing and breakout arrangements of any cable, it is more logical to bring a 12-strand cable to every building. The result is a fiber cross-section of five 12-strand cables at the MPOP location, and, therefore, the size of the entrance duct needs to be changed. The other major negative factor is the inability of the physical star to cope with accidental damages in a cable section. A cable cut can interrupt all transmissions until the cable section has been replaced. For this reason, it is not recommended to use the physical star architecture along public streets. In a campus environment, where more controllable conditions exist, this major deficiency becomes a secondary consideration.

It has been said that the physical star architecture has been promoted by the fiber cable manufacturers to sell more fiber. While this is probably true, the star architecture offers extremely low end-to-end attenuation properties and can be a good solution for both higher speeds in the data network and for RF broadband network continuity.

6.4.2 The Fiber Backbone Single-Ring Architecture

Ring architectures are commonly used in the long-distance segment to accommodate SONET transmission systems. In this arena, the dual ring is essential to provide avoidance routing and to prohibit any failures in the outside plant routing plan. A single-ring architecture can be used on the private property of a campus where outside plant construction events can be controlled and cable damage is minimized. Figure 6.6 shows a single-ring fiber backbone architecture that permits the use of a single fiber strand to interconnect all building entrance locations.

Transmissions leaving any one of the campus buildings are routed in one direction around the ring and through every building. This is not a usable architecture for multimode fiber since all cross-connect losses are additive. However, since the attenuation of single-mode fiber is around 0.5 dB per km and patchcord and connector losses are in the same range, using this architecture for data networking is not a problem. The single ring, however, does not safeguard against cable cuts, and the entire network ceases to function when the continuity is interrupted anywhere in the ring. For the same reasons the single ring is also not recommended for RF broadband technology.

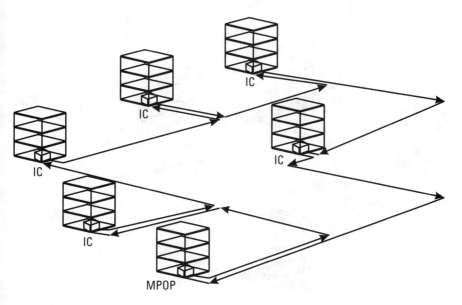

Figure 6.6 Fiber backbone ring architecture—the single ring.

6.4.3 The Fiber Backbone Dual-Ring Architecture

Sometimes referred to as the counter-rotating ring, the dual-ring architecture offers the best solution for present-day high-speed data networking. While the use of fiber is minimal, the QoS of the network can be assured by an automatic routing in both directions. This makes this network forgiving of any outside plant problems and assures avoidance routing of all transmissions. Figure 6.7 illustrates the dual-ring architecture in a campus environment. It is noted that while a cable cut in the ring causes automatic rerouting of the combined traffic, the same cable cut in the building entrance section can be the cause of a major interruption. When in public roads or city streets, it is highly recommended to enter every building at two different locations or to provide a v-shaped entrance facility with a six-foot separation between the building entrance points.

The dual-ring architecture is ideal for any data networking at baseband and can easily support high-speed ATM transmissions as well as multiple loops. The architecture lends itself also for RF broadband LANs in any desired spectrum and between 5 and 1,000 MHz.

Even more flexibility can be obtained by using the ring architecture as the route for direct point-to-point connectivity between all outlying buildings and the operations center. By overlaying the star architecture into the ring, a direct

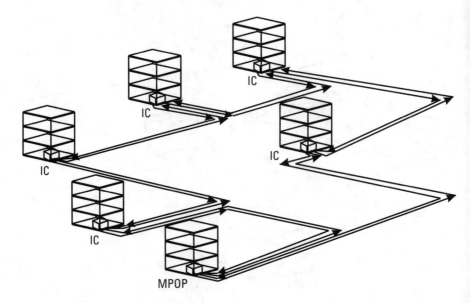

Figure 6.7 Fiber backbone ring architecture—the dual ring.

return path can be provided for each building. The same can be done for the forward path, and each building could have its own 5–1,000-MHz RF broadband spectrum. A gigabit facility operating at baseband can only provide for a gigabit party line. The RF broadband technology permits segmentation of the transmissions and can provide a 750-MHz bandwidth to every participating building location.

6.4.4 The RF Broadband Star Architecture in the Dual Ring

The fiber requirement of the dual-ring architecture is minimal. Figure 6.7 indicates a requirement of two fiber strands in the ring. Assuming that a 24-fiber strand cable is installed between every building entrance location, a star layout can be imposed on the physical ring and permit any future service upgrading or additions. At a cross-section of one-half inch, the cable can easily be installed even in the most crowded duct systems and deliver 750-MHz bidirectional bandwidth to every campus building. The physical star in the dual ring architecture also offers various migration paths for the integration of data, digital voice, and video. Initially, the user may only use the campus network for high-speed ATM data services, and RF broadband is only used to add more capacity in one of the buildings. When telephony and the PBX is converted to ATM-based digital transmission, a larger portion of the campus is activated to RF broadband and integrated voice and data services are delivered. In addition, and at a later time frame when the MPEG-2 video technology becomes affordable, a separate frequency spectrum can be assigned for internal full-motion videoconferencing. Even if these developments do not occur within the next decade, there is a good chance that the integration of all media will be the final stage in enterprise networking. Early preparedness and implementation of a single-mode fiber-optic campus backbone will be rewarded for decades to come. Having some spare fibers available will reduce any new outside plant cabling requirements, except for new construction events. All upgrade sequences are scalable and only require the addition of equipment at building entrance MDF locations.

6.5 Versatility of the Building Network Architecture

Buildings come in all shapes and sizes. Present-day data networks have been installed to fit the most difficult routing conditions. The biggest wiring problems exist in older buildings where riser spaces are nonexistent or improvised. Small buildings are usually wired with just UTP CAT-5 wiring between the MDF or the telephone closet at the cable entrance location and the outlet in the second floor. Larger buildings, over four stories high, require multimode fiber in the riser and fiber/copper translation/routing equipment at some floor levels. High-rise buildings, with a large user population on each floor, have had to install HCs at every floor level to separate the data network vendor equipment from the service distribution segment of the network. The implementation of new high-speed ATM and gigabit services requires a closer look at the changes that may be required in the risers of these larger buildings.

RF broadband technology offers flexible solutions for both the riser backbone and the horizontal distribution segment. Baseband and RF broadband in-building wiring methods are discussed briefly in Sections 6.5.1–6.5.3.

6.5.1 RF Broadband in the Small Building

The term *small building* requires a short explanation. Typically, it is identified by the fact that standard UTP CAT-5 wiring has been installed within specified limits between the equipment closet in the basement and any outlet location. Such a building can be one, two, or three stories high, as long as the total cable length does not exceed 295 ft.

Assuming that the CAT-5 wiring is placed correctly and in good working condition, the user should ask whether the presently shared 10-Mbps service to the desktop will be sufficient for the next decade. If the answer is no, there are a few considerations to be made. Based on the existing equipment complement and the capital outlay, it may be more economical to do the following:

1. Downsize the user population and provide multiple Ethernet service groups;
2. Change all equipment to fast Ethernet and challenge the UTP installation;
3. Add a coaxial distribution network to portions or to all outlets in the building;
4. Use the UTP wiring and upgrade the backbone to RF.

It would be nice if the decision making process could be as easy in practice as it is in concept. In the real world, however, there is historical baggage to consider, such as the investment made, the available budget, and the vendor promises. Let us assume, for this purpose, that the single building is located in a campus environment, and the duct supplying the building has two spare single-mode fiber strands. In addition, an assessment of the workstations' activities indicates that some do not require higher speeds; others require at least low-speed ATM at 25.6 Mbps; and there are a few that have to be able to operate at fast Ethernet speeds.

By using the two spare fibers for an RF broadband backbone and transmitting 25.6-Mbps ATM over a cable modem, the group of workstations not requiring higher speeds could be upgraded to ATM and still utilize the UTP wiring. The stations requiring 25.6-Mbps service individually could be wired using RG-6 coax. Each of these workstations requires its own cable modem. The few stations that require upgrade to fast Ethernet will carry a high price tag, but all other modifications are economical and far less expensive than providing multiple Ethernet loops or upgrading the entire network to fast Ethernet. Even in partial upgrades, the use of RF broadband technology is economical. The data networking vendors have developed a hierarchy of levels to encourage the placement of equipment at the location of the HC. RF broadband makes all rearrangements at the MCC. A cable modem establishes the frequency, and the transmission continues without additional routers or switches directly to the HC or even to the desktop. The only equipment content at the HC is the fiber-optic/electrical conversion equipment.

6.5.2 RF Broadband in the High-Rise Building

The properties of a large horizontal or a high-rise building are similar to a campus layout. The different floors can be looked upon as one-story buildings. The backbone is located in the riser space and the horizontal wiring terminates in an HC in the riser room. There are a number of intelligent buildings that have been designed by architectural firms and feature multimode fiber in the riser and CAT-5 horizontal wiring.

For many reasons already discussed, the use of single-mode fiber in the riser is a desirable upgrade process. While the higher attenuation of multimode cable may not be the driving force for the upgrade, the accumulation of patch-cord and connector losses certainly will be if the riser has not been designed as a physical star.

The type of the fiber route architecture within a high-rise building can follow the same rules that apply to the interconnectivity between buildings. The fiber can be laid out as a single ring, a dual-counter rotating ring, or a physical star. The single ring again minimizes the fiber requirement but arranges all floors in a serial configuration and precludes the use of multiple loop technology.

The dual ring provides for the avoidance route in cases of any cable cuts within the riser, but has the same deficiency as the single ring. If there is more than one baseband service entering the building, the riser needs to be rear-ranged to permit both services to reach different floor levels. Figure 6.8 depicts the various possible riser physical layer architectures. It can be seen that only the star architecture offers the flexibilities of dealing with each floor separately. It is also possible to use the star architecture within a dual redundant ring, as formerly discussed for the campus backbone. Within a building, however, cable cuts are unlikely and increase the complexity of the installation.

Segmentation of data services by floor and by user groups is an important issue in the data network of the future. Currently, the data networking industry is using routers and switches at every floor to separate workgroups, depart-ments, and authorized users. Most commonly, the traffic is a single 10-Mbps Ethernet or 16-Mbps token ring service. The vendors will argue that progres-sion to higher speeds is required and that newly developed faster backplane units should be deployed on every floor. The software-programmable segmen-tation will work as before and by elevating the backbone traffic in the riser to 100 Mbps or FDDI, every workstation can enjoy the higher speed.

There is absolutely nothing wrong with this promotion of new hardware. It is, however, a little like the long-distance service. A monthly review and com-parison with other vendors is in order. The existing cabling in the riser is also worth an analysis. Because the star architecture supports a fiber pair to each floor level, a segmentation of a 10-Mbps service to each floor level can easily alleviate any equipment change-out. Multiple Ethernet loops can minimize the equipment upgrade in the riser and transfer all upgrading to the building entrance location. The upgrade then consists of one additional fast

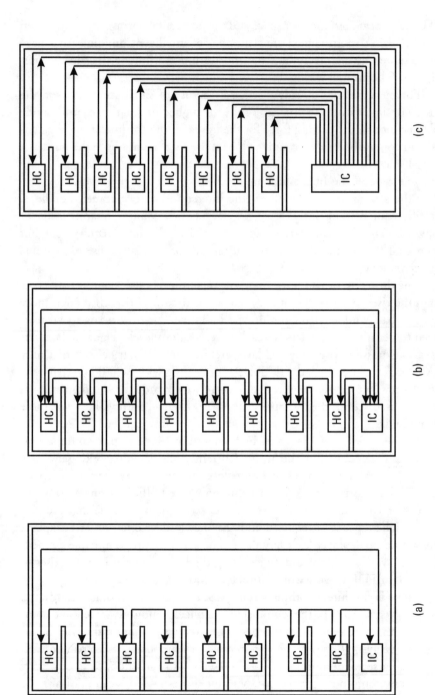

Figure 6.8 Ring and star fiber backbone in high rise buildings: (a) single ring, (b) dual ring, and (c) star.

motherboard at the equipment room and a few more Ethernet cards and avoids changing the equipment at the floor locations.

It can now be argued that it is necessary to have a fiber pair for each floor level and that this requirement may lead to a campus-wide construction proj-ect. Luckily, however, five years ago, the purchase of a combination multi-/single-mode fiber cable in the campus prevents any new construction. Two single-mode strands are all that is required to take the multiple Ethernet traffic to RF broadband and, using a cable modem, convert all individual loops back to baseband at the building entrance intermediate cross-connect (IC) or forward it at RF to the individual floor levels.

RF broadband can also simplify the cabling in the riser. The dual ring architecture from Figure 6.8 is all that is required to make the entire 750-MHz bandwidth available at each floor level. However, in the return direction, a star architecture is desirable to maximize the flexibilities for operation and growth. Figure 6.9 offers two examples of a riser-based RF broadband system.

Figure 6.9 shows a six-story building. Single-mode fiber is brought to floor levels two and five only. Backbone coaxial cable is used to provide service to the next adjacent floors. The RF broadband spectrum is available at all floor levels and the decision to provide a portion of the spectrum, or the entire spec-trum, to the desktop, is yours. The service options are very flexible:

1. Single-mode fiber to each floor level;
2. Single-mode fiber to some floors and coaxial backbone to other floors;
3. One CM at the HC location and UTP to the desktop;
4. Multiple CMs at the HC location and UTP to the desktop;
5. RG-6 wiring to the desktop and a CM at the desktop.

In options (1) and (2) above, the entire RF spectrum is available at each floor level. Option (2) is more economical as it uses less fiber and fiber-optic equipment. Option (3) reuses the existing UTP wiring, requires a bridge between the cable modem and the HC, and requires that all workstations are using the same RF spectrum slot. This can be an Ethernet, token ring, or OC-1 ATM service. In option (4), for example, the UTP wiring is grouped in the cross-connect to form multiple user groups. Each of these groups utilizes an RF spectrum slot through a dedicated CM cable modem. The RG-6 coaxial wiring to the desktop of option (5) permits the selection of every available RF spec-trum slot right at the desktop. The cable modem can be assigned to any RF fre-quency, as desired.

It is noted that this kind of flexibility cannot be achieved in conventional data networks unless the equipment at the HC location features individual

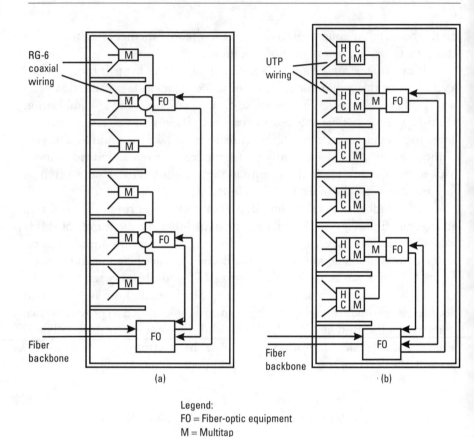

Legend:
FO = Fiber-optic equipment
M = Multitap
CM = Cable modem
HC = Horizontal cross-connect

Figure 6.9 RF broadband in the high-rise building: (a) RG-6 wiring to the desktop, and (b) UTP wiring to the desktop.

boards for each workstation. While the data network equipment providers have populated the HC location with motherboards, station modules, and fiber-to-electrical translation equipment, the RF broadband equipment complement can be reduced to the fiber-to-electrical equipment, some inexpensive coaxial components, and one or more cable modems. The price tag for this equipment is less than one-half of the conventional multimode switch/router and only a small percentage of the faster backplane technology required for the over 100-Mbps service formats.

6.5.3 RF Broadband in the Horizontal Wiring

The horizontal wiring usually consists of UTP cable between the HC panel in the riser room and the outlet. UTP CAT-5 has been used most frequently and is a good conduit for baseband speeds of up to 100 Mbps. Higher speed applications such as FDDI, 155-Mbps, and gigabit transmission systems have challenged the applicability of CAT-5 technology in the future. A large team of UTP and STP vendors are trying to determine CAT-6 and CAT-7 standards that would offer error-free service at these higher speeds. As discussed earlier, there are physical limitations of what can be done with UTP cable. STP wiring is also trying to make a reappearance in the radio-frequency interference (RFI)-conscious European market, while standard and inexpensive RG-6 coaxial cable could permanently solve the bandwidth problem.

It is not evident why gigabit service should ever be provided at the desktop. While there is no workstation on the market or in development that could use this speed, there seems to be the desire to convince the user to upgrade "one more time" to CAT-6 or CAT-7 wiring. Using 2.4 Gbps in the backbone makes sense, but to bring more than 25.6-Mbps ATM (OC-1) or native ATM at 51.2 Mbps to the desktop does not (except for very special needs). CAT-5 UTP can handle these speed ranges well, even if the wiring has not been installed in a perfect manner. The user, however, has been told that higher speeds are coming and that the horizontal wiring system must be changed to accommodate these future developments. A change to coaxial RG-6 wiring, of course, would eliminate any future rebuilding. RG-6 coaxial cables can handle the entire frequency band of up to 1,000 MHz and could also accommodate multiple gigabit transmissions. Figure 6.10 compares the commonly used UTP wiring and RG-6 coaxial wiring.

From Figure 6.10, it can be seen that the wiring of both types of cable is very similar. Instead of a four-pair twisted-pair cable, the coaxial cable has a center conductor and an outer shield. The connector hardware is different and features a conventional F-connector. The installation costs for both wiring solutions are comparable. While the UTP cabling starts in the riser room at the cross-connect panel, the RG-6 wiring is connected to an eight-port multitap. If necessary, a backbone coaxial cable can be used, similar to the multi-user cable to reach a group of outlets. UTP would then break-out to the individual outlets, while a multitap would be used for the distribution of single RG-6 cables to the individual outlets.

RG-6 coaxial cable has been used for baseband video as well as for RF broadband applications. It has gone through repeated redevelopment cycles

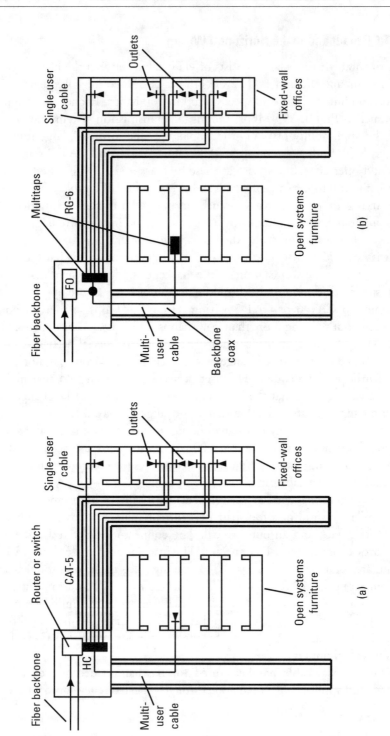

Figure 6.10 Horizontal wiring: (a) UTP CAT-5, and (b) RG-6 coaxial.

to make it work flawlessly in the rugged outdoor environment. It is used by most cable companies for the CATV service drop and is the link between the pole in front of your house and your television set. The life expectancy of the RG-6 cable in the outdoors exceeds 15 years. When used in the ceiling space of a building, the life expectancy can easily be tripled.

6.6 Flexibility in RF Broadband Networking

One of the many benefits of the RF broadband technology is the routing design flexibility. Baseband networks do not feature any routing flexibility, since the signal has to be transported between all node locations of the network. Departmental segmentation is commonly used in VLAN arrangements, but speed assignments require changes of hardware configurations. The RF broadband network can do the following:

1. Route the entire spectrum to any and all locations;

2. Assign different spectral segments to different buildings;

3. Assign different spectral segments to floor levels;

4. Carry any spectrum portion or the entire spectrum to the desktop.

A spectral segment can be any 6-MHz channel that is equipped with a cable or NTSC video modem. ATM-compatible cable modems can presently handle 25.6-Mbps speeds and can be used for data or telephony. Upon finalization of more economical encoding solutions, cable modems will carry digital video in the MPEG-2 format as well. The management of the spectrum will become the future challenge of the integrated multimedia LAN and of the network manager. With capacity to spare, the RF broadband LAN has a simpler and more economical profile than any conventional and future baseband technology. Figure 6.11 is a schematic of a four-location RF broadband LAN installation. The locations have been named A through D for the purpose of this example.

In Figure 6.11, the network control center is located at A. It consists of the ATM switch matrices and the service modules for the entire network. Our example assumes that all service modules are either 10-Mbps Ethernet or 25.6-Mbps ATM. Each of these cards is connected to a cable modem, which upconverts the baseband to an RF frequency. The cable modem bank consists of as many cable modems as there are service modules. The output of the cable modems is combined in a simple coaxial device and forwarded to the

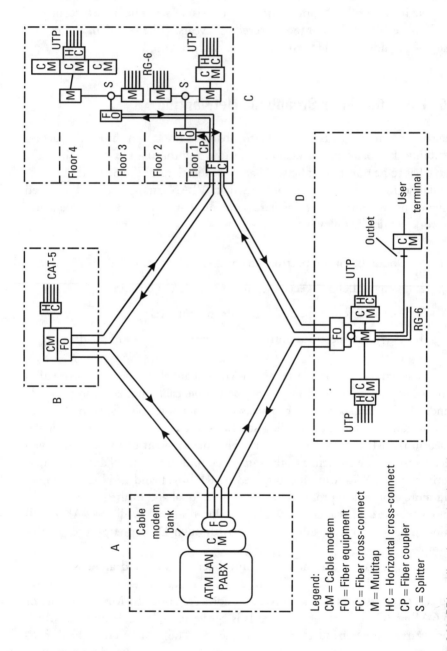

Figure 6.11 RF broadband network schematic.

Legend:
CM = Cable modem
FO = Fiber equipment
FC = Fiber cross-connect
M = Multitap
HC = Horizontal cross-connect
CP = Fiber coupler
S = Splitter

fiber-optic transmitter. The single-mode fiber is illuminated and the entire spectrum is transferred to the other locations via a dual ring connection. The fiber receiver at all other locations converts the optical energy to electrical for transportation on the coaxial medium.

At location B, the fiber transmitter/receiver is connected to a single CM. Since this location is a small building with only 20 users, the most economical solution has been employed. All workstations will be served with 25.6-Mbps ATM. This is easily accomplished by installing a single cable modem to the in/out ports of the fiber equipment and interconnecting the baseband in/out ports to the already existing HC panel. Rack space requirements for this equipment are minimal and UTP CAT-5 wiring is already in place to complete the installation.

Location C is a larger building with special needs for special people. In the basement MDF location of this building an open rack has been placed that contains a fiber cross-connect (FC) panel. The single-mode fiber has also been placed in the riser. This fiber is interconnected with the campus fiber through an optical coupler. Optical two-way couplers have a 3-dB loss. Because the entire campus fiber only has a loss of 1 dB for optical attenuation and 2 dB for patchcord and connector losses, the additional 3 dB of coupler loss is well supported. The total optical attenuation will be 6 dB, and the optical budget between the fiber transmitter and the fiber receiver is between 10 and 16 dB for different lasers.

The FO equipment in the riser is placed at floor levels that permit symmetry in the short coaxial backbone sections. The FO equipment in/out ports are connected to a coaxial splitter that feeds the equal length coaxial backbone cable. Multitaps are installed at each of the floor level riser rooms to provide the desired mix of services. For instance, on floor 4, it has been determined that three different user groups are to be served. Three CMs are placed and interconnected with the HC panel in a manner that separates the three user groups. These user groups do not have to be departmental assignments, they could be an Ethernet group, a token ring group, and a 25.6-Mbps ATM group. Every workstation on floor 4 uses the existing UTP wiring.

On floors 2 and 3, the speed requirement is more critical. These floors house the imaging and CAD design groups that need 25.6-Mbps service to the desktop. The horizontal wiring on these floors has been changed to RG-6 cable. A separate RF frequency is assigned to each outlet location and the workstation connects through a 25.6-Mbps cable modem. The workstations on floor 1, however, are a part of the old token ring network and do not require any change. A single cable modem is used and its baseband input/output is connected to the existing HC. The UTP wiring does not have to be changed and connects to the workstations.

Building D is a medium-sized building with about 50 users. There are five workstations that require high-speed service. These users will be served with coaxial cable through a 25.6-Mbps ATM cable modem. The remaining users are subdivided into two groups of 22 and 23 users and have access to two Ethernet service loops. Conventional UTP wiring is used for these users.

While there are many different HFC architectures for RF broadband networking applications, Figure 6.11 illustrates the operational flexibilities that are available and usable in an economical manner. While baseband data networks only provide a single high-speed connectivity between equipment components and the user, RF broadband offers a new alternative dimension. Individual users and user groups can be supplied with applicable speed ranges. This means that, for the first time, the network manager can make choices and provide the user with the best solution for his or her problem. With the advent of voice and data integration, as well as the future implementation of digital high-quality videoconferencing, flexibility of rearranging the network without the need for expensive rebuilding is an important issue.

Data networking vendors are working on layer 3 equipment for providing the network manager with new upgrade options, including equipment configuration changes at all node locations. While some of these changes are realistic, others are unnecessary and wasteful. In contrast, RF broadband LANs do not need expensive equipment at HCs, ICs, or even MDF locations. The frequency assignment and service structure of the network is determined at the network control center or at the MCC. As a result, fast backplane motherboards can be centralized; all service modules are located at that location as well as at least fifty percent of the cable modem population. This concentration of equipment eases both management and maintenance activities while increasing the operational economy of the network.

6.7 Performance Considerations of the RF Broadband Network

Early RF broadband networks, as used by the cable industry for the distribution of multiple video channels, featured numerous amplifiers in cascade. Every active element introduces noise into the signal stream. This noise is amplified in the next amplification stage, and the noise level increases.

The single-mode fiber-optic transport does not have any amplifying components. The noise content is directly related to the laser noise of the fiber-optic transmitter and the optic-to-electric conversion of the receiver. Since the coaxial tail of the RF broadband network is passive, the total noise accumulation is limited to the fiber segment. Similarly, any modern single-mode fiber/UTP system only accumulates noise in the fiber segment. Ingress and

egress components in the coaxial segment are kept to a fraction of the level produced in the unshielded UTP wiring.

6.7.1 The Carrier-to-Noise Ratio

In a coaxial transmission system, the carrier-to-noise ratio (C/N) is reduced by
3 dB for every three amplifiers in cascade. One of the reasons for upgrading the
cable television networks to HFC technology is to reduce the number of amplifiers in the distribution segment. Current HFC technology limits the number
of amplifiers in the coaxial segment to two and sometimes three amplifiers in
cascade.

In the RF broadband LAN, coaxial amplifiers can be totally eliminated.
The reason for this approach is not the higher noise level that may be caused by
the use of these amplifiers, but the flexibility of the network. For instance, the
riser in even the largest building could easily feature a coaxial backbone without
any performance problems. However, the flexibility of being able to bring
different spectrum portions to different floor levels is lost. In addition, the
5–42-MHz return spectrum is available only once, and the capability to origi-
nate the entire return spectrum from each floor level is lost. The RF broadband
LAN combines the best performance parameters when constructed as a single-
mode fiber trunk and a passive coaxial distribution tail.

The C/N defines the number of decibels that the carrier is located above
the noise floor. Nonlinearities in the electronic circuitry of any active equipment cause the equipment to have a particular noise figure. In the typical RF
broadband LAN, consisting of the single-mode fiber segment and the passive
coaxial tail, there are only four contributors that affect the overall C/N ratio of
the system:

- The fiber transmitter C/N;
- The fiber cable C/N;
- The receiver shot noise C/N;
- The receiver thermal C/N.

The C/N of a fiber-optic transmitter is determined by the noise created
by the laser. Single-mode fiber transmitters use DFB or Fabry-Perot lasers.
While the noise figure changes with channel loading and between these two
types of lasers, a C/N value of about 51 dB can be expected.

The fiber cable noise is directly related to the distance of travel through
the single-mode fiber strand, the optical power of the laser transmitter, and the

type of laser. The formula for a Fabry-Perot laser is 23 log $(2/D)$ and for the DFB laser is 10 log $(2/D)$, where D is the distance in kilometers. Because distances in the RF broadband LAN are short, the fiber cable C/N can be expected in the range of 52–53 dB, depending upon the type of laser and the manufacturer's specifications.

The fiber receiver has two noise components, the thermal noise and the shot noise. The thermal noise C/N is typically 1 dB lower than the C/N value stated on the manufacturer's product data sheet. The C/N depends upon channel loading and is typically about 52 dB. Thermal C/N is 51 dB. The shot noise level is a function of the receiver location. Typically, the shot noise C/N is obtained by adding the optical path loss to the manufacturer's stated noise figure. Assuming that the data sheet states −47 dBmV, an optical pathloss of −13.0 dB results in a shot noise level of −60 dBmV. The resulting receiver shot C/N requires the use of the receive level formula: receive level = $2x$ (optical receive power − 14) − 1. This reduces the shot noise level of −60 dBmV to a shot noise C/N of −59.0 dB.

Combining these C/N values—that is, the fiber transmitter C/N of −51.0 dB, the fiber cable C/N of −53.0 dB, the thermal noise C/N of −51.0 dB, and the shot noise C/N of −59.0dB—results in the system C/N. C/N ratios are combined in accordance with tables that combine two values at a time. The combined value is always lower than the lowest C/N value. In our example, the lowest C/N is the fiber transmitter and the receiver thermal noise. The combined system C/N of the four values is −48.6 dB [1].

In the early days of cable television, the FCC specified a minimum value of −39.0 dB. Visual impairments in television transmission systems become noticeable to the human eye at −43.0 dBmV. A combined system end-to-end C/N value of −48.6 dB offers a crystal clear optical presentation in the more critical analog format. Digital transmissions are more forgiving and do not require these outstanding C/N values. However, it is noted, that whether the system is a gigabit baseband fiber/copper-based data network or a fiber/coaxial-based RF broadband network, the obtainable C/N ratios exceed the operational requirements by a wide margin.

6.7.2 The Composite Triple Beat

The composite triple beat (CTB) is a third-order intermodulation product that contains the additions and subtractions of three RF carriers (F1 ± F2 ± F3)[1]. CTB is the accumulative disturbance of the number of RF carriers when re-amplified through long cascades of coaxial amplifiers. The new HFC technology, using fiber nodes and only two or three amplifiers in the new cable delivery systems of the cable industry, eliminates any possibility of CTB buildup. Likewise, the RF broadband LAN technology, discussed herein, reduces the CTB factor to zero. The subject, therefore, is only mentioned as a reference to the evolutionary problems of the RF broadband industry.

6.7.3 Reliability of Service

The reliability of service in a network facility is directly related to the number of active devices and the interrelated outage rate of these components. A reliable network must have an availability of 99.999% of the time. This amounts to a total outage time of about 0.1 hour/year, or six minutes per year. The RF broadband LAN uses fiber-optic transmitters and receivers as the only active elements that can fail. In the rare occasion that coaxial amplifiers are used, history can point to decades of operation in inclement weather conditions. Additional electronics such as cable modems do not affect the functionality of the entire system and are automatically shut down when not performing to specified values. It is evident that conventional data network equipment components are more complex than RF broadband components. The combined meantime between failure (MTBF) value of cable modems is considered better than that of the more complicated motherboards and service modules.

In addition to percentage availability of the system and the MTBF of the system components, there is the meantime to repair (MTTR). This value is an indicator of the time required to clear an outage and repair or replace a faulty unit. A comparison between the conventional data networking equipment population and the RF broadband equipment population shows that better MTTRs are achieved when the equipment is concentrated at one location. A good example is the PBX installation.

Most repairs and replacements are made at the switch location. The same holds true for RF broadband. Most switching and routing equipment can be centrally located, while data networking equipment is distributed to even HC locations. The resulting conclusion is that the MTTR of the RF broadband LAN will be substantially better than that of conventional data networks.

Class of service (CoS) and QoS are being addressed by the major data networking equipment providers. The translation from gigabit to 100 Mbps down to 10 Mbps speeds in the data networks causes latencies that are produced by the many congestion points in the network. These latencies are virtually eliminated in the RF broadband LAN because of the front-end concentration of most of the equipment. However, with increasing traffic loads it is good practice to ensure maximum latencies for traffic categories as well as control jitter (the variation of latency of a packet through network elements). The use of

ATM in the backbone is the best strategy to assure QoS in the network of the future. ATM- and IP-based cable modems can extend QoS control throughout the LAN by virtue of centralization of most network elements.

6.8 Selecting the Desired RF Broadband Network Architecture

It can be concluded that the RF broadband LAN permits flexible network architectures beyond those of present day data networking concepts. The new dimension of assigning frequency bands to different users adds new design options to the conventional networking principles. While it is not possible to cover all topological and operational options that are available, it is noted that the existence of single-mode fiber in the LAN backbone can provide scalable upgrade and overlay alternatives for new future service requirements. RF broadband offers new opportunities to the network manager to assess the migration path of the future. The decision to implement the first phase of the RF broadband network depends entirely on the perceived longevity of the existing data network; the desire to implement an integrating voice, data, and video network; and the economical constraints. Scalability of implementation can assure a step-by-step development toward this goal.

While it is not possible to prescribe each and every migration path, it is possible to provide guidance in three basic areas of the decision-making process—the selection of the frequency spectrum, the optimization of the system architecture, and economical issues and considerations.

6.8.1 Optimizing the Spectral Capacity

There are three basic choices for the assignment of forward and return transmission frequencies.

1. The subsplit system;
2. The high-split system;
3. The dual system.

In the case of the subsplit system, the forward RF carrier frequencies are selectable between 50 and 750 MHz. In the return direction, only the frequency range of 5–42 MHz is available. This asymmetric choice of frequencies stems from the desire of the cable company to transmit broadcast channel 2 on

54 MHz. While this rule does not make sense in the enterprise environment, it has become the basis of the development of cable modems. There are symmetric and asymmetric cable modems readily available to operate in these frequency ranges. An enterprise RF broadband LAN can readily utilize the availability of these very cost-effective cable modems by using a star architecture for the return system. For instance, a separate fiber strand from every floor of a tall building and from every small building will form a formidable return architecture. The return fiber star will permit the frequency spectrum of 5–42 MHz to be available from every floor or every building of the network, and cost-effective conventional cable modem equipment can be used. The forward architecture can be a single-fiber strand to every one of these locations. In a ten-location system, the resulting forward capacity would be 50–750 MHz or 116 six-MHz channels, and the combined return capacity would be ten groups of 5–42 MHz, for a total of 370 MHz or 30 six-MHz channels.

In the case of the high-split system, the crossover between forward and return is selected between 186 and 222 MHz. Cable modems operating in this band are available and have been developed for the MAP/TOP standard. Applying the return star architecture to the return band of 5–186 MHz, our ten-location model has a combined return spectrum of 300 six-MHz channels. The single fiber strand forward architecture, however, limits the forward transmission capacity from 222 to 750 MHz or 93 six-MHz channels. To make the architecture more symmetrical, it is possible to use fewer dedicated return fiber strands and/or to reinforce the forward direction with more outbound fibers. It is interesting to note that by reducing the return star to every third floor or building, symmetry is achieved. The result would be 93 forward and 100 return six-MHz channel assignments.

The dual system architecture, under (3) on the previous page, requires the duplication of the coaxial tails. If the RF broadband technology is used for backbone purposes only, existing UTP will be used and no duplication of the coaxial tail is required. Assuming a star architecture in both forward and return backbones, the capacity of the network to each floor and each building in our ten-location example would be 5–750 MHz or 124 six-MHz channels in both directions to each floor or building. The combined capacity of the network results in 1,240 channels or 7,440 MHz in each direction. The addition of the dual coaxial service drops is scalable and can be implemented when required. It is noted that the total fiber requirement in this example of 10 locations could be satisfied with a 24-strand fiber cable. The dual system star architecture, by far, provides the highest spectral capacity as well as numerous flexibilities to grow the infrastructure on a scalable basis and to cope with the bandwidth requirements of the next century.

6.8.2 Optimizing the Availability

The equipment availability has already been optimized by the elimination of coaxial amplifiers and the application of only fiber-optic transmission equipment. In addition, the availability of the network can be improved by using a ring architecture interconnecting all campus buildings. The physical ring does not interfere with the recommendation for a star connectivity. The star cabling is simply laid into the ring that interconnects all buildings. A counter-rotating ring is established by providing optical couplers at the building entrance locations and routing of the fibers in opposite directions through the ring. Using this method to avoid outages caused by cable cuts has a price tag. The number of fiber strands in the ring has doubled.

6.8.3 Economical Considerations

Flexibility and scalability are the big advantages of the RF broadband technology. The network manager is in command to set the goal for the infrastructure of the future.

The first step is the implementation of single-mode fiber in the campus and in the risers of large buildings. The migration from the single-mode fiber infrastructure to the completed integrated voice, data, and digital video facility can be accomplished on a step-by-step basis and in accordance with the budget constraints of the organization.

In many ways, the first step, the availability of single-mode fiber connectivity, will cause vendors to develop new and more economical products. For the first time in the evolutionary process of the development of data networking, the user is in charge of the future and can influence the vendors. It is certain that vendors will follow and meet the competitive challenges. RF broadband LAN technology permits every organization to assess the vendor offerings of the future and to expand the network infrastructure in an economical manner and at the desired speed. The end goal, the integrated multimedia network, however, can only be realized with dedication and determination to implement the ultimate telecommunications network and the desire to eliminate any network rebuilding requirements in the twenty-first century.

Reference

[1]   Tunmann, Ernest O., *Hybrid Fiber-Optic Coaxial Networks*, New York: Flatiron Publishing, Inc., 1995.

7

ATM-Based Network Applications and RF Broadband

RF broadband LANs can be utilized by any entity—small or large. RF broadband features versatility that can provide alternative choices in communications networking. Present-day data networks operate exclusively at baseband frequencies and, therefore, need to increase their speed ranges to accommodate the ever-higher traffic loads. RF broadband technology opens a new dimension in the transport of digital traffic. Analog and digital traffic can be stacked one on top of the other when appropriate carrier frequencies are assigned to each of the services.

7.1 RF Broadband—Versatility

The application of RF in the data network enables the user to make alternative decisions. If a network is based on 10-Mbps Ethernet and has been upgraded many times to subdivide the usership between several Ethernet loops to accommodate the increasing transmission loads, the time is right to think about upgrading. The choices of the data network equipment providers are aimed at higher speeds. Upgrade the backbone to fast Ethernet, FDDI, or even gigabit speeds and the problem is solved. Much of the time, especially in campus networks, this simple advice can cause monumental rebuilding requirements for the wiring system due to the bandwidth limitations of new equipment offerings. Checking with competing vendors does not offer a more cost-effective alternative. As the user is considered a valuable customer base, the incumbent vendor will promise a painless migration path in order to retain business.

Both ATM-based networking and RF broadband technology open new alternative approaches to higher speed and integration solutions. In the example of the legacy multiloop Ethernet network, the more economical alternative is to stack the frequencies, use cable modems for transmission, and regroup the user base to accommodate any particular needs. There are some expenses involved, such as requiring a single-mode fiber and additional backplanes with Ethernet cards, but a speed increase is not required. The same speed is available by a factor of ten.

Another frequent problem, the mixing of token ring, Ethernet, fast Ethernet, and FDDI in one set of equipment promotes dependency on proprietary vendor solutions.

These solutions turn into expensive propositions, especially in large systems where workgroup switches appear at every building entrance point and floor level. In RF broadband, the protocol conversion can be concentrated at the data center, and the desired format is delivered on its own carrier frequency to the user.

As telecommunications budgets are usually limited, funds should not be wasted on delivering higher speeds to the desktop without a critical analysis of the present network and future requirements. Five to ten years from now, it is likely that a high percentage of corporate networks will feature integrated ATM-based technology, delivering telephony, data, and high-quality video throughout the corporate LAN, the enterprise WAN, and the public network. Therefore, the network manager is wise to spend the communications budget in increments toward the goal of multiservice integration.

This chapter deals with applications of the RF broadband technology in today's separated telephony and data network environments and in the integration of the multiservice LAN and typical video and telephony applications.

7.2 The SOHO and Small Office LAN

7.2.1 Conventional Choices

The conventional wiring system in a small office installation typically consists of a key telephone system and CAT-5 LAN wiring. Ten Base-T stacks are used to interconnect the workstations with a server. The Internet connectivity can then be established through standard 56K modem equipment or even at lower speeds. There are many different wiring aids available from a number of suppliers that permit reasonable arrangement of the wires and some form of cable management. Figure 7.1 shows such a conventional installation.

Key systems require multipair interconnect cables of a large diameter. CAT-5 requires a four-pair cable to every workstation in a star architecture. The result is a large amount of wires to satisfy just voice and data requirements. In cases where the small office is a branch office of a large company, the interconnectivity in the WAN is normally the first bottleneck. ISDN, FR, and T-1 private lines have been used to increase the WAN connectivity. The latest DSL offerings can operate over ATM, do not require a full-time lease, can serve data and compressed voice, and reduce costs substantially.

Assuming that the small office is growing, both the key system and the LAN are a work-in-progress. Discussions between staff members about upgrading to a PBX and to fast Ethernet will become more frequent. In these talks, the various upgrade options are discussed, and the question about telephony and data integration on one network is raised. Then, headquarters makes it known that any upgrade must include an ATM access concentrator because the private line connections in the WAN will be changed to "low-speed" T-1 ATM. The desire to integrate voice and data into one network changes the equation in favor of an ATM-based system. Product investigations show that the per port pricing of layer 3 equipment has dropped and that a fast backplane ATM

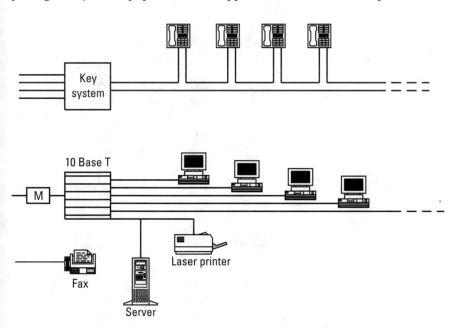

Figure 7.1 Typical small office communication system.

switch/router with a WAN access concentrator, Ethernet, and 25.6-Mbps ATM ports have become more cost-effective. It appears that a CTI solution is more economical than separate PBX and data systems. Local telephone lines can be integrated into the ATM switching matrix, and phones required at non-computer locations can be separated using phone hubs. Such systems look very much like Figure 7.2, except that only existing CAT-5 wiring is used for all workstations.

7.2.2 The ATM-RF Broadband Alternative

CTI can be provided in an ATM-based network as VOIP or as voice over ATM. The subtle difference between the two is the ability to use inexpensive Ethernet cards when using IP. In doing so, however, QoS will be limited to just priority and will not permit various CoS, as this is the case when using ATM. In our example, the ATM switch/router has been equipped with both Ethernet and ATM 25.6-Mbps modules. The old existing Ethernet stacks can be used to further adapt the workstations to the required speeds. Low-speed users can share an Ethernet port; high-speed users can be provided with a 25.6-Mbps ATM port. Every computer requires an NIC card to translate the ATM-voice back to the analog format. This permits the use of standard analog phones.

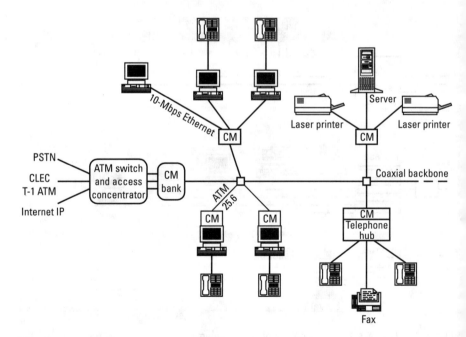

Figure 7.2 The small office RF-ATM LAN.

Phones located in halls and in areas that do not require an associated computer are connected to a phone hub using standard CAT-3 wiring. The phone hub provides the translation to analog voice and powers the telephone sets. Figure 7.2 shows the integrated ATM-based voice and data network with an additional feature—a coaxial backbone has been used to replace some of the CAT-5 wiring.

In this topology, cable modems operating at 10-Mbps Ethernet are used to group workstations in a more flexible manner. The advantages of using RF broadband technology in the small office LAN can only be realized when the workstation population is expected to increase at a fast rate. RF broadband offers the network manager an additional tool in the assignment of authorization, resulting in a finer segmentation of service classes within the organization. In addition, there are the advantages of less wiring, easy reassignment of service classes and speed ranges, and a reduction of rewiring requirements caused by moves and changes. The result is a more streamlined management of user needs and the ability to expand the system to more users without additional equipment. Future expansions do not require wiring changes and are limited to an additional module in the front-end ATM switch/router and a pair of cable modems.

7.3 The Intelligent High-Rise Building

Architectural firms throughout the world have played a role in increasing the efficiencies of new buildings. Today's intelligent buildings feature many innovations to reduce the operating costs of building tenants. Innovative approaches to climate control problems, lighting, acoustics, security, and quality of living have been incorporated in these designs. Intelligent buildings come with an array of tenant-oriented control systems, from solar collectors for heat conservation, outside light transfer to internal sections of the building by fiber-optic cabling, entrance authorization networks, automatic lighting control systems, to "chimney effect" natural cooling systems. Today's high-rise intelligent building requires substantial wiring networks to support these environmental functions as well as to establish monitoring, evaluation, and programming of these technologies from remote locations.

As a result, architects have a good appreciation for the space requirements that are needed to accommodate these support systems. Every modern high-rise building features a spacious arrangement of riser rooms to accommodate vertical cabling as well as wiretrays for horizontal support of wires to the service locations. In most cases, the riser rooms are stacked in a manner that will permit the tenant to install equipment racks and facilitate vertical connectivity from floor to floor through conduit stubs that can support fire stopping in accordance with local regulations. The tenant can now easily establish any horizontal wiring needs by utilizing the provided wiretray system and determine the outlet locations for each room. In cases where the use of the building is predetermined at the time of design, the architect will accommodate the client's wiring needs for telecommunications as part of the electrical design process.

High-rise buildings are usually built for multitenant occupancy, and as the tenants' communication system requirements are different, the architect often refrains from incorporating voice and data wiring schemes in the initial design. This frequently results in additional wiring projects by clients after the building's completion and violations of structured cabling standards and local/regional building codes. Fortunately, the emerging ATM technology provides a solid foundation for inclusion of multitenant communications requirements as part of the initial building design. Figure 7.3 depicts a typical riser

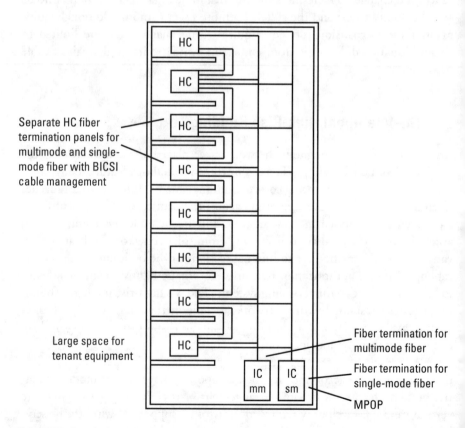

Figure 7.3 A typical multitenant, high-rise riser design.

infrastructure that can be utilized by multitenants simply by installing equipment at the various floor levels.

The riser layout shows single-mode and multimode cabling in a star and dual-ring configuration. In this example, separate fiber strands are available for transmission and reception at each floor level and for the interconnectivity between floor levels.

Open racks are placed in the riser rooms at each floor level and equipped with fiber termination equipment. All fiber management panels are placed by the architect in accordance with the structured wiring recommendations of BICSI, except for tenant-provided patch cords. The availability of both multimode and single-mode cabling offers a tenant flexible choices for single and multiple floor wiring concepts and equipment add-on. The equipment can be installed in the provided open racks and interconnected as desired. The availability of both fiber types in the MPOP in the basement will permit any desired connectivity for WAN and CLEC traffic, whether presently available or anticipated in the future. The MPOP location requires a larger space than floor-level riser rooms to accommodate ATM access concentrators for each client. The architect may consider a full rackspace for every future tenant and provide the fiber termination panels in a centrally located rack. The use of overhead fiber guides for the interconnections between the riser fiber termination racks and individual tenant racks is also recommended. Following the BICSI guidelines, the architect can assure compliance with all wiring codes and offer the client a cabling structure that will allow easy equipment integration and patchcord cable management to acceptable standards. The architect-provided infrastructure also discourages and minimizes the creation of new rewiring projects by client staff and assures that any additional wiring projects, if any, will fully meet any guidelines of structured wiring systems.

The architect can also predetermine the horizontal wiring component with reasonable certainty. The determination of outlet locations and outlet sizes is of key importance to avoid later rearrangement requests. The use of doublegang outlets with modular cover plates is recommended. The wiring standard of CAT-3 for telephony and CAT-5 for data delivery can be used by most tenants without the need for modifications. Bringing fiber or coaxial cabling to the desktop is an option that can always be implemented by the tenant in any later time period, provided that sufficient space for such additions has been provided in the flush-mount doublegang housings. HC panels, which can be placed by the architect at every floor level, will assure compliance with BICSI standards. Whether the tenant desires gigabit, ATM, or RF broadband services, he or she will be able to accommodate any desired equipment configuration.

7.4 Apartment Complexes

7.4.1 Conventional Wiring Methods

Apartment complexes have used RF broadband technologies for many years in an attempt to deliver cable television services to the tenants. Cable companies, when asked to supply cable services to a small number of apartment units, have to obtain permission from the landlord to string their coaxial cables throughout the complex. This permission has often been denied because of routing problems, aesthetics, and a landlord's desire to increase revenues by providing his or her own receiving equipment.

The first wave of apartment house cable systems consisted of Master Antenna Television Companies (MATV) that provided communal antennas on the roof of each building. They featured nonadjacent channel amplifiers and a loop-through coaxial wiring system to provide an outlet in each apartment. Loop-through coaxial wiring systems are substandard: They cannot provide equal levels, deliver a limited number of channels, and have been the cause of frequent complaints by tenants.

The second wave of RF broadband wiring was the result of the locally franchised cable company trying to market its services to the tenants. Some landlords saw the revenue potential of a centralized receiving location for cable television service, refused to give their consent, and opted to approximate the channel lineup of the cable company by installing their own satellite receivers.

Many of these "private cable" systems exist today in the apartment house community. Most of them do not comply with the basic wiring standards required to produce a maintenance-free and addressable network. Apartment complexes, for the most part, have been built without regard for the communications needs of the tenant. They do not feature risers or communication closets. Many telephone and cable television wiring systems have been installed by drilling vertical access routes between floors and using surface mountable molding to hide the wires in hallways and staircases. The idea of a structured cabling system for all communications services has escaped the industry. Energy-efficient building designs are seldom found and the need for tenant communications requirements are recognized only in the more expensive high-rise units in the downtown areas of most cities.

Even the cable companies went through a series of different ill-fated attempts to provide service to tenants until they developed the "home-run" distribution system concept. The home-run system consists of a RG-6 service drop cable that is connected between the building entrance location and each of the apartment units. This wiring system is very similar to a CAT-5 horizontal wiring where the building entrance location or the floor-level riser room provides a centrally located distribution point. A riser room on each floor level represents a service location to the service provider. The tenant can be serviced at this accessible location without requiring the technician to enter the apartment. The service can be connected, disconnected, altered with respect to service categories, and tested for quality and performance of the service.

7.4.2 Multiservice Infrastructure Alternatives

It is the existence of such a home run architecture that can provide new revenue sources to the landlord. Our discussion of multiservice alternatives for landlords starts with the assumption that the landlord already operates a satellite-delivered cable television system to tenants and that the wiring follows the general principle of structured wiring standards. This means that a riser space with a minimum of 5- by 5-ft riser closets are available at each floor level and that all horizontal wiring is "home run" between the riser closet and each of the apartments.

Since LECs or CLECs will enter the facility or the complex of buildings at street level, the wiring concept is similar to that of an intelligent building. All cabling is fed from the MPOP to each floor. Whether the cable television satellite installation is on the roof, at ground level, or in an adjacent building is not important. The satellite derived multichannel television transmission is simply interconnected with the MPOP of each of the buildings using single-mode fiber. In the case of a single high-rise building with satellite dishes on the roof, the combined channels can be connected with the MPOP in the basement using a single strand of single-mode fiber or a coaxial cable with a minimum diameter of 0.5 in featuring a solid aluminum sheath. It is further assumed that the existing horizontal wiring consists of quadshielded RG-6 cable with distances of 250 ft or less between the riser closet and the outlet. This distance limitation permits utilization of the cable for frequencies up to 750 MHz and enables the drop to be used for two-way services. Again, the installation of both multimode and single-mode fiber in the riser is recommended in order to be able to fulfill any future service requirement.

What does "multiservice" mean to the landlord? The Telecommunications Act of 1996 deregulated the business of telecommunications. A provider of telephone, data, or video services is not required to operate as a common carrier. A resale certificate will be issued by the FCC to anybody, after a 45-day notice period, when a #214 application form is filed with the Common Carrier Bureau of the FCC. This means that any landlord can establish reseller status and offer telephony, data, and video to its tenants. Whether telephony, data, or cable television, the landlord has the ability to negotiate telecommunications services at resale price levels and to provide its tenants with many economical communication solutions while accruing revenues never before anticipated.

Figure 7.4 shows the wiring and equipment requirement in the riser to offer multiservices to the tenant community.

The sketch illustrates a mixed fiber/copper and fiber/coax wiring in the riser that can provide standard ATM-based telephone service to the tenant via phone hubs and CAT-3 wiring, cable television distribution through RG-6 wiring, Internet access through cable modems over the RG-6 wiring, and data services at T-1 speeds over standard CAT-5 wiring. CTI can also be provided on the data network.

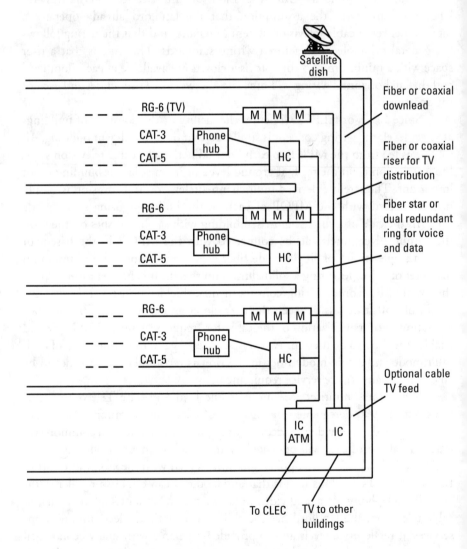

Figure 7.4 Multiservice alternatives in the apartment complex.

The tenants' obligations are to supply the telephone sets, computers, and cable modems. The landlord's obligation is to obtain the best connectivity arrangement for ATM-based integrated voice and data services from the serving public carriers or CLECs in the local loop. It is expected that local loop carriers will multiply during the next decade and that multiple choices will become available to the user. These offerings will include copper-based services like DSL, wireless services like LMDS, and HFC-based voice, data, and video services by the cable companies. The bargaining power of the landlord for discounted rates and the reselling of these services to its tenants can become a formidable source of additional revenues to every landlord who has installed a compatible multiservice infrastructure.

The example of Figure 7.4 is not the only solution. The system functions can be provided exclusively on the RF broadband system, which can avoid the CAT-5 and CAT-3 wiring to the units. The coaxial cable is certainly capable of transmitting all multiservice categories. The deficiency of this alternative is the requirement for cable modems for telephony, data, and Internet access within the apartment unit, which reduces the maintainability of the installation. Another alternative is to only provide CAT-3 wiring for telephony and to leave data services on the coaxial cable. After all, the cable modem used for Internet access can also be used for office connectivity and the need for CAT-5 wiring is negated.

In summary, the provision of multiservices by landlords to their tenants is of great importance in future apartment-house wiring concepts. The utilization of home offices in the conduct of business is increasing at a rapid rate. Telecommuters must be able to rely on voice and high-speed data connectivity with their places of employment. A properly designed RF broadband network in combination with improved data and voice connectivity can provide the landlord with additional revenue potential and the tenant with an array of economical multiservice solutions. Without the intervention of the landlord in providing these services, the problem of additional wiring requirements by tenants will involve the landlord, the tenant, and every new local loop service provider in frequent and unnecessary negotiations over new and ever-increasing wiring requirements of the future.

7.5 The Campus Network

Existing wiring concepts in the corporate campus have been based on the evolutionary development of proprietary vendor solutions. They represent the equipment and the wiring method that was successfully sold by the marketing person with the most convincing story. The proliferation of bridges, routers, and switches in today's campus networks is proof of this development. It has prepared the user to believe that any change in technology can be accommodated by upgrading the equipment. This migration path has been detoured by ATM and gigabit equipment offerings, and a new wave of higher speed backplane equipment is being offered to provide ultimate solutions for in-campus and WAN traffic requirements.

7.5.1 Conventional Wiring Systems

The three-level wiring system is the most common campus network architecture. The voice network consists of heavy multiple-pair copper cables with the PBX being the central location and the access point to the public network. Telephone network managers worry about new extensions, moves and changes, voice mail systems, long-distance service, and WAN connectivity on a daily basis. The expansion of phone service to dormitories in the university environment has caused wasteful expenditures in underground construction for more cables. The typical campus duct system is filled with copper to capacity.

Data networks started innocently in the administrative area of the campus, consisted first of a few isolated Ethernet islands, and grew in only a few years into campus-wide installations. In university campus settings, the student population added to the data network growth by desiring access to academic servers and Internet connectivity in the dormitories. Luckily, there are no copper cable star architecture requirements in the data LAN, and a total reconstruction of underground conduit plant was seldom required.

The third wiring system, especially in university campus environments, is the cable television distribution system. A revenue producer to the administration, the cable television system is usually a coaxial cable-based tree and branch network with the headend located in one of the campus buildings and with service drops reaching every dormitory. Some installations feature a separate system for two-way academic services and can be used to bring on-demand type videos to the lecture halls. This two-way RF broadband application reduces the need to physically move video presentation equipment carts through the campus pathways. In most cases, these wiring systems have been designed and implemented by the low bidder and feature no adherence to installation standards or quality requirements. In order to save cabling, these systems often feature substandard passive equipment and sometimes even amplifiers in manholes.

The life span of the three wiring system communication network is reaching its end. It is time to plan for a transition to a more unified and integrated technology that can serve the ever-increasing need for more voice, data, and video communication in a more economical manner.

7.5.2 Data Networking Alternatives

While the more advanced data networks of today feature FDDI in the backbone and CAT-5 wiring to the desktop, there are data equipment vendors already offering OC-48 switch/routers with 40-Gbps backplanes and OC-12 boards to workgroup switches that reduce the speed range down to fast Ethernet or 155.2-Mbps ATM for connection to the desktops. The offerings of ATM-based equipment and of gigabit-Ethernet will dominate the market during the next decade. However, there is no need to consider total rebuilding. A step-by-step upgrading process over a number of years appears to offer many economical solutions. (Users of this process are well advised to window-shop and research the many options.) Here are some examples of the upgrading process:

1. Reduce private line lease costs in the WAN by applying FR and low-speed T-1 ATM. Insist on QoS multilevel service categories.

2. Consider 10-bps voice compression technology to reduce long-distance telephone expenses to branch offices. Every private line (56K or 64K international) can carry five compressed good quality voice conversations.

3. Consider outsourcing of WAN lease line services to service providers offering ATM technology with all levels of QoS. VPN solutions over the PSTN network and tunneling through the Internet are available.

4. Refrain from using telephony over the Internet solutions until the next-generation Internet assures low latencies and QoS through IP over ATM.

5. Integration of data and voice services in the WAN is the first step in the multistep integration process and reduces the cost of communications.

6. Use the savings from the new WAN communication services to build up a LAN upgrade fund.

7. Consider the installation of a few single-mode fiber strands for backbone connectivity as the first upgrade activity within the campus.

8. Consider packet over SONET (POS) if your business only requires high-speed data traffic.

9. Consider ATM-based upgrading of the data network backbone first, even when considering voice/data integration and videoconferencing. The ITU-H.323 standard, however, is not yet perfected. Videoconferencing over IP networks is still a work-in-progress.

10. Consider using the new single-mode fiber infrastructure for RF broadband-based campus extensions, and overlay networking with telephony, data, and video in different 6-MHz frequency assignments.

11. The merging of the net management for voice and data is a good way to start the upgrade process. The lightweight directory access protocol (LDAP) permits the administration of policies and access rights for both data and telephone users. LDAP combines the directories of both networks and the foundation for a convergence to multiservice networking.

7.5.3 Computer Telephony Integration

CTI will be the topic for many years to come. There are many vendors and products that can fulfill the desire to integrate voice and data services in small LANs. CBXs are becoming a tool of choice for call centers. VOIP in a LAN with an ATM-based backbone is a feasible solution. QoS for IP is still in the standardization process. Voice over the Internet telephony (IP telephony) is usable through tunneling but not yet a perfected technology. There are too many routers in the Internet that work perfectly in a one-directional mode, but when required to organize a two-way voice conversation cannot cope with the delay and routing requirements. Latencies of more than 300 ms are offensive to the ear and do not permit a fluent conversation. While jerky 56K video sequences to the eye look like still picture sequences, the latency-rich voice exchange appears like a series of sound sequences. The connectionless Internet may route one direction through a different number of routers, and the voice quality appears good in one direction but becomes unacceptable in the other direction. Despite these problems with the status of the Internet technology of 1998, IP telephony will soon work fine when the new Internet is completed and uses an ATM/SONET/DWDM infrastructure with assured QoS levels.

Within a campus LAN, the problems of latency are minimized. Do not hesitate to try to integrate VOIP in short sections of the LAN, but convert to ATM-based integrated voice and data on the backbone. Java telephony application programming interface (JTAPI) from SUN, Microsoft's TAPI, and Novell's TSAPI are not yet recognized standards. ATM-based CTI will assure multilevel QoS within the LAN, the WAN, and through the outside world. However, VOIP will work fine at the desktop and on the established IP links. Conversion to ATM cell transfer on the backbone may offer the best compromise and speed the integration of CTI.

7.5.4 Multiservice ATM and RF Broadband

There are many reasons for a migration to ATM in the backbone. LANE permits the integration of legacy protocols within the LAN. The control of speed ranges becomes a tool to the network designer. The initial backbone speed may be OC-3 at 155.2 Mbps and can be increased to OC-12 and OC-48 on a scalable basis.

Servers can be interconnected with NIC cards in any format desired. The interconnectivity with service providers uses standard interfaces. While IP is connectionless and uses broadcasting to reach the desired server, ATM can reduce the load on the network by a connection-oriented transmission. While this technology is not easily accepted by IP experienced personnel, it is more easy to understand that a workstation requires less bandwidth if it only has to receive messages that are directed to that workstation. As a result, and with ATM in the backbone, a workstation does not require fast Ethernet or 155.2-Mbps ATM speeds. The old 10-Mbps Ethernet or a 25.6-Mbps ATM connection can accommodate the most stringent operational requirements.

The geographical layout of the campus and the already existing equipment are governing factors in the planning for ATM. These factors also form the basis for the integration of RF broadband technology into the network. The first step is to determine which of the backbone cable routes will use single-mode fiber. The decision can be made simply by estimating future service requirements and by assessing the campus geography to determine areas that may be better served by centralizing the data center. The physical locations of the PBX, the data center, and the servers are also important considerations in the determination of single-mode fiber routes. Other factors are the condition of existing outside plant facilities and the rate of the desired integration of voice and video into the network. It is good practice to, first, develop a planning document that permits the assessment of "what if" assumptions to assure flexibility for future requirements. Figure 7.5 shows a simple campus layout that utilizes ATM and ATM-RF broadband technologies side-by-side.

The schematic representation shows a single-mode fiber-optic backbone in a star architecture within a dual redundant ring interconnecting the PBX, the data center, and all buildings. Buildings with medium data speed requirements and CTI technology are served by ATM-based cable modems over RF broadband. Other buildings are served with ATM switch/router equipment in an OC-12 and workgroup switch hierarchy without CTI, but contain VOIP segments. The equipment layout is representative of a network upgrade in progress. The implementation of both high-speed ATM and ATM-RF broadband offers a good opportunity to monitor performance differences, study

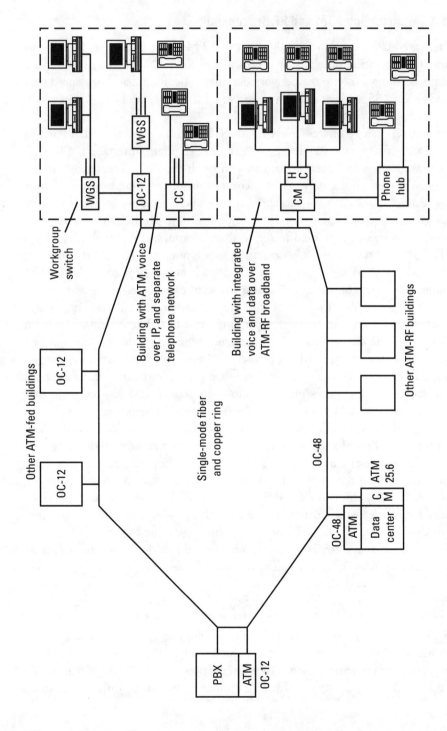

Figure 7.5 An ATM and ATM-RF broadband campus.

network management and QoS functionalities, and compare advantages and disadvantages of the two backbone technologies. Having high-speed ATM and RF-ATM working side-by-side will make it possible to better assess the differences of VOIP versus voice over ATM as well as to determine the benefits of voice and data integration and the various types of legacy system interfaces.

7.6 Ancillary Applications Using RF Broadband

RF broadband lives in the environment of the single-mode fiber and can fulfill any network application. Whether the signal is in an analog or in a digital format, RF broadband utilizes the bandwidth of the fiber to the fullest. The advantage of RF broadband is the ability to transmit a multiplicity of services within 6-MHz bandwidth increments.

7.6.1 Cable Television

One obvious application is the distribution of multichannel video. Cable television in analog form will soon be replaced with MPEG-2 encoded digital formats. Up to 12 digital television channels can be accommodated in one 6-MHz channel. The need for cable television, however, only exists in the college and university campus environment where video delivery is a revenue-producing commodity. RF broadband offers the only solution for the simultaneous distribution of many video channels. The inclusion of cable television in a multiservice network requires the selection of a block of frequencies for analog video distribution. The RF spectrum is wide enough to accommodate analog cable television as well as digital voice, data, and video frequency assignments. Campus buildings requiring cable television service must use coaxial service drop cables to the outlet locations. While the single-mode fiber backbone may carry all services side-by-side, the distribution segment may use RG-6 cables for cable television and CAT-5 wiring for voice, data, and digital video services.

7.6.2 Video on Demand

VOD, on the other hand, can be used not only for entertainment purposes but also in an educational setting. A hospital campus may opt for VOD to offer training material for paramedical, nursing, and medical staff. The ability to order a video lesson, at any time and whenever desired, makes VOD a desirable technology for the continuous education of all employees in any corporate setting. Internal company personnel training is now conducted during working hours and requires the scheduling and attendance of staff, thereby reducing the productivity of the company. VOD training can be conducted by any individual at any time and scheduled for slow business hours or after-hour time periods without affecting the productivity. VOD courses can also be provided with interactivity. By using the data network for ordering, the workstation for viewing, and the keyboard for responding to questions and problem solutions, VOD can offer automatic evaluations on the students retention and development. The new DVD technology is expected to aid the development of corporate VOD services through easier and truly interactive handling of the training material. Studio as well as storage and playback requirements can be accommodated in smaller spaces and assure a more efficient production.

The use of RF broadband offers the opportunity to make VOD available to individuals, groups, and departments by assigning RF channel frequencies of cable modems. In an OC-12 system, the use of multiple VOD transmissions of 8 Mbps each would affect the throughput. In the RF network all VOD traffic occurs on a separate RF frequency with no affect on other traffic requirements. The special issuance of cable modems for VOD amplifies the importance of the learning process and heightens the trainees' interest.

7.6.3 Videoconferencing

The incorporation of videoconferencing into day-to-day business activities has not progressed on the fast track as expected. The presentation of "talking heads" on the computer screen has not been considered a valuable tool to increase productivity. One of the many reasons for this rejection is the quality of video at 56K. The jerky picture presentations do not agree with our childhood experiences in front of full-motion television sets. Uncompressed video requires 140 Mbps to properly translate all artifacts into a digital format. High-quality compression systems used by long-distance service providers apply DS-3 at 45 Mbps. The new MPEG-2 compression technology is often compared to the NTSC analog standard but requires a VBR of up to 8 Mbps and is not yet available at reasonable prices. The most common transmission format used today is MPEG-1 using a T-1 bitstream at 1.544 Mbps. This format is used in long-distance applications and provides marginal motion quality. While it is not suited for full-motion television, it has become a useful tool for slow-motion picture sequences with good definition. The acceptance of videoconferencing in our daily business activities is directly related to picture quality. (See Chapter 11.) Whenever economically priced MPEG-2 encoding and decoding equipment is available, videoconferencing will flourish in both the LAN and WAN environments. The multiservice network provides a solid platform for the integration of videoconferencing into every corporate setting.

JPEG and MPEG-2 can readily be transported over ATM, fulfill the requirements for VBR transmission, and offer connection-oriented transmission between participants at multiple QoS levels. The use of RF broadband technology in the LAN assures efficient segmentation of MPEG-2 transmissions and workstations. A 25.6-Mbps cable modem can carry multiple MPEG-2 videoconferences with VBRs of 3–8 Mbps within a 6-MHz channel assignment without affecting the simultaneous transport of data.

7.6.4 Cable Data

The expression "cable data" has been used frequently to describe the ability to transmit data on cable television systems. The cable industry is in the process of upgrading the distribution systems to HFC-based technology. The first "cable data" offering by the cable industry has been the use of cable modems for fast Internet access. Some of these products only concern themselves with downstream support and use the telephone for upstream communications. Cable modems, which have been developed by a large number of companies and come in many sizes and shapes, will be the subject of Chapter 9. Cable modems simply convert the data stream to an RF frequency (6-MHz) assignment for travel over the HFC network. They can be used in the same manner in the LAN environment and in the multiservice network of any corporate entity.

7.6.5 Cable Telephony

There are a number of companies that have developed cable modems for the explicit purpose of telephone transmission on the HFC networks of the cable industry. While the industry's goal is to compete with the established LECs in providing telephone service to single residencies, the introduction of telephone service has been slow due to the lack of finalized standards and operational considerations. Cable telephony also uses cable modems for the transmission of voice. The new cable modem standard accommodates an Ethernet port and two telephone connections, ideally suited for residential subscribers. For businesses and apartment complexes equipment is available for massive voice interconnectivity. Using standard 56K compression technology, up to 240 telephone transmissions can be conducted within a single 6-MHz channel assignment. The status and the capacity of these cable modems is the subject of a more detailed discussion in Chapter 10. Cable telephony can be used in the LAN environment as well and offers the intriguing possibility of eliminating the complex copper wiring systems in today's corporate networks.

7.6.6 Personal Communication Services

The introduction of cellular telephones and digital PCSs has incorporated mobility of communications into our daily lives. It is certain that wireless communications will expand in the future. The FCC conducted auctions in the 1994 to 1996 period for a frequency band between 1.850 and 1.990 GHz. This band was auctioned by 15-MHz blocks for all major trading areas (MTAs) in the United States.

Commonly referred to as blocks A, B, and C, these frequencies cover the major metropolitan areas. Separate auctions for 5-MHz spectrum slots followed. These blocks of frequencies were called blocks D, E, and F and comprise areas in accordance with 493 basic trade areas (BTAs). In addition, the FCC has set aside an unlicensed frequency range in the 1.910–1.930-GHz transmission spectrum. This frequency range can be used by corporate entities for internal wireless PCS. A campus-wide wireless telephone system has a number of advantages to the user. It can improve the response times to react to emergencies, locate personnel, and increase productivity. A good example is the ability to reach medical personnel in a large, multibuilding hospital campus.

The multiservice network can be planned with the integration of a future internal PCS system in mind. Equipment available today features a strand-mounted transmitter/receiver unit that can be placed in the coaxial distribution network of a cable television system. The same technology can be used in the campus LAN. Figure 7.6 depicts such an installation.

In Figure 7.6, the transmitter/receiver units are placed in a coaxial section directly adjacent to a fiber receiver location. This may be a riser room at any floor level in a multistory building. Each of these PCS units can handle 19 simultaneous telephone conversations. When additional transmitter/receiver units are needed because of increasing traffic requirements, they may be placed in the riser on every second or even at every floor level. The installation is scalable but requires RF broadband technology for transmission over single-mode fiber strands. With new cellular digital packet system specifications (CDPD) in the standardization process, it will be possible in the future to use the PCS technology to transmit wireless data. One proposition suggests utilizing silence times and two-way conversational differences during voice transmissions to transmit data. The CDPD architecture can be used for TCP/IP and become a wireless access to the Internet. PCS used over the RF broadband system in a campus may be the answer for wireless extensions of the corporate LAN. There are many wireless LAN systems available, most of them applying point-to-point technologies to establish wireless data transfer. The unlicensed frequency band of PCS may offer an alternative to point-to-point

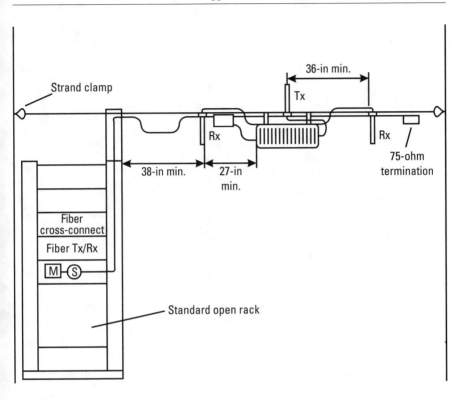

Figure 7.6 Typical campus PCS installation.

extensions and provide omnidirectional coverage for the integration of campus-wide wireless telephony and data.

8

Planning the Multiservice Broadband LAN

The process of planning a multiservice network is the most important first activity in establishing a multiservice broadband LAN. In the past, the network has grown in segments to fulfill the desires of users with either large budgets or political power. Now, however, an integrated network development plan must be conceived. A detailed planning document can unify the various infrastructure segments and address any future requirements. The first activity in this planning effort is familiarization with existing facilities. Second, a good understanding of structured wiring requirements will pave the way toward detailed identification of routing and implementation problems. Nothing is more frustrating than to complete the planning only to find that implementation is delayed because the costs and timelines for major construction activities have been overlooked. The evolutionary development of LANs and telephone networks has, in most corporations, taken separate paths. The reasons for this parallel development are organizational boundaries, departmental and reporting structures, budgetary differences, political considerations, and reliance on vendors. The planning effort for the multiservice network requires a fresh look at the organization and an integrated managerial structure. This does not mean a restructuring of every department, but it requires the application of a program management solution to network management. Program management techniques require the formation of a small action-oriented group of experienced people assigned to the planning process and exempted from their day-by-day operational duties for the duration of the planning phase. The program management group operates across departmental lines, collecting the departments' future requirements, analyzing every requirement into an integrated planning document, reporting directly to the executive branch, presenting findings of infrastructure alternatives to management, and obtaining approval to conduct the agreed-upon planning effort. This chapter deals with the sequences of such a planning effort.

8.1 Existing Plant Records

Locating existing plant records, and finding out that they are not complete, is the most frustrating part of the planning process. Many organizations retain detailed records of their infrastructure through as-built documentation packages provided by vendors. Because the submittal of as-built information by contractors is subjective and because contractors only concern themselves with the work that they have just completed, it is easy to detect deficiencies in content and continuity. Often, there are excellent network records prepared in Visio 5.0 that show every component of the network but lack cabling types and routing details. Other times, the facility management department maintains explicit records of cable and routing details but was not privy to a number of departmental additions during the last five years. Planning for a multiservice infrastructure requires the integration of all existing records and a physical inventory of the entire plant: every wire, cable, and piece of equipment.

8.1.1 Outside Plant

A detailed physical inventory of the outside plant segment is an integral part of the planning process. In the campus outside plant, a physical survey may reveal differences between the CAD drawings on record and the actual continuity. Planning for a nonobsolescent multiservice LAN requires an accurate record of the cabling continuity.

8.1.1.1 Routing, Manholes, and Distances

If the physical survey confirms the conduit routing information, credit goes to the facility department. Often, the continuity cannot be confirmed, especially when electrical and telecommunications conduits are in close proximity or even crossing each other. To determine conduit continuity in an accurate manner, it may be advisable to (1) install pullines in a spare duct between each of the manholes and (2) measure the distances between manholes and building entrance points. Manholes have been installed to various standards. It is a good practice to record the conditions of each of the manholes and to note the conduit entrance facilities. Innerduct did not exist a few years ago and is a good tool to separate cable runs from each other within the same conduit. The planning effort for the multiservice LAN may offer the opportunity to equip an empty duct with innerduct. The installation practices used for the installation of telephone cables did not include any mounting specifications unless the cable had to be spliced. When installing single-mode fiber cables it is helpful to apply observations of the deficiencies and to begin making notes describing the desired installation quality.

Innerduct can support two multipair fiber cables, when pulled at the same time. Fiber cables pulled at different time periods should occupy separate innerducts. Fiber cables destined to serve a campus zone can reside in the same innerduct. Depending on the system architecture, both fiber cables continue to each of the buildings in a particular geographical zone or split to form a ring between the buildings. When routed in separate directions, the use of separate innerducts is desirable. In any case, the cables require a service or expansion loop in each of the manholes. Such a service loop can be formed using both innerduct and cables in cable trays and using standard support hardware. While devising such a superior fiber cable support arrangement, make sure that the cable or the innerduct is not mounted in front of the ladder entrance. There are a surprising number of manholes that contain cable feeds at the entrance ladder and are frequently used as foot rests. While this condition is not harmful to heavy telephone cables, the use of fiber cables for human support should be avoided.

8.1.1.2 Conduit Occupancy

Conduit occupancy is the foremost problem in every campus outside plant. While plant records show available space, a physical survey will almost always reveal the presence of unknown cables. A two-inch conduit with a single cable can present a major problem to the accommodation of a new fiber cable. Metallic telephone cables have the tendency to crisscross and twist within the conduit envelope and can obstruct any further use. Ideal for planning purposes is the empty four-inch duct that can be subdivided into four innerduct routes. However, if this ideal condition does not exist, it is good practice to survey the existing cables, determine their use, and search for the unused cable that can be removed to make room for the new fiber infrastructure.

Since the purpose of the multiservice LAN is to eliminate the various wiring systems, there will come a point in time when most copper cables can be removed. It is left to the ingenuity of the network planner to determine the best and most economical approach toward this goal. This may include (1) avoiding a congested route by using longer distances, (2) replacing a congested segment,

(3) using less than a ring architecture in the beginning and devising a logical sequence of events to upgrade at a later date, or (4) use the existing pulline and monitor the pulling tension during the installation phase.

The completed outside plant network planning document should be based on accurate record keeping containing (1) manhole locations, (2) manhole continuity, (3) distances between manholes, (4) sizes of manholes, (5) condition and geographic location of conduit entrance stubs, (6) conduit occupancy, (7) determination of suitable conduit space, (8) suitability for innerduct installation, (9) availability of pullines, (10) room for cable extension loop and mounting methods, and (11) requirements for innerduct splicing to accommodate different cable routes.

8.1.1.3 Building Entrance Locations

Conduit sections between the serving manhole and the building entrance are usually congested. New cables for data and voice services have been added over the years, and construction practices have produced questionable routing solutions. In some cases, electrical contractors have been employed and, because entrance sweeps were not specified, 90-degree pull boxes have been installed.

To maintain pulling tensions and observe the minimum bending radii of fiber cable, the entrance facility requires a more detailed study. While it is often impossible to sweep through the foundation, a realistic practice is to use an S-curve sweep for building entrance at ground level and to sweep upward inside the building. Sufficient inside space or a 90-degree orientation of the incoming conduit to the building wall is needed; when available, they will provide satisfactory entrance routes. The network planner is well advised to budget priority funding for new conduit in the building entrance segment. The number of options that are available in the campus ring are generally not available in the entrance route. The physical survey of the entrance facility also includes an assessment of the space requirements for equipment. Usually referred to as a building, main, or intermediate distribution frame (BDF, MDF, or IDF, respectively) the building entrance equipment room will become the backbone interface location for the multiservice network and justifies special attention.

8.1.2 Inside Plant

The inside plant includes all wiring systems that have been installed during the evolution of telecommunications. Wiring systems between the building entrance location and the outlets exist for telephony, data, and sometimes video. Often this cabling has been installed in different time frames, by different contractors, and for various purposes. Accurate records are seldom available.

While in the past the use of different components, routings, and installation practices did not alter the network performance, in the age of higher digital speeds there cannot be any compromise. New wiring systems must be developed on a system basis and follow the established standards of BICSI. The application of structured wiring standards and methods as well as a system approach to wiring can assure that the new cabling will not have to be replaced. Cables and compatible connecting components have matured to assure long life and maintenance-free operation.

8.1.2.1 MDF Equipment

An accurate record system of existing equipment at the MPOP, at alternate points-of-presence (APOP), and at MDF building entrance locations is essential for the development of the multiservice LAN. A physical inventory of existing equipment, space availability, and electrical service is required to assess any expansion construction requirements that may be needed. Some of the major concerns are listed as follows.

- *Powering:* When the electrical service wiring is provided as a single circuit, it is wise to establish a multicircuit feed and a secondary connectivity to a standby power source consisting of battery backup and emergency generator. New ATM equipment features two independent power supplies. If one of the two power supplies is connected to standby power, the reconvergence times due to a power failure can be in milliseconds.

- *Space:* The available space for new equipment is often limited, and it is prudent to utilize unused rack space for new equipment. Single-mode fiber termination equipment as well as transmission equipment requires special attention to cable management details and deserves its own open rack with properly installed AC strips, side guards, and horizontal and vertical cable management guides.

- *Cleanup:* A good assessment of all existing wiring and a redressing and cleanup of telephone and data network terminations are desirable. Multimode fiber cables that have been terminated in cross-connect panels may use various types of patch cords and connectors that do not meet present structured wiring standards. Existing horizontal wiring cross-connect panels require attention as well. Check all cable markings and record the source and destination points of all wiring. Performance testing of fiber and horizontal wiring systems may also be required to assure performance of the existing wiring for higher speed ranges.

8.1.2.2 Vertical Risers

The days of multiple riser installations without a stacked riser room at each floor are numbered. Developing a record system for separate telephone and data risers may be difficult and may offer ideas as to consolidation and a vertically stacked line-up of all equipment closets. While the new installation may require single-mode fiber, coax, multimode fiber, and CAT-5 to run side-by-side, the separation of these cables in the vertical riser should be done in an orderly manner. Check for the existence of fire stops between floor levels and make a list of all cleanup and construction activities that are necessary.

8.1.2.3 IDF Equipment

IDF equipment at the various floor levels is used for the termination and cross-connect of telephone and data wiring. Existing installations sometimes feature wall-mounted telephone cross-connect panels, lack of markings, and wires in disarray.

To reuse the CAT-3 wiring system for phone hubs in a multiservice environment, it is good practice to inventory all incoming and outgoing wires. The interconnection of telephones with computer terminals and of stand-alone telephones require detailed cut-over sequences.

8.1.2.4 Horizontal Wiring Systems

Recently constructed CAT-5 wiring systems require performance testing and a good determination of length. Any cables found to be over 90m should be exchanged to CAT-5-enhanced (CAT-5E), which means using the cable with enhanced hardware. Cable lengths under 90m may prove to be an asset, as they might meet CAT-5E performance specifications. Check all terminations and connecting hardware at the cross-connect panel. By changing this hardware and patchcords to the enhanced type, the overall performance of the existing wiring can be improved. The record system is not complete until all source and destination points are identified, tested, and properly marked.

8.1.2.5 Outlet Installations

Outlet installations exist in many variations caused by ever-increasing requirements for communication. An exact record of outlet types and location almost never exists. There are many installations that feature separate telephone, data, and video outlets. Modular outlet plates offer an opportunity to combine these services.

When planning for integrated service outlets, double-gang housings and plates are recommended. The additional room behind the cover plate can become very useful in the cable termination process. The avoidance of overbending is an important consideration in CAT-5, fiber, and RG-6 cable terminations. The usual practice of the contractor is to terminate the cables with enough room to use the tools of the trade.

After completing the terminations, the outlet plate is simply forced to fit the housing, and the excess cable has to adapt by bending and kinking. The use of double-gang housings with a minimum of two inches depth will reduce this problem and provide for the mounting of up to eight modular connectors on one face plate. In cases where dual RG-6 connectors are required for multiple users or for the separation of transmit and receive direction, the dual housing offers sufficient room to locate a miniature splitter.

While recording outlet types and locations, it is advisable to remove a number of face plates at random and to survey the physical condition of the wiring. Mishandled CAT-5 wiring terminations with reduced twist will not meet recommended performance requirements. If doubtful conditions are found, the scheduling of performance testing and retermination of the connectors is recommended.

8.1.3 Computer Center and Server Farms

The planning process for the multiservice LAN should include a survey of all server locations to determine whether changing the location may provide economical advantages. Since gigabit or OC-48 connectivity is desirable between servers and backbone ATM switching equipment, the assembly of many servers at a common location permits more economical interconnectivity. The planning process may also include an assessment of a server area network (SAN) to improve data accessibility.

After recording the locations and type of servers and when a new location for the server farm has been found, it is important to consider cut-over procedures that will minimize outage times. Other planning considerations are:

1. Powering (the development of a stand-by power service feed for all server units is more easily accomplished for server farms than for decentralized servers in a multibuilding environment);

2. Cabling (gigabit or OC-48 interconnectivity may require the use of single-mode fiber because of the required distances).

However, whether single-mode or multimode fiber is used, bringing the fiber cables to terminations and cross-connect locations in close proximity of the equipment is recommended. The use of CAT-5, CAT-6, or any copper wiring should be limited to short footage or avoided entirely by direct fiber terminations.

In planning for the server farm location, it is important to consider the various campus cabling architectures that are available. Figure 8.1 shows a typical campus architecture using a double redundant source interconnect ring and a star cabling campus delivery network.

The dual redundant source interconnect network is a good solution for high-speed interconnectivity between all major traffic sources (see also Chapter 15).

The MCC serves as the gateway to the campus. ATM switching equipment at the source locations can provide reconvergence times in milliseconds in

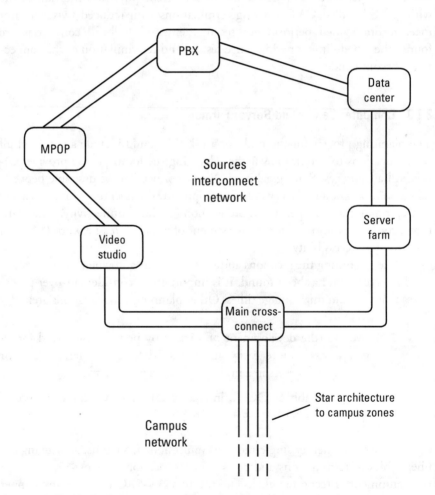

Figure 8.1 The dual redundant source interconnect network.

case of damage to the ring continuity. Locating the server farm within the ring interconnecting all source locations enhances the robustness of the network.

8.1.4 PBX Equipment

Single-mode fiber continuity to the existing PBX equipment location is a basic consideration for the future multiservice network. Even in the case that there is no immediate requirement for voice and data integration, the PBX is a source location and deserves to be included in the dual redundant source interconnect ring.

When updating the record system, it is good practice to inventory the existing PBX equipment and to study the extension numbering arrangements, the conditions of the termination panels, the availability of cables and equipment for PTSN interconnectivity, the MPOP location, and the cable routes used to the individual campus buildings.

This information will be helpful in the planning process for the multiservice LAN and will identify problems and shortcomings. The findings can then be used to establish a step-by-step conversion toward voice and data integration.

8.1.5 Data Network Center

A good inventory of the existing data center equipment is the foundation for planning the multiservice network. If the new campus backbone is planned for ATM-based services, the existing layer 2 and 3 switches may serve an important role for upgrading the data network in larger buildings. Dependence on known IP technology can be perpetuated; VOIP can be added. The functionality of VOIP is excellent in the restricted environment of a building or even a group of buildings. The use of ATM-based technology in the backbone, however, increases the robustness and usefulness of the network by adding QoS parameters and faster circuit setup and tear-down intervals. With reconvergence times reduced to milliseconds, ATM cell switching increases the resilience of the network for the multimedia requirements of the future. The ATM-based dual redundant source interconnect backbone optimizes the flexibilities in the network emanating from the MCC. A mixture of workgroups can be established and can be IP-, ATM-, or RF broadband-based.

8.2 Conformance to Structured Cabling Standards (BICSI)

Over the last 10 years, Building Industry Consulting Services, International (BICSI) has developed international cabling standards for all structured cabling systems. BICSI is a nonprofit telecommunications association focused on low-voltage wiring. By the end of 1997, BICSI exceeded 12,500 members residing in 60 nations. An international cabling standard was initiated in 1990 and developed by ISO/IEC committee SC25/WG3. ISO/IEC 11801, the generic premises cabling standard, was published in August 1995. Fifteen nations—Australia, Belgium, Canada, Denmark, Finland, France, Germany, Holland, Italy, Japan, Norway, Spain, Sweden, the United Kingdom, and the United States—participated in the project. The existing U.S. cabling standard, ANSI/TIA/EIA-568, was used as the basis for this international standard but was further modified and amended to include other inputs. The ISO/IEC 11801 standard defines cabling infrastructures with a life expectancy of more than 10 years. The standards apply to cable manufacturers, suppliers, end users, building professionals, and application groups. The standard is designed to support LAN and telecommunications services within a building and a multi-building campus area. ISO/IEC 11801 covers all aspects of cabling and connector hardware between equipment of any kind and the telecommunications outlet. While work area cabling is beyond the scope of the standard, some useful guidance is provided.

ISO/IEC 11801 offers two different approaches for implementation. One method is based on standard components, the other on link performance. The standard defines two performance parameters, the *link* and the *channel*. The link is commonly defined as the work that the contractor performs. This work incudes the installation of cable and the terminations of the cables. The channel defines the performance of the entire path from the equipment through the cross-connect, the patchcords, the link, and the work area cabling. The development of BICSI cabling standards is ongoing and similar to that of the ATM Forum, which is organized in task groups. The recommendations of the task groups are covered by task recommendation (TR) numbers. The more important task groups are listed as follows:

- UTP systems task group;
- Connector task group;
- Editorial task group;
- Fiber-optic task group;
- STP-A task group;
- ScTP task group;
- Harmonization task group;
- Next generation task group.

There are various standards in existence, and new standards are under development. The user of campus structured cabling systems is well advised to participate in BICSI to be able to understand the reasoning for the more important standards. BICSI presentation summaries are available for every area of interest to the network designer. They include items such as horizontal telecommunications cabling support systems and fastening techniques, characteristics of coaxial cables in HFC systems, premise wiring and cable management, the relationship between infrastructure and management, coping with continuous change, near-end crosstalk (NEXT), far-end crosstalk (FEXT), return loss, and performance testing.

8.2.1 Single-Mode Fiber

The link performance specifications in ISO/IEC 11801 define various classes. Copper cabling links are covered by classes A through D. The optical class defines links above a bandwidth of 10 MHz and addresses both multi- and single-mode fiber connectorization.

8.2.1.1 Fiber Termination and Cable Management

The standard specifies two optical connectors, the SC and the ST connector. If ST connectors are already in use, then their continued use is allowed. New installations shall utilize the SC connector for all multimode and single-mode applications. The SC connector is the only connector that is available in a duplex version for forward and return transmissions.

Fiber termination equipment as well as cross-connect panels shall use the paired SC connector versions for all single-mode fiber installations. To meet the link performance specification of the fiber backbone cabling, it is important to consider matching components for the entire channel. This means that rack enclosures, termination panels, paired SC connector modules, and patchcords are compatible in design, manufacture, and deployment.

BICSI is working on the standardization of a single dual-fiber SC connector. This product, which has been introduced by AMP (a fiber-optic component supplier), would reduce the connector space requirements by 50%. In this architecture, the fiber connector panel could have twice the density of standard copper wiring connector panels. In the BICSI meeting of November 1997, the TR 41.8.1 working group did not reach the two-thirds majority required for the introduction of the AMP- MT-RJ connector, but it is possible that this new connector standard may soon become effective.

8.2.1.2 Equipment Interconnection

In the planning effort for the single-mode fiber backbone, it is important to determine (1) the rack space requirements for fiber cable termination, (2) the equipment space requirements, and (3) the cable management configurations. Cable management is an important factor in the dressing of patchcords used for interconnection of equipment and cabling. The use of single patchcords should be avoided. Paired patchcords with SC connectors can be routed and managed more easily. The selection of appropriate vertical and horizontal cable guides that permit separate routing of patchcords is recommended in order to be able to identify each cable without manual handling and to avoid excessive pulling tensions or gravitational loads on the cords.

8.2.2 UTP Wiring

Link performance specifications have been standardized for the following:

- *Class A:* Bandwidth to 100 kHz for speech and low-frequency applications;
- *Class B:* Bandwidth up to 1 MHz for medium bit-rate applications;
- *Class C:* Bandwidth up to 16 MHz for high bit-rate applications;
- *Class D:* Bandwidth up to 100 MHz for very high bit rates (CAT-5).

A new class E specification is under development for CAT-6 UTP. The European community also requires a class F specification to cover CAT-7 STP, shielded twisted pair, which is required because of more rigid radiation standards.

The key performance parameters of the class D-CAT-5 cabling are the attenuation of a two-pair system, NEXT, and the attenuation-to-crosstalk ratio (ACR). These performance parameters have been standardized by TIA/EIA-568-A for CAT-5 cable and by ISO/IEC 11801 for class D cabling. An addendum of the standard covers the four-pair nature of the cabling. It adds additional performance parameters, such as the power sum NEXT (PS-NEXT), the power sum equal-level FEXT (PS-ELFEXT), power sum ACR (PS-ACR), return loss, and propagation delay (skew). In the four-pair architecture, the PS-NEXT represents the accumulated NEXT from three pairs when measured on the fourth pair. The PS-ELFEXT represents the power sum of the FEXT from three pairs carrying equal-level signals to the fourth pair. The return loss is a measure of the impedance mismatch of each pair and a measure of electromagnetic interference (EMI). While NEXT can be somewhat compensated for by digital signal processing, FEXT and EMI cannot be reduced by any signal processing method. Propagation delay or skew increases with frequency and attenuation, and limits the performance of UTP wiring.

8.2.2.1 UTP—Enhanced CAT-5

While it was assumed that the original performance specifications for CAT-5 cabling would provide a satisfactory transmission up to 100 Mbps, additional testing has proven that the originally installed CAT-5 cabling may not meet the increased requirements for attenuation, return loss, and skew. The new CAT-5 performance requirements, which are commonly referred to as enhanced CAT-5, currently comprise the only version being sold. The user should remember that the difference between the two specifications is not so much caused by the cable itself but often a factor of insufficient care taken in the installation process as well as the use of substandard hardware. In many cases, it will be possible to meet the channel and link specifications of the CAT-5E cabling by testing the installation and by identifying the problems with workmanship and hardware. The use of CAT-5E hardware and adherence to quality of workmanship can eliminate most performance problems, except in cases of cable lengths that are longer than 90m. Table 8.1 presents the link and channel performance specifications of the enhanced CAT-5 cable.

When using UTP cabling, it is important to consider matching hardware components. The use of individual cable components is not a viable option. The cabling systems should be a structured system of performance tested components. While UTP has come a long way, it is still based on a telephone

Table 8.1
Enhanced CAT-5 Cable Performance

Performance at 100 MHz	Channel (dB)	Link (dB)
Attenuation	24.0	21.6
Pair-to-pair NEXT	30.0	32.0
PS-NEXT	27.0	29.3
ELFEXT	16.0	18.0
PS-ELFEXT	16.0	18.0
ACR	6.0	10.4
PS-ACR	3.0	7.7
Return loss	10.1	12.1

connector standard that, instead of transporting 3 kHz, is asked to transfer 100 Mbps.

8.2.2.2 UTP—CAT-6

The proposed standard for 200-MHz performance UTP cable is called CAT-6/ class E. The ISO/IEC 11801-A standard has been released to the IEEE and the ATM Forum to assist in new networking technology. Whether this standard will be used is a question of whether higher than 100-Mbps speeds will be required at the workstation level. It appears today that there is no need for 155 Mbps at the desktop and that gigabit Ethernet has no functional requirement. All other speeds such as 25.6-Mbps and 51.2-Mbps ATM as well as 100-Mbps fast Ethernet can be handled with the enhanced CAT-5 standard. However, new installations may be well advised to use the CAT-6 wiring system to cope with any future high-bandwidth requirements that are not known at this time. Table 8.2 outlines the link and channel performance specifications of the CAT-6/class E cable.

It can be seen that the performance of UTP cabling has reached its technological limitation at a bandwidth of 200 MHz. Figure 8.2 presents a graph of the NEXT loss and attenuation as a function of decibels over megahertz.

The ACR specification is 3.1 dB minimum for CAT-5, 6.0 dB minimum for CAT-5E, and 18.3 dB minimum for CAT-6. It is obvious that the CAT-6 wiring system requires perfectly matched components (cable, twisting, jacks, jackfields, connectors, and patchcords) to be able to meet the performance requirements.

Table 8.2
CAT-6/Class E Cable Performance

Performance at 200 MHz	Channel (dB)	Link (dB)
Attenuation	31.8	27.3
Pair-to-pair NEXT	34.9	36.9
PS-NEXT	31.9	34.3
ELFEXT	17.2	19.2
PS-ELFEXT	14.2	16.2
ACR	3.0	9.6
PS-ACR	0.0	7.0
Return loss	9.0	12.0

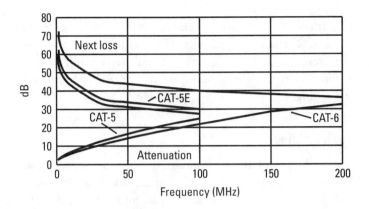

Figure 8.2 ACR for CAT-5, CAT-5E, and CAT-6.

8.2.2.3 STP—CAT-7

In the rest of the world, the use of shielded cables is quite predominant. The CAT-7/class F standard originated in Germany under E DIN 44312-5 and was recently proposed by ISO/IEC 11801-A to accommodate speeds for ATM service at 622 Mbps. STP cables have many advantages over UTP. The overall foil shield restricts radiation of energy due to impedance mismatches, and the individual shielding of each pair improves the NEXT performance substantially. Final performance specifications have not been adopted, but CAT-7 is used frequently in new installations on the European continent. The reason for accepting this higher cost solution in workstation wiring is related to the laws covering EMI transmissions. In many European countries, the user is liable for any damage caused by radiation of electrical energy. While the shielding properties of CAT-7 cables can widen the bandwidth performance, the length of the cable is identical to the 90m specification of UTP because of the unchanged attenuation value.

8.2.2.4 Testing Guidelines

Since the frequency range of twisted cables extends to 300 MHz, test equipment requires performance in a swept frequency domain. There are many products on the market that meet a full suite of autotest functions and that can be used for link and channel measurements. The latest edition from Microtest® is the Omniscanner, which can be used for measurements at 100, 200, and 300 MHz and provides for a variety of test functions. Every aspect of the wiring can be tested, inclusive of NEXT, return loss, ELFEXT, attenuation, ACR, PS-NEXT, PS-ACR, and PS-ELFEXT. In addition, the unit can be used to find miswires, opens, shorts, and crossed or split pairs as well as to measure the exact length of the cable to a 1-ft resolution and the impedance to a resolution of 0.1 ohm. Testing of all existing UTP installations to the enhanced standards is recommended. Planning for the multiservice infrastructure for the future requires a good understanding of the capabilities and limitations of the existing wiring.

8.2.3 Coaxial Wiring

Coaxial wiring does not have the limitations that become evident when assessing UTP cable performance. The RG-6 quad-shielded cable has an outside diameter of 0.25 in and costs only about 15% more than CAT-5E cable. The advantages of using a coaxial wiring system for the last 250 ft of service footage are plentiful. There are no bandwidth limitations, no NEXT or FEXT, no ACR, and no radiation. Coaxial cable has been proven in the outdoor environment for years and its only performance parameter is attenuation. While UTP cables in a 90m wiring system have an attenuation of 21.6 dB at 100 MHz (CAT-5E) and of 31.8 dB at 200 MHz (CAT-6), the RG-6 coaxial cable has an attenuation of about 10 dB at 1,000 MHz. Figure 8.3 shows the attenuation over frequency of a 250-ft-long RG-6 coaxial cable and a comparison to CAT-5E and CAT-6 cables.

In Figure 8.3, the cable terminates in a simple F-connector at both ends that provides radiation-proof connectivity with multitaps and outlets. One of the major deficiencies appears to be the lack of a telephone-type connector, but the screw-on F-connector has the advantage of a more robust connection. Coaxial cabling to the desktop is not a requirement for baseband transmissions of speeds up to 500 Mbps, but should it be required to bring 622-Mbps ATM, gigabit Ethernet, or RF broadband frequencies to the workstation, the RG-6 coaxial cable offers an attractive and economical alternative solution.

BICSI recognizes the need for multiservice wiring on a single network and has issued guidelines for hybrid fiber/coaxial and coaxial installations; it describes the RF technology as to its importance in campus and premises installations. While it is recognized that the coaxial cable is a robust medium for wide bandwidth performance, it is stressed that connectorization has to be performed in strict accordance with the supplier's recommendations. In many ways, the installation skill level required to install an F-connector is less demanding than the skill to properly terminate four pairs of UTP in the RJ-45 connector. However, while the mangled RJ-45 connector still works, the improperly prepared F-connector will totally impair the electrical performance.

The testing of the coaxial service drops does not have to include sweep testing. A simple level test of the link can determine continuity and transfer of the signal at a predetermined RF frequency. A signal generator tuned to

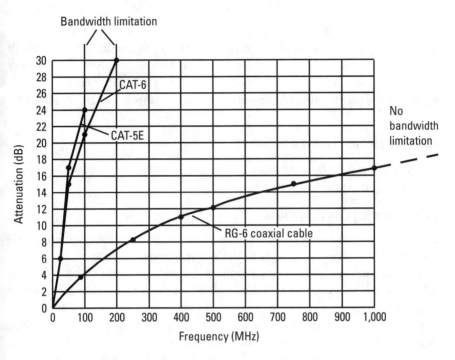

Figure 8.3 The 250-ft RG-6 service drop attenuation.

1,000 MHz can be connected to the input port of the first multitap at the HC horizontal cross-connect location. The level is set to +16 dBmV. A signal level meter is taken to every outlet location, and the received level is measured. A measurement of better than +5.0 ± 1.0 dBmV proves the continuity and quality of the installation. Every unused service drop shall be terminated with a 75-ohm F-terminator to maintain the proper impedance of the network.

8.3 Network Architecture Preferences

The architecture of a network concerns itself with the planning of the frequency spectrum, the physical routing of the backbone fiber system, and the interconnectivity between communications sources. While there are multiple choices in each of these areas, it is important to relate these choices to the existing plant, the geographical layout of the campus, and the user population. Because the records of the existing communications network elements are documented, it is a good practice to take a bird's eye view of the campus and to assemble several "what if" scenarios relative to user topology, future demand,

integration of voice and data, integration of videoconferencing, WAN requirements, campus zoning, and sequencing of the steps required to transform
the network into a multiservice facility. The brief description provided by
Section 8.3.1 cannot give justice to the numerous factors that require consideration in the preplanning phase; only a few basic components are mentioned.

8.3.1 Preplanning Considerations

Before starting the planning process there are a number of preplanning activities that require consideration.

8.3.1.1 Routing Considerations

The multiservice network must be self-healing and robust enough to limit connectivity and equipment failures to the shortest possible outage times. A good
measure toward this goal is to interconnect all source locations with self-healing
ATM switching equipment in a dual-redundant ring architecture. This source
ring becomes the headend to the campus backbone and the gateway to the outside world. The campus backbone emanates from the source ring at a strategic
location that can be chosen in the most economical manner based on the existing underground plant geography. This location will be referred to in the following discussion as the MCC location. While the MCC may be collocated
with one of the communications sources, it is the network's origination center.
Figure 8.1 indicates the source's interconnect ring and the campus backbone.
Campus geography, existing conduit facilities, zoning, and service redundancy
dictate whether the campus backbone should become a star, ring, or star within
a ring architecture. Figure 8.1 shows a simplified presentation of the ATM
high-speed self-healing interconnection between all source locations. The campus backbone is shown as a star but can also become a dual-ring connectivity to
protect every campus building with routing redundancy.

8.3.1.2 Zoning Considerations

To assess the various service requirements and the existing particularities within
each of the campus buildings, it is advantageous to subdivide the campus into
planning zones. As these planning zones later will become implementation
sequences, it is a good practice to identify the more pressing service requirements into the low zoning numbers. A campus zone can consist of groups of
adjacent buildings or be developed in accordance with business interests. In an
educational campus, the zoning may address the grouping of academic buildings and student residencies. The alternative architectural choices can be
applied in accordance with the desired service robustness. A hospital campus
may utilize a star within a redundant ring architecture for all medical

emergency and surgical departments and apply simple star architecture to administrative and material management groups. Chapter 15 and Figures 15.1 and 15.3 deal with the subject of campus zoning in more detail and address baseband and RF broadband solutions.

The use of single-mode fiber permits a new level of routing ingenuity. The distance factor is eliminated even for future multigigabit service requirements. As a result, the network designer is not limited by geographical zoning considerations but can develop the assignment of zoning numbers in accordance with business functionalities. The result may be a meshed interconnect fiber connecting high-priority zones with full redundancy and using single-star architectures for less essential business units.

8.3.1.3 RF Spectrum Planning and Management

The self-healing, ATM-based source ring does not require any spectrum considerations. Even at OC-64 or higher, the transmission is still at baseband. RF spectrum planning is only required for backbone continuity that uses the single-mode fiber for multiple RF services. The spectrum of an RF broadband transmission system ranges from 20 to 750 or even 860 MHz, depending on the laser transmission capacity. The single-mode fiber is a unidirectional transmission facility and can be divided in 6-MHz channels for forward and return signal transport. In cases where existing or new UTP wiring is used for desktop connectivity, it is only necessary to plan the utilization of RF frequencies in relation to the location that is served by one cable modem. User areas at floor levels and smaller buildings can each be assigned a set of frequencies for symmetric or asymmetric forward and return transmission.

Only in cases where coaxial distribution to the desktop is used, the RF spectrum planning must consider the two-way capability of the coaxial cable. For these cases, it is necessary to assign the cable modem frequencies in accordance with the subsplit or high-split frequency standard. Also, when using coaxial amplifiers to extend the vertical riser, RF spectrum planning is required to be able to use readily available amplifier architectures. Should the decision be made to build a dual system, the entire passband of 20–860 MHz becomes available for both directions.

8.3.1.4 Bandwidth Planning and Management

A closer look at the present and future user population may reveal a large number of workstations that will never require more than 10-Mbps Ethernet or 25.6-Mbps ATM service. These user groups that may be located in the outlying campus buildings can be served by the application of RF broadband. Using RF frequencies on a single fiber strand will reduce the number of fiber cables required in the campus network.

Other user groups, such as CAD departments and radiology or imaging centers, are better served at higher speed ranges and require high-outbound speeds to connect to servers. A survey of the user population can be used to establish user classifications such as the following:

1. High-inbound and -outbound user-OC-3 scalable to OC-12;

2. High-inbound and low-outbound user-OC-3 scalable but asymmetrical;

3. Medium-inbound and medium-outbound user-25.6-Mbps symmetrical;

4. Medium-inbound and low-outbound user-25.6-Mbps asymmetrical;

5. Low-inbound and -outbound user-10-Mbps symmetrical;

6. Low-inbound and lower outbound user-10-Mbps asymmetrical.

Having determined existing users in this manner, and by mapping the user population categories within the geography of the campus, network planning can proceed in an intelligent manner. It is noted that the above user classifications can be made solely on the basis of data traffic. Voice only adds a fraction to the data load and can be considered as overhead. Videoconferencing, in contrast, adds substantial load requirements in both directions. A future videoconference user should be classified as in (3) above. A 25.6-Mbps bidirectional continuity will even satisfy future MPEG-2 video streams running at VBRs of up to 8 Mbps. VOD services are typically a (2) requirement.

The network designer's ingenuity can now be applied to develop economical network alternatives. Bandwidth planning is one of the most important considerations in the preplanning effort for the multiservice network. The designer must make many decisions relative to route redundancy, developing service alternatives, considering fiber routing alternatives, and determining alternative equipment complements, keeping in mind the costs of all network components and the goal to determine the most economical solutions.

8.3.1.5 Traffic Management and QoS Considerations

QoS levels can be used to enhance the traffic engineering of the network. As stated before (in Chapter 5), ATM-based cell switching products can apply a contractual transmission relationship for any user. For instance, a videoconferencing user can be assigned rt-VBR with a higher priority than a distance-learning user. Each of the QoS service categories, including VBR, rtVBR, ABR, CBR, and UBR can be assigned with four priority levels. Manufacturers of ATM equipment have taken these classifications to even finer detail. For example, FORE and others offer network management software that permits a total of 256 sublevels. Some of these sublevels can be programmed by the user;

others are activated automatically in the cell header of the information. The selected QoS level and sublevel assures minimum traffic speeds to be transmitted and received by every user terminal.

8.3.1.6 Powering Issues

The preplanning effort includes an assessment of the present power service. In locations where self-healing ATM equipment with dual-power supplies is to be installed, stand-by power service is required. Connecting the equipment power supplies to another phase of the main power service is not a viable solution. Once the routing of the single-mode fiber and the major equipment installation locations are determined, critical power service locations can be identified and the cost of a second independent or stand-by power feed established. Stand-by power is an important consideration in the design of the multiservice LAN. Often, it is argued that the telephone system is the network of last resort. A failure of the data network requires the existence of a separate telephone system. At least, it is argued, there is voice communications when the power is down. These statements were true for the age of the "best-effort" data network but do not apply to resilient ATM-based multiservice networks as long as a second power source is provided at all backbone equipment and at phone-hub locations.

8.4 The Planning Document

The planning document is an assembly of proposed activities to be conducted in a sequential manner. A hierarchical presentation of all steps and events is best suited for this documentation as it permits the inclusion of a timeline for construction activities, development of RFPs, network design activities, implementation, and operational readiness of the new facility. It is good practice to develop the planning document in phases in order to account for cut-over events and an upgrading to the new multiservice network on a step-by-step basis. Because campus geographies, building layout, and network requirements are very different, the information in Section 8.4.1 and for all network components of Sections 8.4.1–8.4.6 is presented as a checklist of the most important tasks only.

8.4.1 The Single-Mode Fiber Interconnect Facility

8.4.1.1 MPOP, Data Center, PBX, Server Interconnectivity

1. Fiber routing requirements:
   - Inside routing plan for redundant ring;

- Outside plant routing for redundant ring;
- Recommended location of MCC;
- Accessibility of MCC to campus outside plant network.

2. Fiber cable requirements between source locations and MCC:
   - Interconnection with PBX;
   - Interconnection with servers;
   - Interconnection with data center;
   - Interconnection with MPOP;
   - APOP considerations.

3. Alternative speed requirements in the sources ring:
   - 2.488-Gbps OC-48 routes;
   - 622-Mbps OC-12 routes.

8.4.1.2 Campus Zoning Requirements

1. Rationale for zone 1:
   - Number of present users and service speed requirements for data;
   - Number of future users and service speed requirements for data;
   - Percentage of computer telephony voice integration required;
   - Percentage of videoconferencing users;
   - Percentage of VOD users for continuing education;
   - Fiber routing requirement;
   - Dual-ring redundancy or star architecture;
   - Number of buildings in zone;
   - Number of buildings with riser requirements;
   - Building numbers with fiber termination equipment at MDF;
   - Buildings with fiber or coaxial riser.

2. Rationale for other zones—repeat entries from zone 1 above.

8.4.2 Outside Plant Requirements

1. Zone 1—outside plant requirements:
   - Fiber routing from MCC—manhole continuity;
   - Redundancy considerations;

- Conduit construction requirements;
- Estimated cost of conduit installation;
- Alternative routing solutions;
- Number of buildings;
- Number of fiber strands;
- Recommended fiber cable, type, number of strands;
- Total estimated length of fiber requirement.

2. Other zones—repeat entries from zone 1 above.

8.4.3 Power and Space Requirements

8.4.3.1 Power and Space Requirements at Source Ring Locations

1. Space and powering requirements for data center network equipment:
   - Number of racks;
   - Size of fiber termination panel;
   - Estimated rack space for equipment;
   - Present powering availability;
   - Additional emergency powering requirement and cost estimate;
   - Special entrance cabling and wiring recommendations;
   - Equipment room modifications and cost estimate;
   - Estimated completion schedule.

2. Space and powering requirements for PBX center—repeat as in (1) above.

3. Space and powering requirements for server locations—as in (1).

4. Space and powering requirements for MPOP and APOP locations—as in (1).

5. Space and powering requirements for the MCC campus gateway—as in (1).

8.4.3.2 MDF Space and Powering Requirements

1. Space and powering requirements for MDF locations of zone 1 buildings:
   - Existing conditions and space;
   - Required equipment room modifications and cost estimate;

- Number of racks required;
- Size of the fiber termination panel;
- Rack space requirement for new equipment;
- Fiber or coaxial riser;
- Connection location for horizontal wiring;
- Special cable entrance construction requirements;
- Cable and wiring routing and installation particulars;
- Existing powering availability;
- Additional powering requirement and cost estimate;
- Estimated completion schedule.

2. Space and powering requirements for MDF locations of other campus zones—repeat for each building and all zones as in (1) above.

8.4.4 IDF and Vertical Riser Requirements

1. Space and powering requirements for IDF locations in zone 1 buildings:
   - Existing riser conditions and space;
   - Required routing and riser space modifications and cost estimate;
   - Fiber and coaxial riser requirements per floor per building;
   - Fiber termination panel requirements;
   - Coaxial cable and component requirements;
   - Rack space or wall-mount requirements;
   - Power requirements at IDF locations, existing and additional;
   - Estimated completion schedule.

2. Space and powering requirements for IDF locations of other campus zones—repeat for each floor, each building in all zones as in (1) above.

8.4.5 Outlet and User Requirements

1. Outlet and user requirements for zone 1 buildings:
   - Existing outlet count per floor, per building;
   - Additional outlet requirements per floor, per building;
   - Count of user classifications (1) through (5) per floor, per building;

- Count of future CTI users;
- Count of future videoconferencing users;
- Count of future VOD users;
- Summary matrix of outlet and user requirements per floor, per building.

2. Outlet and user requirements for all buildings in other zones—repeat for each room, floor, and building in all zones as in (1) above.

8.4.6 Horizontal Wiring Requirements

1. Horizontal wiring requirements in zone 1 buildings:
   - Count of existing horizontal wiring per floor, per building;
   - Count of existing CAT-5 wiring subject to performance testing;
   - Count of new CAT-5E, CAT-6, or RG-6 wiring requirements;
   - Cost estimate for testing of existing CAT-5 wiring;
   - Cost estimate for new installations of horizontal wiring and outlets;
   - Summary matrix of horizontal wiring system requirements.

2. Horizontal wiring requirements for buildings in other zones—repeat for each floor, each building and each zone as in (1) above.

8.5 The Multiservice LAN Project Plan

The planning document can now be sorted into the various implementation activities, and a timeline can be established for budget and project flow as well as completion. The timing of the project implementation activities can be chosen as required; however, there are many activities that form the critical path of the project and that have to follow in sequence. Section 8.5.1 briefly describes activities that must be performed in sequence. Again, the zoning concept within a campus will help to organize the interdependency of the activities in a manner that will permit positive scheduling control throughout the project. It is good practice to apply conventional project management techniques to identify a logical sequence of the various construction activities and to identify problem areas that require long time periods for an early start.

8.5.1 Facility Construction Requirements

8.5.1.1 Outside Plant

An overview of the conduit problems and additional construction requirements may show that some work is required in many parts of the campus. A multiservice network is implemented from the network center to the outlying areas. Any outside plant construction required to complete the sources' interconnect ring has to be scheduled on a high-priority basis. Outside plant alterations and additions for the various campus zones can then be scheduled in accordance with the priorities of the service requirements. Activities such as the rodding of conduit sections to determine available space or the installation of pullines is usually the first activity and determines the scope and the sequencing of the outside plant upgrade project. The selection of an experienced local contractor can save time and get the upgrade started. While many organizations desire the services of a turnkey implementor, it is more expedient and economical to break the project activities into manageable segments. Outside plant, in-building alterations, and electrical services fall into this category.

8.5.1.2 Building and Riser Alterations

In-building alterations can usually be scheduled in parallel with any outside plant activities. However, the sequencing of the activities must be determined in accordance with the accepted implementation schedule and progress from the inner ring through the various campus zones. Minor in-building alterations can usually be handled internally, while others require the use of experienced local contractors. The network manager/designer should have the authority to make "make or buy" decisions on the basis of safety and protection of the existing communications network components. Nothing is more annoying than an outage caused by craftspeople who do not know the purpose of existing cables and the affect of an outage in an ongoing business. Any cleanup of cabling and redressing of wires and patchcords is in better hands with a staff person than a hired contractor.

8.5.1.3 Power Requirements

The facility department is usually in a good position to plan and provide any additional electrical circuits. While the members of the staff have a good knowledge of circuits and loading requirements, they are usually overwhelmed with the daily chores. Assurance to stay within the scheduled performance periods must be obtained, or the work, when considered beyond the department's capabilities, is better augmented by an RFP and outside electrical contractors, especially in cases where the project includes the installation of emergency generators and batteries.

8.5.2 Implementation of the Multiservice LAN

The planning document can now be used to stage the various cabling and equipment implementation tasks. Implementation activities can be grouped into measurable activities, each containing its own timeline and requirements. Equipment and performance specifications are as important as detailed task descriptions. The result of this effort can be a single RFP document or consist of multiple RFP documents that clearly identify every task and milestone of the project. The development of an RFP and the details of the implementation requirements are discussed in more detail in Chapter 14.

9

The Status of Cable Modem Development

9.1 The Evolution of the Cable Modem

The term *cable modem* is used for a two-way transmission device that uses conventional RF broadband channels. The cable industry has used one-way modulators and demodulators for entertainment television services for many years. The definition of a cable modem is the integration of downstream and upstream RF frequency transmissions for digital transmission over the HFC cable network. Cable modems are designed to transmit within the 6-MHz spectrum of a standard television channel assignment. In the United States, the downstream spectrum is usually in the RF frequency range of 50–860 MHz or even 1,000 MHz. In contrast, the return spectrum has been limited to 5–45 MHz in order to accommodate off-air broadcast stations on channel 2 (54–60 MHz) in the forward direction. While Chapters 4 and 6 explain that corporate LANs do not have to follow these bandwidth restrictions, it may take some time until VOIP and ATM-based products will become available for the 20–1,000-MHz band in both directions. In the meantime, good use of the available products can be made within every corporate LAN. An application of standard cable modems in a campus can eliminate the need for additional fiber cables because the services can be stacked using discrete RF frequencies on the same fiber strand. Sections 9.1.1–9.1.3 briefly discuss past, present, and future cable modem developments.

9.1.1 Analog Video Cable Modulators and Demodulators

The cable television industry has used analog RF modulators and demodulators since the beginning of the distribution of television over cable. At the headend of every cable television network there are numerous analog RF demodulators that receive the off-air and satellite transmissions, demodulate the signals to baseband frequencies, and, using modulators, up-convert all baseband frequencies to the desired RF channel for transmission to the subscriber. The set-top converter demodulates the RF channels to channel 3, or in systems without converters, the television set tuner demodulates the signal for viewing.

9.1.2 Digital Video Modems and Set-Top Converters

The advent of digital video brought about the development of digital video modems and set-top converters. Their introduction is directly related to the beginning of digital broadcasting. The beginning of the digital broadcast era is scheduled for 1999. For the first time, ATV and HDTV will be transmitted by the existing broadcast stations in addition to the present analog signal. Because the cable operators are mandated to carry every broadcasted signal, whether in analog or digital form, the stage is set for the deployment of digital video demodulators, modulators, and set-top converters. The subscriber should not, however, expect that the cable company will exchange the set-top converter to the new analog/digital model without request. When the first affordable digital HDTV television set has been bought, the cable company will oblige and change the set-top converter. Viewers will then be able to watch both the old analog format and the new digital HDTV over the RF cable network. The new digital video demodulators, modulators, and set-top converters have the same functionality as their analog counterparts but include numerous innovations. Set-top converters are designed for entertainment purposes and are intended for one-way downstream and upstream transmissions.

The big advantage to the cable operator is the ability to transmit a number of digital video transmissions over one 6-MHz channel. In the case of MPEG-2, the VBR requirement can accommodate a minimum of five digital transmissions in a single 6-MHz assignment. HDTV requires a wider bandwidth, but it is intended to transmit at least two HDTV transmissions in a 6-MHz channel.

9.1.3 Digital Cable Modems

The development of two-way cable modems for use on HFC-based cable television networks has been under way for a few years. Driven by the desire to fill the market void for faster Internet access, numerous manufacturers have pro-posed a number of different approaches for both the physical and the MAC layer. The various directions and proposals made by the industry are briefly described in this chapter, as it shows that technological advances can be achieved in many different ways. However, the more options there are, the more difficult the task of consolidating the beliefs and agreeing on standardiza-tion. At the time of writing, it appears that a consolidation of opinions has been reached, at least among North American cable television operators. The agreed-upon Multimedia Cable Network System (MCNS) standards offer cable modem interoperability and assure that the residential subscriber can pick up a plug-and-play cable modem at any retail outlet.

However, it must be noted that the MCNS standard is not a global stan-dard and does not promote high-volume, two-way voice, data, and video communications within a cable television service area. The just-released MCNS standard only satisfies the urgent demand for high-speed Internet access and the desire to establish a new and profitable revenue base for the cable industry. The future demand for multiple QoS and ATM-based multilevel services will address the need for business-to-business communications in the local loop and lead to more advanced technologies and standards.

9.2 Cable Modem Manufacturers and Market

The list of cable modem manufacturers is long. There are many that specialize in the business of high-speed Internet access over standard coaxial-based cable television networks in the downstream direction and offer the user telephone dial-up access through the PSTN for the low-speed return traffic. These ven-dors are not addressed in the following discussion, as they have no mission in the corporate high-speed data transfer environment.

Because of the different speed requirements between downstream and upstream data and to accommodate the 5–45-MHz return spectrum, the uni-verse of cable modem providers is also divided into two classes, asymmetric and symmetric cable modems.

9.2.1 The Cable Modem Market

While the deployment of two-way cable modems was slow in 1997, the agreed-upon MCNS standard will bring compatible products to market in the latter part of 1998. An exponential increase of deployment and user popula-tion is expected from a level of 200,000 units in 1998 to about 1.3 million in 2000. These market forecasts do not include the deployment estimates for foreign countries. Already, cable modems are being used by many international cable television organizations. They have been deployed in Argentina, Australia, Belgium, Chile, China, Columbia, England, France, Finland, Germany, Israel, Japan, Korea, Mexico, Norway, Romania, Scotland, Singapore, and Switzerland.

The cable modem universe has not been clearly defined at this point in time. Various technological changes are in progress to provide IP- and ATM-oriented designs with a large number of different modulation schemes. Also, the chip set technology is in a state of constant change. The major cable modem technology suppliers are Broadcom, Stanford Telecom, Libit Signal Processing, Lucent Technologies, LSI Logic, Rockwell Semiconductor Systems, Terayon, Ultracom Communications, and international giants like Alcatel, Siemens, Philips, and Toshiba. While it is impossible to provide information on every new development, it is obvious that the special need for high-speed point-to-point and point-to-multipoint in the corporate LAN will be satisfied within a short period of time. A few examples of the rapid technological development are briefly described as follows:

- Broadcom has introduced a MAC chip set compliant with a DOCSIS and MCNS recommendation. The chip set permits a 40-Mbps downstream and a 20-Mbps upstream speed. Contracts are in place with major cable modem vendors such as 3-Com, Bay Networks, Com21, Cisco Systems, General Instrument, Hewlett-Packard, and Scientific Atlanta.

- Stanford Telecom has developed a new single-chip solution to handle 16QAM/quadrature phase shift keying (QPSK) modulations upstream and 64/256QAM modulations downstream for physical layer (PHY) transmission functions.

- Libit Signal Processing Ltd. has introduced a 64/256QAM demodulator chip to European digital video broadcast (DVB)/DAVIC standards for set-top converters as well as a single-chip solution for MCNS-compliant cable modems.

- Lucent Technologies has introduced a single-chip PHY solution that meets MCNS upstream and downstream specifications.

- Terayon has developed spread-spectrum technology for upstream transmissions in the interference-rich return band and is backed by Cisco Systems, Inc. and Sumitomo Corp.

- Ultracom Communication specializes in VC/MTM™ technology that is similar to orthogonal frequency division multiplexing

(OFDM). VC/MTM can run at full speed delivering 64 or 256QAM of 5–8 Mbps upstream per 1 MHz of RF spectrum compared to 1.5–3 Mbps for standard QPSK/16QAM solutions.

The above examples have been picked at random and are shown only to illustrate the pace of the technological changes that are taking place in the development of new and ever-more economical cable modems.

9.2.2 Symmetric and Asymmetric Cable Modem Providers

With improvements in modulation methods accommodating more and more bits per hertz transmitted over RF frequencies, justice cannot be given to the many vendors that develop new products. The integration of new chip sets will change the offerings on a month-by-month basis. The following is a list of cable modem vendors that either manufacture their own products or have their products manufactured on an OEM basis. (All products are frequency agile and provide downstream transmissions at speeds up to 40 Mbps in the range of 50–860 MHz. In the return direction, speeds up to 20 Mbps can be provided in a single 6-MHz channel in the 5–42-MHz band, and in some cases in the range of 5–186 MHz.)

- 3Com;
- Alcatel;
- Alexon;
- Along Technology;
- Bay Networks;
- Cabletron;
- COCOM;
- Com21/3Com;
- General Instrument;
- Hybrid Networks;
- Motorola;
- NEC;
- NetComm;
- NetGame;
- New Media Comm;

- Panasonic;

- Phasecom;

- Philips NV;

- Pioneer;

- Scientific Atlanta;

- Siemens;

- Sony;

- Terayon;

- Toshiba;

- US Robotics;

- Zenith.

While this list probably does not include some entries, it provides insight into a burgeoning industry that hopes to serve the residential cable network subscriber. Soon, interoperable products will appear on the shelves of the retail stores at prices that may compare with 56-Kbps telephone modems. While the computer workstation still needs a 10-Mbps Ethernet card, the rest is plug-and-play connectivity through the RG-6 coaxial drop wire. Current products are aimed at faster Internet access for the average residency. Later products will provide higher speed, two-way, IP-based data services between residencies and the SOHO community. The next phase will address the transmission of telephony as well. This phase will start in the VOIP mode but will advance to multilevel QoS utilizing ATM-based technology right to the desktop. The migration path is a matter of supply and demand. If fast Internet access is the initial revenue producer, telephony, videoconferencing, and the interconnection of business establishments will follow. While the present cable modem technology is well suited for the campus LAN environment, it is predicted that symmetrical high-speed and ATM-based cable modem products will be available within a short time period and enhance the choices available to the business community.

9.3 Cable Modem Technology

The HFC reference model sets the basis for the development of cable modem technology. Figure 9.1 illustrates the peer communication relationship of the HFC network.

The cable modem layers defined by IEEE 802.14 are the channel, the PHY layer, and the MAC layer. The PHY layer consists of the downstream and

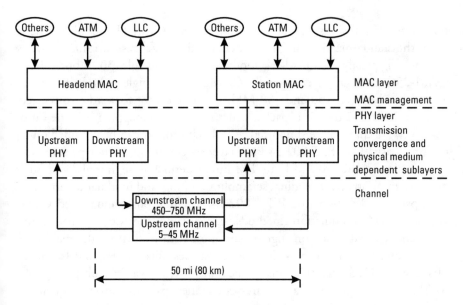

Figure 9.1 HFC peer relationship.

upstream path requirements and the convergence specifications. The development of MAC layer specifications are the most complex because of the shared nature of the HFC network architecture. Considerable development time has been devoted by numerous companies to the standardization effort of both PHY and MAC layers.

9.3.1 The Physical Layer

The PHY layer contains two sublayers. The transmission convergence (TC) sublayer is responsible for HEC processing and performs encryption/decryption functions and the alignment of the data into the PHY layer. The second sublayer, the physical medium dependent (PMD) layer, permits synchronization and ensures bit timing of construction, correct reception and transmission of bits, and the modulation coding of transmitter and receiver operation.

Major industry participants like IBM, Alcatel Corta, Zenith, Lucent, General Instruments, Stanford Telecom, Aware, and Scientific Atlanta have provided inputs into the standardization process relative to QAM/QPSK modulation methods, ATM-based PHY, spread spectrum, and discrete wavelet multitone (DWMT) PHY layer structures.

After lengthy considerations, IEEE 802.14 adopted QAM-64 and QAM-256 for downstream transmission and QPSK and QAM-16 for upstream transmissions. While there were proposals with a better bit-per-hertz ratio, the standardization of QAM offers good transfer rates within a 6-MHz channel. Typically, a QAM-64 modulation can handle 30 Mbps, and a QAM-256 modulation can handle 40 Mbps in a single 6-MHz frequency assignment between 150 and 750 MHz. Parameters for forward-error correction and MPEG-2 transport packet structure and easy mapping into the European 8-MHz channel are supported. ATM cell support can be provided on an optional basis. The IEEE 802.14 PHY layer specifications include performance specifications related to AM hum, FM hum, thermal noise, impulse and burst noise, framing structure, coding, scrambling, timing, and modulation errors. In the upstream direction, the IEEE 802.14 specifications standardized QPSK and QAM-16 modulation methods, recognizing the varying bandwidth requirements and limits on ingress, common path distortion, thermal noise, impulse and burst noise, as well as nonlinearities, phase noise, and frequency offset. While QPSK modulations can be transmitted in various subspacings of a 6-MHz channel, the standard addresses channel spacings, carrier frequencies, timing and synchronization, modulation impairments, modulation bit rate, and coding.

Most standards have been finalized, but the final issue of the PHY specifications is not expected before the end of 1998. The problem with the final release is caused by the ITU recommendation J.83, which consists of an annex A and B. Annex A is derived from the DVB standard that originated in Europe. It incorporates Reed-Solomon block coding of an MPEG-2 video transport stream. Annex B uses concatenated Reed-Solomon block coding, which is widely accepted in North America. While both annex A and B have been adopted, there are concerns about interoperability and international standardization. In the end, the world may have to live with two different PHY layer standards.

9.3.2 The MAC Layer

The MAC layer protocol has been the most challenging standard development and is not complete at the time of writing. The MAC committee of IEEE 802.14 is trying to establish a cable modem interface standard that not only accommodates current Internet access requirements but also assures interoperability with future demands and services like MPEG-2, ATM, and VOIP. The goal was to provide a universal standard for cable modem multimedia services. Many problems exist in the return path, including ingress and propagation delay. The return path requires the transmissions originating from many users to arrive at a single receiver in sequence. Because the propagation delay for a nearby station is less than for a distant station, the near station will obtain priority service. The concept of QoS also must be an integral part of the standard, so that CBR, ABR, VBR, and UBR service levels can be offered.

Some of the MAC objectives are listed as follows:

1. Both connectionless and connection-oriented services must be supported.

2. QoS service must be on a per-connection basis.

3. Bandwidth access must be provided in accordance with the ATM requirements for CBR, ABR, VBR, and UBR.

4. Interoperability with ATM must be assured.

5. The MAC must be built with inexpensive electronics and assure inexpensive implementation.

6. Arbitration for shared access must be assured within every level of service.

7. Usage monitoring must be assured.

8. Protection against a faulty terminal is required.

9. The MAC must be able to perform in the tree-and-branch architecture of the network without any regard for the physical location on the tree.

10. Reaction measures to prevent congestion collapse must be incorporated.

11. Cell delay variations and latency in voice interconnections must be minimized.

Since 1995, the major industry members have provided proposals to further the MAC technology. There are two major MAC areas: access and the management considerations. The same companies, listed above for the PHY layer, participated in the process. In addition, Philips Research and National Tsing Hua University and others offered solutions. A most interesting proposal by Lucent Technologies covered the ADAPt™ protocol and incorporated it in its HFC 2000 product.

Lucent supported the advanced multimedia HFC network installations with SNET and proved that multiservice RF broadband could succeed. SNET, however, used a special high-voltage powering method within the sheath of the fiber/coax cable that did not pass the outside plant safety standards for voltage in communications cables, and the project was terminated.

Many useful features that have been combined into the IEEE 802.14 convergence agreement form the basis for the MAC working group draft specifications. The main goals of the MAC protocol are the ability to support both ATM-cell transfer and variable packet transfer while providing multilevel, high QoS. The standardization process is ongoing and concentrates on issues such as:

1. Station addressing consisting of the conventional 48-bit 802 addressing plus a local 14-bit identifier;

2. Upstream bandwidth controls defining the types and structures of bandwidth minislots;

3. Upstream PDU formats for both ATM and variable-length packets;

4. Downstream data flow of ATM cells and variable-length packets with bandwidth information, allocation information, and request for minislot allocation and contention;

5. Encryption and decryption algorithms.

9.3.3 MAC Management

MAC management consists of configuration management, performance management, accounting management, fault management, and security management. Configuration management deals with addressing, access control, and levels of service parameters. Performance management deals with the RF signal quality and traffic statistics and migrates to fault management for error statistics, testing records, and substandard RF signals. Some of these areas are not under the jurisdiction of IEEE 802.14, which is mainly concerned with the management of the layer. Figure 9.2 shows the complexity of the tree-and-branch architecture and the resulting cable modem deployment structure.

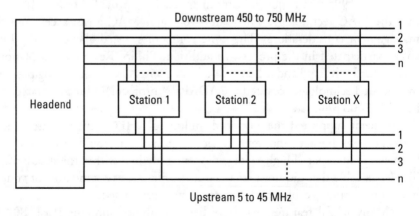

Figure 9.2 Cable modems in the HFC network.

While the HFC network uses a star architecture in the fiber segment, coaxial branches emanating from the fiber node can be numerous. The number n signifies the complexity that can be expected.

9.3.3.1 Access Control Parameters

A cable modem requires initialization before it is taken into service. These access control parameters consist of synchronization, authorization, local address assignment, level setting, the assignment of default upstream and downstream RF frequency assignments, and any required encryption information. Each one of these initialization processes require a series of dialogs consisting of multiple scenarios between the headend and the cable modem. There are numerous command messages consisting of transmissions from the headend to the station and from the station to the headend. The entire initialization process is fully automated to permit the user to purchase the cable modem and connect it to the system without the need for technician service.

Some of the more important message exchanges are listed as follows:

- *Headend-to-station messages:* Invitation to join, address for the new station, request to tune to an available frequency, the headend's current time, the synchronization time-base message, poll request messages, power-on collision messages, local address verification, switch downstream channel message, message to stop transmission, channel reassignment messages, retransmission requests, address verification requests, test execution requests, SNMP statistics requests, encryption assignment and confirmations, network busy message, cancel network-busy message, and decommission messages;

- *Station-to-headend messages:* The station response messages include station identification, identification codes, repeat identification, station deactivate requests, changed channel status, station activate requests, built-in test responses, test result messages, SNMP statistic responses, encryption confirmations, station-alive messages, and dialog messages to other headend command messages.

9.3.4 QoS Considerations

ABR control is more complex in an HFC system than in standard ATM network configurations. In order not to load the HFC network with resource-consuming data streams to establish ABR traffic, it is proposed to control the ABR from the headend. The ABR cells destined for the ATM core network are buffered and spaced in accordance with the flow control rate. The resulting QoS can be classified as follows:

1. A high-priority CBR class of service that can also cover rt-VBR traffic;
2. A class for VBR traffic with multiplexing gain;
3. A class with lower priorities; covers ABR traffic with shaping at the headend;
4. UBR service that is not mixed with ABR but has its own minislot assigned for all UBR traffic and carries all UBR service.

9.3.5 Security Considerations

Security considerations are also a part of the MAC layer. The functions of a secure transmission are access control, authentication, data confidentiality, and data integrity. While the traffic within a corporate LAN has a lower security requirement, security in a public HFC network is of considerable importance. The handling of security by both the ATM Forum and by IEEE 802.14 is a part of the MAC layer. The standard, whose details have not yet been released, intends to specify a large number of security mechanisms to make the shared HFC medium comparable to that of nonshared media access networks.

The IEEE 802.14 requirement document is a guideline for security mechanisms. Some of the minimum requirements are described as follows:

- *Authentication:* Every terminal must be validated as an authorized user with a service profile and accounting record. Authentication must be accomplished with a minimum number of exchanged messages and minimize the content to a small number of bits.
- *Encryption:* The user data between the terminal and the core network must be made confidential.
- *End-user convenience:* The security functions must be automatic and not require an intervention by the user.
- *Repeat protection:* Retransmissions of messages are to be avoided to limit intrusion by any third party.
- *Transparency:* Security is considered transparent and does not affect the contracted level of service.
- *Exportability:* Shared media access will be used worldwide and must adhere to import and export regulations.
- *Bit error tolerance:* Single transmission bit errors should not affect blocks of data or corrupt messages.

- *Reliability:* The system must recover synchronization immediately after transmission interruptions.

- *Upgradability:* Any future security mechanisms must be includable without hardware modifications.

9.3.5.1 Encryption and Decryption

Encryption and decryption should use proven cryptographic techniques that employ the digital encryption standard (DES) algorithms as specified in ANSI X3.92-1981. Both hardware and software implementation is desirable in the encryption process to keep bandwidth intensity to a minimum. There are *secret key* and *public key* cryptographic solutions that require complex and expensive equipment additions but that are not suitable for the shared medium security requirement of the cable modem. Secret key is a traditional solution based on two parties using the same coding. Public key is the concept of using a pair of keys, one called the private key, the other the public key. The private key is kept secret and never revealed. The public key is used for the communications path. Public key cryptography increases both security and convenience but requires complicated mechanisms in the device.

Some have argued against any security precautions within an HFC network, their main argument being the uselessness of the device in case of a failure and the lack of communications. Cable television operators are fearful of expenses incurred in connection with service calls that are caused by the malfunction of equipment or user ignorance. DOCSIS specifications have addressed these concerns and determined that privacy is the basic minimum security component. The new approach to security is a low-cost and simple mechanism called baseline privacy (BPI). It uses *cipher block chaining* algorithms with the 56-bit DES. This security mechanism is believed to provide sufficient security for the users of the shared medium, especially when the DOCSIS cable modem is used mostly for Internet access purposes.

While DOCSIS has found a "light security" solution to get the deployment of cable modems under way, IEEE 802.14 is seeking further improvements of the security mechanism for ATM-based cable modems. In addition, the U.S. federal government has an input into the debate. The Department of Commerce is responsible for evaluating cryptographic products, and technical assessments are performed by the Office of Export Control at the Department of Defense and the National Security Agency. A concern, often voiced by security product vendors, is that in this global community the U.S. advantage can rapidly disappear, because these regulations do not apply to companies outside of the United States. The determination of IEEE 802.14 standards for export-able ATM-based security products will take additional time.

9.4 Present Cable Modem Standards

Frustrated by the slow progress of the IEEE 802.14 cable data standards com-mittee, the MCNS, which consists of major cable companies such as Comcast Cable Communications, Cox Communications, Tele-Communications Inc., Time Warner Cable, Rogers Cablesystems Ltd., Continental Cablevision, and the Cable Television Labs (Louisville, CO), decided to speed the timeline of standardization. With the help of the DOCSIS and under the guidance of Cable Labs, the industry representatives concentrated on the development of an interoperable cable modem design that can be deployed in 1998. The mission of the DOCSIS project was to reach agreement on standards for badly needed faster Internet access within the shortest possible time frame.

9.4.1 The MCNS Interoperable Cable Modem

The reference model of the MCNS cable modem combines all functions between the data network and the user terminal. The intended service will allow IP traffic to achieve transparent bidirectional transfer between the cable modem termination system–network side interface (CMTS-NSI) and the cable modem to customer premises equipment. Figure 9.3 shows the MCNS cable modem reference model.

The communications specifications identify the IPvs.4 standard at the network layer, but it is upgradable to IPvs.6 (IETF RFC 1883) when it becomes an accepted standard.

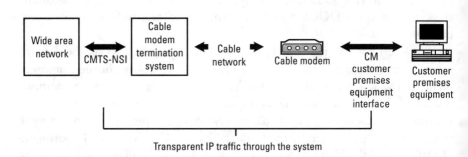

Figure 9.3 MCNS cable modem reference model.

9.4.1.1 The Universal Network Interface

At the network side, data link and PHY combinations are required to carry the IP traffic from the HFC system to the network in the following formats: ATM over STS-3c, ATM over DS3, FDDI, 802.3 over 10Base-T, 802.3 over 100Base-T, Ethernet over 10Base-T, and Ethernet over 100Base-T. Figure 9.4 shows the architecture of CMTS-NSI.

While using IP over the HFC network segment, the network interface system can be interconnected with ATM-based or Ethernet-based backbone LANs and WANs as well as with the PSTN network and the Internet. The DOCSIS network side interface specification indentification number is SP-CMTS-NSII01.

9.4.1.2 The Customer Premises Interface

The MCNS cable modem can be provided as an external unit or as an internal PC configuration and supports every hardware platform such as IBM, Apple, DEC, HP, Sun, and others. The supported operating systems cover every option ever conceived, inclusive of Windows 3.1/95/98/NT, MAC system 7.0 or higher, OS/2 WARP 3.0 or higher, and UNIX. The customer premises equipment interface can be Ethernet 10Base-T and the universal serial bus (USB). The communications software is a TCP/IP stack software capable of supporting DHCP/BOOTP, SNAP addressing, and multicast. The user remains responsible for the required Ethernet card and software. The DOCSIS cable modem to customer premise equipment interface specification identification number is SP-CMCI-102. Figure 9.5 shows the block diagram of the Broadcom Corporation cable modem design which was released in May 1998 and will be used by many cable modem vendors in upcoming product deliveries.

The cable modem board combines all control, stream processing, and baseline privacy functions in addition to supporting an external high security device. The DOCSIS cable modem specifications support most of the standards developments of IEEE 802.14. The exception that has been made is the IP-based, packet-based transport. This selection had to be made to start up the deployment of cable modems and to satisfy the demand for Internet access. While the IEEE 802.14 standards committee faces many delays, DOCSIS required a solution that interfaces with existing and economical CPE equipment. The standard 10Base-T computer board is the most economical and familiar interface. In order to incorporate QoS into the standard, *packet fragmentation* has been used to provide VOIP and future video compatibility. Packet fragmentation assists in the application of MPEG-2 transport streams to move IP data. MPEG-2 encapsulation will allow a single cable modem to

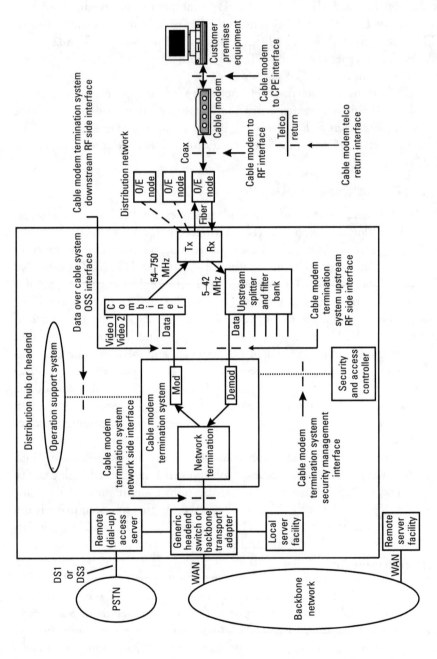

Figure 9.4 CMTS-NSI.

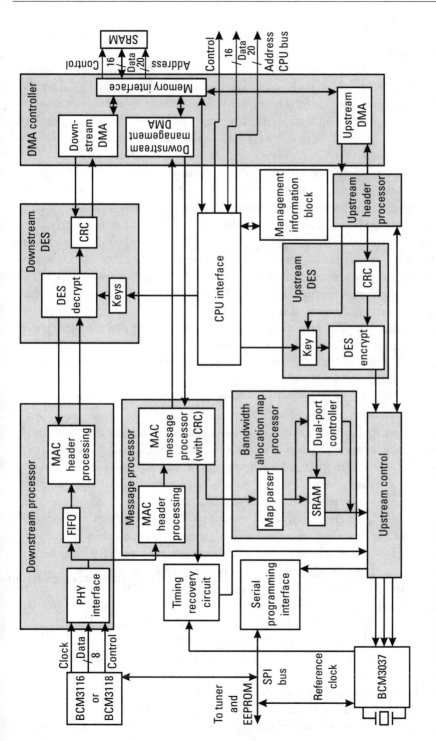

Figure 9.5 The BCM3220 MAC chip from Broadcom Corporation.

support multiple sessions and multiple users as well as delay-sensitive multimedia streams for VOIP or video over IP.

The MAC chip utilizes a standard serial program interface (SPI) bus master interface, which is used to configure and control both the QAM receiver and burst modulator chips. This control allows setting operating values such as tuner frequencies, gain, and levels. The upstream processing section passes the data to the headend. It transmits MPEG-encapsulated data within a series of time slots but has to share these time slots with other modems. In order to set up the transmission, the unit places requests to the headend and is assigned a time period. For extended data, the unit requests a series of prescheduled piggyback time slots to improve throughput. There is also a provision for concatenation where the MAC is allowed to build up several packets into a super-packet. This is accomplished by a single request for sufficient bandwidth and the data is sent with a single concatenated header all at once. The headend can make available dedicated request regions in the upstream bandwidth to assure service contract requirements for higher level QoS users. Referred to as headend request polling, additional bandwidth can be made available outside the usual contention mechanism. The improvement of the cable modem performance through polling, prioritization, and fragmentation can maintain packet delays within 20 ms. This means that CBR and rt-VBR QoS level requirements can be met over the HFC network.

As mentioned before, BPI has been included in the basic architecture of the MCNS cable modem. The 56-bit DES cipher block chaining algorithm is considered applicable for normal user requirements. Additional security can be added by a separate modular insert. DOCSIS specification SP-SSI-101 covers the dual-key security offered by inserting the security module. The security module is securely identified by the authorized manufacturer as an authentic security module as part of the factory production process. It contains a critical authentication mechanism that is activated by the cable operator's network and identifies the terminal as an authorized user. When initialized, it requests and receives symmetric keys and establishes the services for which the subscriber is authorized. The cable system reviews and compares the keys periodically with the list of services that the user is entitled to obtain. When the subscriber moves to a different cable network, the security module is reset and the user can establish service within the new network. The conditional access system is scalable from a small to a large number of users. The functionality of the security module standards are based on the public key infrastructure (PKI) of the IETF. Security of the conditional access is based on public key and symmetric key cryptography and the content security is based on symmetric key cryptography. The algorithms of RSA are used for public key encryption. DES algorithms are used for symmetric key encryption. The level of security offered is comprised of the most technologically advanced processes and will provide the required protection against intrusion and offer a solid base for fully secured Internet transactions.

While the MCNS cable modem standard has been approved, every vendor and product is subject to an extensive compliance confirmation. Equipment vendors have to answer a 2,300-item questionnaire called the protocol implementation conformance statement (PICS) and comply with a series of test sequences called the acceptance test plan (ATP). In addition, every vendor must produce an interoperability test report (ITR) demonstrating that its equipment has coexisted with other vendor devices on a network. The formal seal of approval is granted by a certification board made up of representatives of Media One, TCI, and Time Warner. High importance is placed on a specification-compliant introduction of the new cable modems. The stakes are high for both manufacturers and cable operating companies. Nevertheless, the digital age is finally coming to the RF broadband industry.

9.4.2 ATM-Based Cable Modems

Fragmentation of packets is technologically very close to ATM-cell transmission. Cells or short frames are required to establish QoS service levels and to be able to reduce circuit assembly, tear down, and reconversion times. While the existence of economical Ethernet boards has moved the direction toward IP-based cable modem technology, the development of ATM-based products and standards continues.

9.4.2.1 IEEE 802.14 Standards

Cable modem standards based on IEEE 802.14 are expected to be finalized before the turn of the century. ATM-based cable modem products will provide ATM cells directly to the desktop and will provide a multitude of ATM-based services to the business community. The recent merger of AT&T and TCI and other long-distance/cable operator agreements will offer competitive multimedia services in the local loop while bypassing the local copper-based telephone infrastructure.

9.5 Cable Modem Applications

Final standardization of the MCNS cable modem indicates the beginning of two-way services over HFC-based RF broadband networks. Cable television networks will interconnect with AT&T and other long-distance providers at the headend of their networks and offer an alternative to current local providers. HFC networks will not just be used for faster access to the Internet but also for telephony, data, and video two-way services to residences and business establishments within the local service area. It is not possible to correctly estimate the entire universe of cable modem applications at this time. Sections 9.5.1 and 9.5.2 describe some of the enormous number of possible applications.

9.5.1 Internet Access for Residences and Businesses

The availability of MCNS standardized cable modems will not just fill the demand for faster Internet access; it will also provide the foundation for rapid development of the e-commerce business. The security features of the cable modem permit the evolution of buying and selling on the Internet to begin, mature, and become a formidable force in changing our lifestyle. The sequence of this development may follow the following migration paths:

1. *Residential:*
   - Deployment of cable modems to residential cable subscribers in TCI service areas; AT&T long-distance discounts and unified billing for long-distance, cable, and Internet access services;
   - Deployment of cable modem access to business establishments for e-commerce functionality; service by serving cable companies and AT&T as ISP at reduced rate structures;
   - Deployment of cable modems to residential subscribers in non-TCI cable service areas;
   - VOIP, MPEG-2 video, and e-commerce over the Internet.

2. *Business establishments:*
   - Deployment of cable modems in university and college campus systems to student residences over already existing RF broadband-HFC facilities;
   - The implementation of special HFC infrastructures in corporate networks to partake in e-commerce developments on the Internet; a separate Internet access network for secure transactions;
   - Integration of legacy LANs into ATM-based multimedia service LANs with VOIP and video WAN communications over Internet and compressed ATM-based voice services over PSTN;
   - Integration of CTI and ATM-based PBX technology into the corporate LAN and WAN with VOIP over ATM and, as desired, over Internet or PSTN long distances.

9.5.2 Applications in the RF Broadband LAN

The application for cable modems within business establishments is discussed in Section 9.5.1. The aforementioned business establishment examples will become common to enhance Internet access and to provide secure access for e-commerce transactions and multimedia services. In addition, the cable modem is helpful in large campus environments and wherever single-mode fiber is at a premium. The cable modem can be used to stack the traffic in the RF format over single fiber strands and offers economical networking solutions. Some of these solutions are described in more detail in the cost modeling of Chapter 16. IP-based cable modems can be used immediately to provide data and low-volume telephone service. ATM-based cable modems, after completion of the IEEE 802.14 standard, can enhance connectivity and integrate multiservices to the desktop.

10

The Status of Cable Telephony

10.1 Cable Telephony in the Cable Industry

The term *cable telephony* is commonly used to identify voice transmissions over an HFC-based RF broadband network. Whether the transmission is data or voice, a cable modem is used to up-convert the information into a frequency of a suitable RF channel. Cable telephony, therefore, relies on equipment quite similar to that described in Chapter 9. Cable modems used exclusively for the transmission of voice are designed to work within the 6-MHz spectrum of a standard television channel assignment. The RF frequency assignment follows the spectral capacity of the cable industry with frequencies between 50 and 860 MHz used for downstream assignments and between 5 to 45 MHz for the upstream direction.

The cable industry is concentrating on serving the residential cable subscriber with both data and telephony. Because the residential requirement for voice and data is limited to a small number of telephone lines and maybe an Ethernet data connection, a combination voice/data cable modem appears to have become the standard of the industry.

A considerable number of trial projects are under way, and offering fast Internet access and one or two telephone connections to the cable subscriber will become a common marketing effort by all cable companies that have upgraded their network to HFC technology.

10.1.1 Synchronous Transfer Method

The transmission of voice starts with a dialing process to establish a two-way connectivity between the parties. The connection is established and must be maintained until the call is terminated. In the beginning, all voice transmission was in the analog format, and connectivity was established by a number of sequential switching functions referred to as line finders and selector switches. These switches were mechanical contact closures and crossbar designs. The transition to voice in digital form has changed these concepts radically. Digitized voice is packetized and can be forwarded over different routes. In order to keep latencies to a minimum, the telephone industry has used the STM. In local and long-distance telephone networks, voice transmissions are multiplexed using time division multiplexing (TDM), connected using the SS7 signaling system, and controlled using STM. STM permits the multiplexing of several circuits over transmission links and switches by dividing the transmission into synchronized time slots. Each time slot is 125-ms-long. It is interesting to note that the same time slot length is used in the SONET structure. Having a common sample time permits telephony to be mapped at the native ATM speeds. Voice communications is a constant bit rate traffic and does not allow bursty conditions. The digital transmission hierarchy extends from DS-0 at 64 Kbps through T-1 (DS1) at 1.544 Mbps, T-3 (DS3) at 44.736 Mbps, to the SONET speeds of OC-3 to OC-192. It is, therefore, not surprising that ISDN and primary ISDN (PRI) are readily available service offerings of the serving telephone companies that will form the bulk of the cable companies' offerings. Figure 10.1 shows the telephone service connectivity in the outside world and the interconnection with cable television company headends as well as business establishments.

The existence of two communication infrastructures, the Internet and the PSTN backbone, allows routing flexibilities that never before existed. A cable company can directly interconnect with a long-distance company or with an ISP. The same applies to businesses. VPN technology is a typical example.

10.1.2 Voice Over IP and Over ATM

The cable industry has opted to standardize the MCNS cable modem standard, which uses packet fragmentation to transmit constant bit rate traffic. Most of the new cable modem designs feature standard ports for Ethernet and telephone service. The manufacturers of cable telephone modem equipment are identical to the listing provided in the chapter dealing with cable modems. In addition, there are global entities that are promoting business telephone requirements. These products are discussed in more detail in Section 10.1.3.

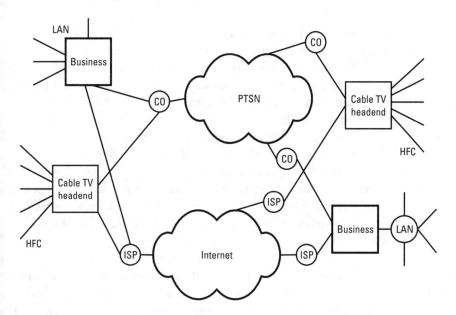

Figure 10.1 Global telephone service.

Both the MCNS standard and the yet to be completed IEEE 802.14 standard have been used to manufacture the most likely combination for the standard U.S. household. On the data side, the service offerings range from 450-Kbps to 10-Mbps Ethernet, and on the telephone side from one to four lines. The telephony offerings extend from a standard plain old telephone service (POTS) line to lines with various features and ISDN connectivity.

Powering problems have plagued the cable industry and have delayed the introduction of cable telephony. The residential telephone has always been powered through the copper cable. The telephone still works during a power failure. To replicate this condition in an HFC network is not an easy task. Stand-by power service has to be provided at the fiber node; power-passing multitaps have to be used in the coaxial segment; and the service drop has to transfer the power to the residency through the cable modem and to the telephone set. The problem is identical to CTI on the data network. A telephone that is connected to a computer will only work when the computer is turned on. The alternative in a cable television-provided telephone service is to power the telephone in the home, which would mean that the service would be interrupted during a power failure. Conventional telephone service provides stand-by power for about eight hours and is fully independent from the power grid. Manufacturers of cable telephony equipment are considering shorter stand-by periods in the range of one to two hours.

Features must be identical to the features now offered by the telephone companies. These features include *call waiting*, which requires a tone to indicate a second call; *distinctive call waiting* to identify the source of the call; *multiple directory numbers per line* to permit the sharing of a line by different parties; *hunting*, which permits a call to find the next line when the called line is busy; *automatic callback*, which allows the subscriber to return the last call; *automatic recall*, which calls again after a busy condition; *call tracing*, the ability to identify the calling party; *call forwarding*, which allows the subscriber to forward selective calls to a forwarding number; and *selective call rejection*, which routes selective callers to an announcement. The solutions to provide these features are critical in the competitive arena of cable telephony, but when accomplished can be directly transferred to the development of multiservice LANs and CTI in the business environment.

Packet cable is the continuation of the cable modem standardization process by MCNS and the Cable Labs to address cable telephony in the new HFC environment of the cable industry. The standard will incorporate latency limitations and QoS levels necessary for quality telephony over the cable system. Working models of the data/voice combination cable modem will be available by the end of 1998 and introduced during 1999. The menu of issues still in the standardization process include packetizing dual tone multifrequency (DTMF) dial tones, developing protocols for initiation, call setup, extension of PBX and Centrex features, and network management, control, and billing. Interfaces with the H.323 multimedia protocol must be perfected. When complete, the cable telephone modem will connect analog telephone sets via a standard RJ-11 jack, convert the voice into the compressed digital format, transmit the voice over the RF broadband network to the headend, and interconnect with both the public telephone network and the Internet. The identical functionality can be used in the LAN environment as well. PBX extension lines can be interconnected with data services and transported over the RF broadband network to workstations using standard Ethernet cards. It is expected that the availability of cable modems will soon provide an economical CTI alternative solution for the corporate network.

10.1.3 Business Telephone Systems

Connecting business telephone service over the HFC-based cable television network is not very complicated. Cable telephony is a symmetrical service. This means that the downstream bandwidth must equal the upstream bandwidth. This brings us back to the asymmetrical cable television passband of 5–45 MHz in the return direction and the ample 50–860 MHz or more available bandwidth in the forward direction. Several modulation algorithms have been developed to provide multiple voice transmissions within the 6-MHz channel assignment. The winning modulation method is the OFDM, which permits the transmission of 240 voice channels in one RF frequency assignment. Consider that each voice channel is based on the DS-0 format of 56 Kbps. The number 240 can easily be multiplied by five when the recently approved ITU standard of 10 Kbps is applied. This formidable ability to carry massive voice loads over a single RF frequency assignment makes cable telephony a viable tool in both cable television networks and the corporate LAN.

Manufacturers of cable telephony hardware are numerous. The list includes ADC Telecommunications, Nortel, ANTEC Corporation, Philips Broadband Networks, ALCATEL, Aware Inc., and a number of the cable modem manufacturers mentioned in Chapter 9. In every case, the cable telephony system consists of the host digital terminal (HDT), which resides at the headend or the CO, and the customer access device (CAD) at the customer end of the RF broadband network.

The Nortel offering *Cornerstone,* a cooperative effort with ANTEC, provides a versatile arrangement of voice ports from two to twelve lines. The Philips *Crystal Exchange* emphasizes that the HDT is compatible with concentrated and nonconcentrated digital switch interfaces that are used worldwide such as VF, TR08, TR303, V5.1, E1/CAS, V5.2, ISDN, FR, and any ATM interface. Most of the other manufacturers offer the same flexibilities as well as advanced HFC domain managers, network management software, and operational support software.

Service to multiple dwelling units (MDUs) is a target application of many suppliers of cable telephony equipment. The owner of a multidwelling unit has the opportunity to provide revenue-producing telephone service to lease holders. As a reseller of communication services, he or she may also include data services and reduced long-distance rates via the Internet. Some MDUs may opt for LEC-provided DSL service, and others may obtain a better rate structure and wider margins through the serving cable television provider. The landlord only has to file for a reseller license with the FCC and, upon routine approval, can become the authorized telecommunications service provider for voice, data, and cable television throughout this establishment and even for new generation wireless services. This subject is discussed in more detail in Chapter 7.

PCS wireless services can also be incorporated into every cable television network and as an adjunct to cable telephony. As mentioned in Chapter 7, a number of field trials have proven the viability of extending the reach of digital PCS services into the far corners of residential neighborhoods. The transmitter/receiver equipment is available in ruggedized pole-mountable housings and can be placed at selected locations in the coaxial segment of the HFC distribution plant. Especially in hilly terrains where the path propagation is frequently interrupted, the addition of PCS repeater stations may become a necessity to widen the coverage area. The availability of a uniform cellular coverage and of a universal numbering plan may serve to reduce the need for wired telephony, which would relieve the stand-by power requirement in the single residency. Uninterruptable PCS service can easily fulfill the eight-hour stand-by power requirement.

10.1.4 Network Management

HFC networks require a fully integrated network management solution. Whether the HFC network resides within the confines of a corporate LAN or in the local loop, the network management solution has to be able to monitor and control the HFC network equipment and RF modem equipment. Most of the available network management solutions use a Windows NT client-server platform with graphical interfaces that visually represent the health of the network elements.

Cable modem management is one of the network management levels. The main function of cable modem management is to monitor the performance of HDTs at the headend location and at subscriber terminals. The main functions of the cable modem manager apply to all cable modem installations whether used for data or telephony. These functions include (1) inventory control of equipment, (2) provisioning and initialization, (3) fault management, (4) remote operational testing, (5) user authentication and security management, and (6) the monitoring and control of bandwidth used for telephony and/or data transmission.

HFC network management is the expanded level of an integrated network management system. Also called the status monitoring system, HFC network management is responsible for the monitoring of all HFC elements. This may include fiber-optic transmitters/receivers, coaxial amplifiers, and even incorporate end-of-line performance reporting. Alarms from all devices are reported on a fault management screen that compares the measured level at discreet RF frequencies with specification levels. HFC network management is considered important by the cable industry and will serve to reduce subscriber complaints and increase the effectiveness of customer service. In the corporate LAN, the smaller number of HFC elements may prohibit additional expense by using redundant equipment complements.

Integrated network management combines both cable modem and HFC component management. Faults can be isolated across multiple network element types and services. Performance can be analyzed with considerations of the interrelationship of HFC elements and cable modems for data and voice transmissions. The common database permits a better comparison of node service area performance, traffic flow, and usage with other nodes and permits the forecasting of element failures.

10.2 Telephony in the Corporate LAN

While the old familiar PBX is evolving into a CBX, the corporate network will undergo the transition from the triple wiring plant to the multiservice single wire facility. Both the evolution of the CBX and the transition to a single facility will not be accomplished by the beginning of the millenium. Present market forecast studies concentrate on the global perspective and predict that enterprise communications will become multimedia-oriented within a decande. The reason is the enormous growth of the Internet and the massive capital infusions that have already been made in the long-distance segment and will be made in the local loop. In the corporate LAN, the transition is subjective and directly related to the cost savings that may be achieved through increasing productivity. It is, however, estimated that a large percentage of corporate networks will undergo a step-by-step upgrading process that will include higher data speeds, the integration of digital telephony, and the inclusion of full-motion videoconferencing and VOD employee education into the daily business.

It is obvious that the investment that has been made in the telecommunications facilities must be protected, but not to the point of losing the competitive edge. On the other hand, there will be savings in the long-distance and in the enterprise segment that can be earmarked for incremental transitional changes within the corporate network. In addressing only the telephony sector in a corporate LAN, the following developments are inevitable over the next decade.

10.2.1 Computer Telephony Integration

CTI has been an explosive development for many vendors of call-center technology. This area, however, is only the humble beginning of a total integration of voice and data in every corner of the industry. At present, switches are circuit-based, and data routers are packet-based. The PBX switching matrix is about to change and become a computer. It does not matter whether a switching function is used to switch a packet, a cell, or a circuit; the switching function requires the intelligence of the computer. It also does not matter whether the switch is located in the CO or in the PBX room; the switching functions require computer technology. Digitized VOIP packets can be mapped into ATM cells and switched as required. The new digital CO (DCO) switching products that are used by the long-distance service providers have been developed by traditional giants such as Nortel, Siemens, Alcatel, Lucent, Ericsson, Rockwell, and Motorola. Mergers and alliances have been formed by these companies with the data networking vendors, and the cross-fertilization of ideas, processes, hardware, and software developments are directed toward total convergence. There are, of course, another three dozen companies that are in the process of selling small computerized PBX products for call-center operations to integrate VOIP or fax into the computer. There are numerous Internet access products that even address G.723.1 and G.729a compression standards, feature H.323-compliant protocol stacks, or follow the ITU T.Ifax2 standard. All these developments may appear as enormous innovations but cannot compete with the quite subdued development effort of the established suppliers to the telecommunications industry.

The Enterprise Computer Telephony Forum (ECTF) has a member staff of over 100 companies worldwide. The ECTF framework consists of an open architecture of modular hardware-software elements and multivendor/multiresource application components. These will include voice, fax, text-to-speech, automatic speech recognition, network interfaces, a specialized TDM bus, and both high- and low-level APIs. This standard, referred to as S.100, is in the process of being upgraded to S.300, which promises interoperability between all vendor products. There is also TAPI, Microsoft's API, which needs to be taken into consideration. As a result—at the time of writing—interoperability between all vendor products has not been achieved. However, one can be assured that upon finalization of the CTI standards, many new computer-based PBX products will be available.

10.2.2 ATM-Based and RF Broadband Cable Telephony

Most existing PBX installations are large and have been upgraded partially to support digital telephone sets. A substantial percentage of the communications budget has been invested into the old and proven circuit switching technology. Because of this investment and the experience gained, it is not possible to simply exchange the PBX for a new and unfamiliar CBX, even when features and functionalities appear to be superior. To begin this convergence, it is an absolute requirement that existing equipment can be upgraded and gradually replaced and that new functionalities can be added to the existing equipment. The integration of telephony into the data network is viewed as a gradual process starting with the telemarketing organization or the student body in a college dormitory and expanding into the general user population on a step-by-step basis.

ATM-based telephony is available as well as IP-based technology to integrate the telephone transmission into the data network of the corporate

LAN. In this scenario the PBX is left in place. Extension lines or T-1 formatted lines, interconnected through a concentrator hub with the ATM core switch at 25.6- or 155-Mbps speeds, can now be brought to other campus locations via the backbone network. The ATM stream containing the telephone information is directly brought to the desktop, which carries an NIC card that translates the cell-based digital voice back into the analog format for connection to a standard telephone set. Phones not requiring computer interconnectivity, such as those located in hallways and public areas, are connected using existing station wiring to a phone hub. This unit can provide service for 24 telephones and is connected to a 25.6- or a 155.2-Mbps module in the workgroup switch and includes power service and battery backup for telephone sets. Similar technology exists for 10/100 Ethernet-based systems. Figure 10.2 shows the wiring arrangement.

The same arrangement can be used when connecting to the public network. In this case, the outside lines or an incoming T-1 format are connected to the concentrator hub, and the cell stream is interconnected with the ATM core switch. While these products either extend the PBX or bypass the PBX, they are not capable of replacing or converting the PBX into a CBX. However,

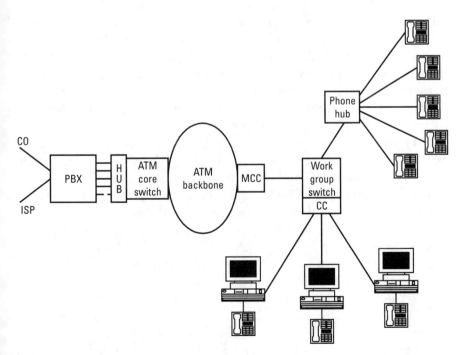

Figure 10.2 An ATM-based telephony integration in the LAN.

within a short time frame, S.100- and S.300-compliant CBX products will become available and true convergence can begin.

RF broadband cable telephony is available on the station side and on the line side of the PBX to extend the reach of the PBX or to add outside lines to a new building location. Assuming that a new building has been constructed in the campus, this requires service for 200 new telephones. The conventional method would be to upgrade the PBX and construct a 200-pair copper cable to the new building entrance location, install a cross-connect panel, and connect CAT-3 station wiring to the new telephone locations. The use of RF broadband technology will negate such a wasteful construction effort. A pair of single-mode fiber strands can provide for the continuity to the new building and the telephone traffic to the new building can be routed over a single 6-MHz channel frequency assignment, using a cable modem. There are two different options to make the interconnection: (1) as a new CBX installation in the new building or (2) as an extension from the existing PBX. In scenario (1), new outside lines can be digitized as individual outside lines or prepared in the T-1 format and connected directly to the HDT cable telephone unit. The HDT is connected to the fiber network, and the cable modem is placed at the new building together with a cross-connect panel. A new CBX or PBX can be placed at the MDF location to serve the user population. In scenario (2), the existing PBX is upgraded to provide 200 new extension ports. The station lines are connected to the HDT unit and interconnected via fiber to the station wiring in the new building. In both cases the telephone traffic of the new 200 stations is transported over a single 6-MHz frequency assignment in a transparent manner and without the need for cable construction. Figure 10.3 illustrates the transparent and cost-saving RF transmission of up to 240 voice channels over fiber in a 6-MHz spectrum.

Figure 10.3 illustrates the use of cable telephony as an extension of an existing PBX and without the use of the data network. While this arrangement can transport massive numbers of telephone transmissions, the cable telephone modem can also be used for CTI-integrated voice and data service to individual workstations, as discussed in the beginning of this chapter.

The integration of telephony into the data network can be a gradual process, but eventually the campus LAN will consist of an ATM core with IP- and RF-based campus zones that provide integrated voice, data, and video services. Not all telephones will be computer-based since there is a requirement for hallways and common rooms where a standard telephone is the only requirement. For these installations, phone hubs like the Spherecom product shown in Figure 10.2 will have their place and reuse the existing horizontal wiring system. At the front-end of the multiservice LAN, the ATM stream will split and

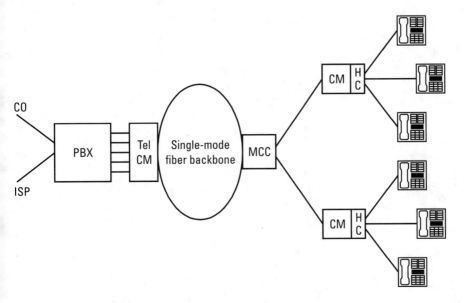

Figure 10.3 RF-based cable telephony in the LAN.

feature cell-based switching for telephony and for data. WAN and long-distance connectivity will provide ATM-based traffic in both ATM and IP formats to interconnect with the Internet and the many new CLECs of the future.

11

Videoconferencing and Video-on-Demand

While there are numerous desktop videoconferencing systems on the market that are trying to transmit video information over standard telephone lines, and at speeds that are insufficient to produce coherent video sequences, there are new technologies that can provide quality motion video. Videoconferencing and VOD within the corporate environment are believed to be an important supplement to the conduct of business. Videoconferencing and VOD are becoming increasingly essential in the competitive business environment to integrate the exchange of visual information into the daily conduct of business, to increase productivity, and to involve employees in continuous education and convey the ever-changing technology. This chapter aims to provide a brief overview of existing and emerging video transmission technologies.

The visual presentation of human activities was first established by the movie industry. When television came along, the desire for more and different programming was satisfied by cable television providers, by numerous new movies, and the video rental industry. We are all familiar with the entertainment value of video but are reluctant to embrace the technology into daily business transactions. We are also used to full-motion presentations and cannot find much usefulness in jerky four- by four-inch moving still pictures on our computer screens. The problem of the jerky picture is a direct result of the limiting transmission facilities in the local loop and in the public network. There are many video applications that require full-motion video quality in the business community. From imaging transfer to distance learning, interactive television, educational presentations, and telemedicine, there exists the need for high-quality full-motion video. It is estimated that the narrow bandwidth restrictions in the local loop and in the public network will be with us for another decade. Until then, any video transmission within the WAN will remain in the "talking head" format. In the corporate LAN, however, the integration of quality video into the daily business can proceed without further delay.

11.1 Analog Videoconferencing (NTSC)

In a campus equipped with an RF broadband coaxial network for cable television delivery, analog videoconferencing and VOD can easily be added. The network needs to be upgraded to two-way operation to accommodate multiple bidirectional video channels, but when configured to provide for video origination from multiple lecture halls, it can enhance the learning environment without major problems. There are numerous colleges and universities that have installed instructional television (ITV) networks and utilize the old-fashioned analog NTSC format for two-way conferences as well as for VOD requirements from dormitories. The technology of analog video lacks modern requirements for automated channel control and switching, and camera operation relies on the telephone or a preassigned schedule to set up, but it is usually far better than running video consoles and tape carts over college pathways. VOD is usually a telephone request by a student to the media center and a confirmation that the requested lecture will be scheduled for a particular time and transmitted on a particular channel. The lecturer can request tapes related to his or her curriculum in the same manner. NTSC video works, is economical, and increases the retention rate of the lecture.

11.1.1 Automation Equipment

Over the years, numerous automation accessories have been developed for distance learning, videoconferencing, classroom command, and control of video sources and cameras. Many vendors offer complete console installations with integrated automatic cameras, echo suppressors, dual television sets, and picture-in-picture features. Remote control equipment to call up videos from the media center is offered by AMX and Crestron and others. Videoconferencing connections can be provided with the ability to adjust pan, tilt, focus, and zoom from remote locations. Transmit and receive channels can be selected in real time or on a prescheduled basis by the use of programmable matrix switching equipment at the systems headend.

All these methods deliver NTSC quality video but require media personnel to feed the right tape into the selected VCR and are too complex to establish videoconferencing and VOD as a business tool. The major problem of all analog video systems is the familiar VCR. It cannot provide for random access, takes a long time to fast forward and rewind, and is subject to frequent failure. To succeed in education and business, videoconferencing and VOD must rely on digital transmission, storage, and control.

11.2 Digital Videoconferencing

Translating an analog video signal into a digital bit stream requires the reading of gray shades and colors of each pixel at enormous speeds. A full-motion video picture has 30 different frames per second, which means that the scanning process of all pixels has to be repeated 30 times per second. It is not surprising that such a digitization requires about 145 Mbps to identify every single picture artifact. To lower this speed requirement, many different compression technologies were developed—some only offering still picture sequences, others not representing the background with the same detail as the motion, others minimizing the number of frames for slow changing background information, and others digitizing every second horizontal line only. Bit rates of these digitizing methods range from 56 Kbps to 45 Mbps and feature fixed and VBR transmission algorithms. As a result, the number of proprietary compression schemes proliferated, and users refused to accept these products for lack of interoperability.

11.2.1 Videoconferencing Standards

Because video and television is a global phenomenon, and because the quality of a full-motion video picture is a very subjective quantity, the development of universal standards has taken the better part of a decade. The number of recommendations issued by the ITU for the transmission of nontelephone signals is plentiful. In 1990, ITU recommendation H.320 was issued to accommodate desktop videoconferencing products for ISDN traffic. Because the standard allows products to communicate with each other, H.320 has become a proven standard throughout the world. Non-ISDN products, developed before 1996, used proprietary coding approaches and lacked interoperability. The ITU issued and adopted H.324 for analog telephone lines and H.323 for IP packet networks in 1996. In addition, H.310, which covers ATM-based video products, was adopted. Each one of these standards includes individual standards for video, audio multiplexing control, data, and communications interfaces.

The video transmission standards follow H.261, H.262 (MPEG-2), and H.263 and are aimed at providing a high degree of interoperability for videoconferencing equipment in both IP- and ATM-based networks.

The major problem with standards is the assurance required by the user that the equipment is truly interoperable between the various vendors and products. To integrate videoconferencing into both the WAN and the LAN in a seamless manner, total interoperability is required. The International Multimedia Teleconferencing Consortium (IMTC) is a nonprofit corporation comprising more than 150 member companies worldwide. IMTC members conduct interoperability tests and make submissions to the standards bodies to enhance the interoperability and usability of multimedia teleconferencing products. The gap between standards and interoperability is caused by the incompleteness of the standards. Refinements are in process but may require more time to solidify interoperability.

First, there is the bandwidth requirement. While videoconferencing in the WAN uses typically 384 or 786 Kbps through the ISDN-switched network, every Ethernet LAN could deliver 6 or 7 Mbps directly to the desktop. In the case of an ATM-based system, 25.6-Mbps and higher bit streams are available. In addition, digital television is moving toward MPEG-2 standards requiring variable bit rates between 3 and 8 Mbps for full-motion video. The second most important technical requirement is latency. This is not an issue for an ISDN connection over the switched network but becomes the predominant issue for high-speed compression algorithms and Internet traffic. The total latency is the addition of the two endpoints and the network latency. Latencies of 50 ms are barely perceptible. An ISDN connection can keep network latency below 10 ms and assure good QoS. In a packet-switched LAN, the latency can exceed 80 ms, and when transmitted over the Internet where 5 to 15 hops between endpoints are likely, latencies of more than 200 ms can be experienced. In addition, compression-decompression latencies of high bit-rate hardware products can add in excess of 200 ms to the connection and cause intolerable picture quality impairments.

Within the LAN network environment it is easy to overcome both the bandwidth and the latency problem by using ATM technology. The short 53-byte cells will organize the transmission in a latency-free manner and provide full-motion videoconferencing throughout the corporate network. When conducting videocoferencing through the WAN or the public network, the established solution is ISDN over the switched network until the Internet features ATM-based controls and perfects the MPOA routing shortcuts. In established IP networks, ATM over IP can be used as well to reduce the problems caused by latencies. Examples of high-quality, point-to-multipoint videoconferencing within the corporate LAN are provided in Sections 11.2.2–11.2.6.

11.2.2 A Comparison of JPEG and MPEG-2

The Joint Picture Experts Group (JPEG) and MPEG have developed different compression methods to identify necessary picture artifacts suitable for high-quality presentation of motion video. The MPEG compression technology uses a comparison of a number of frames and pixel groups within adjacent frames to determine the number of pixels to be transmitted. As a result, the encoder algorithms are voluminous and the bit rate must be variable. When the background in a sequence does not change, the bandwidth requirement is low or around 1–2 Mbps. When motion and camera movement changes the pixel content in each adjacent frame, the bandwidth requirement is high and can reach 15 Mbps.

MPEG-2 is an ideal encoding mechanism for one-way video delivery services such as movies, VOD, HDTV, and ATV. While encoder costs are high and only become affordable for multiple video transmissions, the decoder can become simple and inexpensive. JPEG encoding is a simpler encoding standard and is, therefore, more economical. While both MPEG and JPEG compression algorithms are based on the discrete cosine transformation (DCT) technology, JPEG compresses each frame individually and can provide better bit rate control at the expense of higher decoder costs. JPEG compression permits the predetermination of desired bit rates. If a 640×480 resolution and 30 frames per second are required, the Q-factor adjustment can optimize picture quality and produce a high-quality, full-motion video at a fixed bit rate of about 5 Mbps. If full-motion is not a requirement, a 320×240 resolution with 10 frames per second may be used in conjunction with a higher Q-factor and provide a satisfactory picture quality with a bit rate of under 1 Mbps. The JPEG compression concept is predominantly used in video editing, videoconferencing, and any two-way video connectivity that may be required.

In summary, the advantages of JPEG are:

1. The ability to select and optimize picture quality;
2. The ability to display multiple video streams like quad pictures and picture-in-picture (PIP);
3. The ability to perform multiple videoconferencing functions;
4. The ability to change the size, Q-factor, and resolution at any time.

MPEG-2 advantages are:

1. The ability to use less bandwidth for frames with limited motion content;

2. The ability to combine multiple one-way video streams into a single transmission;

3. Lower decoder costs.

11.2.3 JPEG Over ATM

There is no reason to delay the installation of high-speed videoconferencing equipment within the corporate LAN. The gateway to the WAN or to the public network, however, requires special attention and adaptation to the lower transmission speeds. An example of uncompressed or motion-JPEG-compressed, high-quality, full-motion videoconferencing systems is FORE Systems' StreamRunner product. Standard NTSC video is encoded at the source location into ATM cells and can be multicast to one or more receive locations for display on Unix or Windows 95/NT workstations equipped with an ATM card. The transmission can also be decoded to NTSC analog video and displayed on standard television sets. The system uses Windows NT software for circuit setup and control and permits videoconferencing between up to four participating locations. Because NTSC video is composite video with audio, JPEG transmission is truly multimedia; transmits video, audio, and data through the ATM network; and permits remote camera control of pan, tilt, iris, focus, and zoom functions. The data port supports an RS-232 serial interface, and the audio ports can provide for as many as six mono or three stereo inputs. QoS support can be UBR or CBR as required. Figure 11.1 indicates a typical architecture for videoconferencing.

The translator codec is the turnaround location for the videoconferencing system. It transfers and directs the transmissions to the desired receive locations. The digital video is transmitted in the ATM-format to the workgroup switches and routed to the participating workstations. The transmission can also be extended through an RF broadband HFC network with the aid of a cable modem. Using a JPEG codec at the workstation can translate the video to the NTSC format for display on standard television sets.

An analog camera or more complex studio equipment can be used through the codec for ATM-based transmissions. The videoconferencing system can be used by participating workstations equipped with an ATM card without any codec. Videoconferences can be computer-oriented and interconnect with up to four participants.

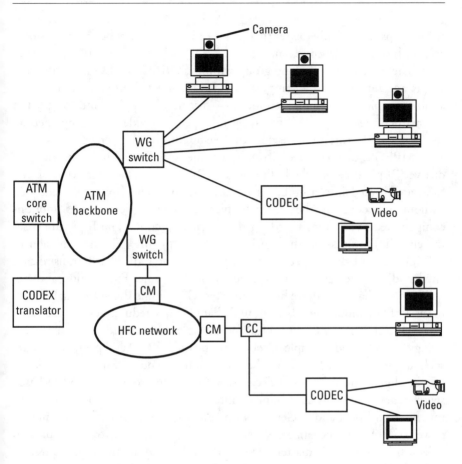

Figure 11.1 Videoconferencing in the LAN.

11.2.4 MPEG-2 Over ATM

MPEG-2 encoders and decoders for videoconferencing are currently a work-in-progress. Designers of MPEG-2 encoding equipment have concentrated their efforts on the broadcast, cable, and satellite industries. There are many products on the market that feature encoding algorithms for multiple video transmissions. Since it is possible to transmit five to eight MPEG-2 bit streams within the 6-MHz channel assignment on an HFC cable system, these encoders can handle multiplexed VBR traffic and are expensive. The goal is to reduce the price level at the decoding end (set-top converter). For the broadcast industry, first in line in the analog-to-digital conversion process, MPEG-2 video encoders are offered for studio-to-transmitter links and electronic news gathering (ENG) mobile applications. Some of these MPEG-2 encoders feature VBRs up to 50 Mbps and promise better than DS-3 or NTSC quality but are not affordable for business videoconferencing.

The MPEG is a working group within ISO/IEC. MPEG-2 specifies the syntax and semantics of a compressed video bit stream and parameters—such as bit rates, picture sizes, and resolutions—that may be applied and decoded to reconstruct the picture. MPEG does not specify a video-encoding method to allow manufacturers the freedom to develop improved encoders.

MPEG-2 defines three picture types, the intraframe (I) picture, the predictive (P) picture, and the bidirectional (B) picture. The three types of pictures are formed into a group of pictures (GOP). The I-picture is encoded without reference to the previous picture. The P-picture uses motion-compensated predictions based on the previous picture. The B-picture bases the encoding process on the previous and next picture. The transmission of MPEG-2 encoded video streams can use VBR or CBR. VBR transmission always adjusts the rate to the complexity of the frame. Full-motion and frequent scene changes require higher bit rates. CBR limits the video quality and, in cases of fast-motion and background changes, may reduce picture quality.

MPEG-2 transmission is ideal for multiple one-way video streams where a single encoder and multiple decoders can be used. VOD and pay-per-view are such applications. MPEG-2 will be the defining standard for video transmission in digital form for many decades. MPEG-2 networks using ATM-based switches are already installed in many interactive video presentations in various museums (at Epcot and at Getty). Numerous kiosk systems use the technology as well. Getty uses an Optibase Videoplex MPEG card, a Matrox Mystique 220 video card, an ATI bus-mastering network card, 20-in monitors with a resolution of 1,024 × 768 pixels, and Beyerdynamic headphones. The MPEG-2 transmission is at 6 Mbps, which is encoded from a Sun Ultra server configured with 400 Gbps of storage. The network uses 100-Mbps TCP/IP from the server to the ATM backbone, 155-Mbps ATM in the backbone, and 10-Mbps Ethernet to the desktops.

11.2.5 ATM Networking and RF Broadband

Philips Communications and Security Systems has introduced a fully digital real-time closed-circuit television (CCTV) system that uses FORE Systems' ATM networking products for the transmission of video, audio, and data. While these type of systems use the JPEG compression algorithm, the transmission of digital video in any format is assured by the cell structure of the ATM technology. Switching times, circuit assembly, and tear-down intervals as well as reconvergence times are reduced to milliseconds and eliminate the latency problems of packet-based network architectures. This does not mean that the entire network must use ATM in an end-to-end configuration. However, the introduction of an ATM backbone component will greatly enhance the multi-service capabilities of the LAN.

For extended campus networks that have a limited number of single-mode fiber strands available, the ATM backbone can be extended using cable modems and RF broadband. High-resolution video/stereo audio can be brought to the desktop display or decoded to conventional NTSC analog video for presentation on standard television sets.

With the many advances made in the transmission of high-resolution video, it is expected that videoconferencing in the corporate network will emerge as a useful tool for intracompany communications. The many hours spent traveling from a campus building to the meeting location and in face-to-face meetings can easily be reduced by the application of JPEG compression and future MPEG-2 technologies over the multiservice network.

11.2.6 Videoconferencing in the WAN

Videoconferencing through the WAN enterprise network is still a work-in-progress. While ISDN over the switched network reduces the latency problems, the lack of bandwidth in the outside world is still a curtailing limitation. Transmissions over the Internet do not provide usable two-way video presentations. The problem is the existing IP routers in the system, each producing additional latencies. However, as in the case of VOIP, many long-distance service providers are upgrading the Internet. The solution is to use DWDM in the core of the long-distance network. DWDM uses multiple optical transmissions over a single-mode fiber and can provide multigigabit transmission speeds in the center of the network. At the tributaries, existing SONET multiplexing and ATM switching assures the network latency considerations and the circuit-oriented functions. The DWDM optical pipe in the center transports data and TDM circuits at high speeds but has no routing and switching functions. When this new high-speed infrastructure becomes available for both Internet and long-distance telephone providers, it is certain that videoconferencing in the WAN will mature to the bit rate and quality that it deserves.

11.3 Video-on-Demand in the Corporate Network

Many corporations have television studios and media personnel on staff to produce educational videos for the employees. In many companies a separate coaxial network exists to disseminate the videos to common quarters and conference rooms throughout the campus. Corporate television has been used for training in the various business processes, for familiarization with company goals, and to increase the productivity of the staff. The benefits to the corporation have been mixed because corporate television does not contain interactive participation and because any such training involves the scheduling of the personnel and the related loss time and productivity. In addition, a substantial number of media personnel must be available to compose, record, schedule, and air the training material using manually operated VCRs.

With the advent of digital video transmission, interactivity and random access can provide the basic requirement toward more personalized employee development. Normally referred to as VOD, the corporate television can be transformed into a dedicated learning experience. The transition from analog to digital video will establish high-quality broadcast video over the corporate ATM-based data network. Distance learning and VOD can then become an integral part of an interactive multimedia learning process.

11.3.1 MPEG-2 and the Digital Video Disk

The development of digital interactive video programming requires the translation of NTSC video into the MPEG-2 format. The easiest way to accomplish this is by using an MPEG-2/DVB encoder. Philips, Viewgraphics, and others have developed such compression systems that provide flexible bit-rate bandwidth handling and optional scrambling. Any analog video information can be compressed into the MPEG-2 format. Streams of different bit rates can be spliced together, and the authored product can then be recorded on a digital video disk (DVD). There are a number of postproduction DVD vendors—like Minerva, Daikin, Optibase, and Sonic Solutions—that offer complete desktop production systems, mostly Windows NT-based software for the authoring of video and stereo audio. MPEG-2 and the DVD disk will replace the familiar tape within a short time period and provide video recording, editing, and authoring inclusive of the interactivity so badly needed to reform the distance learning and educational VOD applications.

While it is not expected that corporate media personnel will develop costly MPEG-2 authoring stations, there are a number of MPEG-2/DVB authoring companies that can be used to prepare the interactive multimedia presentations on DVD disks.

11.3.2 VOD Technology

Whether the application is to select a movie at will, as may be the case in hotel-based or cable television-based distribution systems, or whether it is an educational video sequence required for distance learning or in the corporate LAN, the foremost requirement is full user control. While a VOD movie may only require standard VCR functions such as pause, rewind, fast forward, and play, the interactive VOD learning process requires client interaction with the learning material. The client responding to questions must be able to advance the learning process to a more difficult level by an automatic evaluation of the answers. The VOD technology will, for the first time, provide the tools that are necessary to learn and retain information at the rate of the student's ability.

VOD technology focuses on the ability to multicast a number of programs to a large number of users at the same time and within a limited bandwidth. In the corporate LAN, the transmission format of VOD can be MPEG-2 over ATM, MPEG-2 over IP, MPEG-2 over ATM, or any of the above over a cable modem and RF broadband. In the HFC cable television network, MPEG-2 over ATM or over IP/ATM will always use the cable modem and the set-top converter at the subscriber end. Two terms are usually used to describe VOD: *near-VOD* (NVOD) and *true-VOD* (TVOD). The difference is the availability of the programming to a user. In the TVOD case, the program must be available immediately when requested. TVOD requires the availability of a dedicated video stream from the disk and a dedicated communications channel to the user. NVOD permits the selection of time intervals such as a 10 or 15 minute scheduling.

In order to permit multisession TVOD, there are bandwidth-conserving technologies that reduce the number of disks in the library and provide for bandwidth sharing in the network. Hodge Computer Corporation has developed a multisession VOD concept that permits one or more users to share the same stream for viewing. When the user opts to stop, pause, or change the frame of the video, the user is removed from the shared stream and given a dedicated resource from the disk library. When the user has completed the desired functions, he or she is assigned back to the closest existing stream that corresponds to the frame that he or she was viewing before. If such a stream does not exist, a new stream is created and may be shared by another user.

Servers for VOD are becoming available in many varieties. The Diva Systems Corporation has introduced Sarnoff Server, a powerful video-streaming computer that provides MPEG videos in a massive parallel processing computer on the banks of industry-standard SCSI hard drives. The technology, which is aimed at the pay-per-view cable television market and uses standard DCT-100/1200 set-top converters from General Instruments Corporation, is already in use by a number of cable operators.

For corporate television and training applications, the use of DVD disk libraries appears to be a more versatile solution. As in the case of the CD-ROM tower for storage, the DVD tower can provide flexible storage for interactive videos in a space-saving architecture.

Bandwidth control of the transmission over the corporate multiservice network is of importance. Native MPEG-2 requires bit rates between 1 and 10 Mbps. A 10-channel VOD system can easily utilize a bandwidth of between 10 to 100 Mbps if not managed. Luckily, the VBR content varies between the various streams and is only additive when CBR is applied in the transmission segment. There are also statistical multiplexers available to manage the composite video streams. DiviCom, a member of the C-Cube group of companies, has developed DiviTrack, a statistic multiplexer solution. DiviTrack uses two MV40 encoders: The first is used to overview the complexity of the video streams and the second to adjust the encoding functions with real-time statistics. At full resolution, the output of 10 video streams can be transmitted in a total bandwidth of 20 Mbps.

Content management functionality is required to allocate the optimized number of programs in the disk library. In an educational setting, as in the selection of movies, there will be popular and less popular programs. A shelf-space allocation algorithm (SSAA) is required to specify the minimum of shelf space at the outset and to adjust the requirement to the statistical record of usage. Frequently used video streams are then provided in accordance with user statistics. In cable television networks and corporate networks, programs must be available to users at all times. In the cable television world, this feature is commonly referred to as a barker channel. In corporate and educational VOD systems, an icon on the screen can activate a listing of video streams with their purpose, difficulty level, credit classification, lecture sequencing, as well as grade level and accomplishments of the student under pin number control.

Interactivity in multimedia changes the learning experience. VOD, for the first time, combines crisp and clear video with stereo audio, permits viewing on a full computer screen, enables zoom functions as well as individual interactions, promotes learn-at-your-own-speed individuality, and increases the rate of retention.

11.3.3 VOD Network Architectures

VOD is a multimedia service that can be transmitted over a data network or an HFC cable network. VOD can easily be installed in an existing cable television facility of a college or a university. In a campus environment, VOD can be combined with fast Internet access through the same cable modem. In corporate networking where high-speed facilities are available, VOD can be added to the data stream or use a separate coaxial transmission system that may be available for corporate video. Figure 11.2 illustrates the basic VOD network architecture.

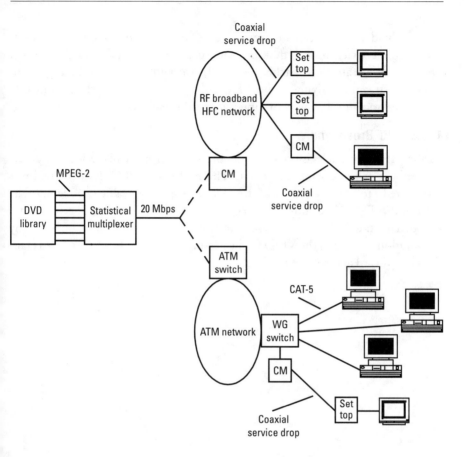

Figure 11.2 The VOD network architecture.

The video material in the DVD library can be accessed on a random access basis from any participating workstation. The video material in MPEG-2 format is fed through a statistical multiplexer that organizes the VBR of each program into a CBR. The transmission through the network in the ATM format can flow over the ATM backbone or over an RF broadband HFC network using cable modems.

11.3.3.1 ATM and IP

The integration of ATM content into the network is required. Video with its VBR requires exact timing for each frame. IP technology and packet transmission systems cannot handle these requirements at high speed. In the corporate LAN, the DVD server can be directly connected to the ATM backbone using a 155-Mbps ATM module.

Individual users can be connected to the backbone using 25.6-Mbps ATM connectivity to the desktop equipped with an ATM card. Another option is to use an ATM backbone with IP workgroups. In this arrangement, the video stream is arriving in the ATM cell-based format over the IP network. At the desktop an ATM network interface card is required.

11.3.3.2 RF Broadband

In the HFC cable television network, the server is most likely located at the headend or at the operations center. The multiple MPEG-2 video stream is modulated to a desired RF frequency using a cable modem. The MPEG-2/ATM RF channel is then transported through the network and encoded in the user's set-top converter for viewing on a television set. Any workstation can access the VOD by addressing the video library through a standard ATM card and a cable modem.

12

Fiber-Optic Transmission Equipment

This chapter discusses the properties of single-mode fiber-optic transmission equipment. Whether the backbone of a campus data network is serving OC-12, gigabit Ethernet, or RF broadband, the use of single-mode fiber has the advantage of unlimited bandwidth and low optical attenuation. Many manufacturers have developed single-mode fiber cables and offer termination hardware and service cabling to the workstation. Fiber-optic cables to the desktop and in the service wiring are not needed in a hierarchical network architecture. Even the workstation of the future does not require speeds in excess of native ATM or fast Ethernet. Existing CAT-5 UTP standards satisfy these requirements. The use of coaxial cable in the service segment can be a desirable alternative for the selection of many RF frequencies when using RF broadband technology. This treatment of fiber-optic cables and devices is therefore restricted to backbone uses in a campus-wide multiservice network.

12.1 Single-Mode Fiber-Optic Cables

The differences between the various types of single-mode cables are due to mechanical features. The electrical specifications of the single-mode glass fiber are identical for all models and types. The fiber strand is a very small optical waveguide that will propagate the transmission of light with minimal optical attenuation. In order to perform in this manner, it is important to keep the core diameter of the fiber strand at 8.3 microns and to use a cladding of no more than an additional 125 ± 1.0 microns. While the cladding is protected with additional coating, it is the core diameter and the cladding that assures the waveguide effect and the superior performance of the single-mode fiber. The common electrical specifications of the single-mode fiber strand are listed in Table 12.1.

The coating around the cladding is the first mechanical strength member around the glass fiber core. To isolate the fiber from external forces, there are a number of protective solutions that relate to stresses during installation, environmental conditions, and bending limitations. Mechanical properties such as tensile strength, impact resistance, and resistance to moisture, chemicals, freezing, flexing, and bending are important considerations during the installation process and have led to a wide variety of models and types.

12.1.1 Loose Tube Cables

In the loose tube buffer construction, the fiber is contained in a plastic tube that can accommodate a number of fiber strands. The interior of this plastic tube is filled with a gel material that permits a controlled movement of the fiber strands. Loose tube cable usually features 6 or 12 fiber strands in a bundle. Multiple bundles are then combined in a larger diameter aramid yarn layer and filled with flooding compound. A jacket or armored jacket is used for further protection. Figure 12.1 depicts the buildup of the loose tube cable.

Loose tube cable construction isolates the fiber from external mechanical and environmental forces. The number of fiber strands within a bundle is predetermined. Should a lesser count of fiber be required, the bundle will contain a filler dielectric. If the entire bundle is not required, it will be substituted with a blank filler of equal diameter. Loose tube cables are available for fiber counts from 5 to 288 strands. The cable can also contain a mixture of multimode

Table 12.1
Electrical Specifications of the Single-Mode Fiber Strand

Parameters	Values
Wavelength	1,310/1,550 nm
Attenuation	0.35/0.25 dB per km
Core diameter	8.3 nominal microns
Cladding diameter	125 ± 1.0 microns
Coating diameter	245 ± 10.0 microns
Cut-off wavelength	1,250 ± 70 nm
Zero-dispersion wavelength	1,310 ± 10 nm

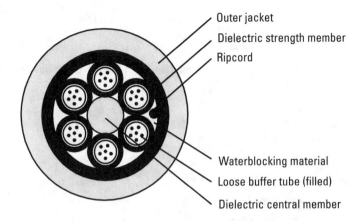

Outer jacket

Dielectric strength member

Ripcord

Waterblocking material

Loose buffer tube (filled)

Dielectric central member

Figure 12.1 Loose tube cable construction.

and single-mode strands and is available with a steel or dielectric center member.

Loose tube cables are recommended for outside plant installations but can also be used for in-building installations when installed in conduit and in accordance with NEC Article 770. The cable is the ideal choice for interbuilding backbone networks. It has good mechanical properties, permits pulling tensions to 600 lb, can be direct-buried or used in innerduct and conduit, withstands the pressure of frozen water in conduit sections, and has a crush resistance to 500 lb. When used indoors without a protective conduit, the length between building entrance and termination is limited to 50 ft. Table 12.2 lists specifications for loose tube cables with fiber counts that are typically used in the campus backbone.

12.1.2 Tight Buffer Cables

There are tight buffer cable types that are suited for indoor and outdoor installations. An example is the tight buffered cable from Optical Cable. Originally a product developed for the military market, it features a 100-kpsi tested fiber, buffered with a 500-μm acrylate elastomeric thermoplastic material. The cable is ideally suited for campus installations where termination locations are at distances exceeding the 50-ft limitation. Figure 12.2 illustrates the makeup of the indoor/outdoor tight buffered cable.

In addition, there are many riser-rated tight buffer cables on the market especially designed for riser applications. The riser-rated jacket makes this cable ideal for all indoor installations. It can be run in innerduct for identification and does not require a conduit enclosure. Tight buffer cables meet UL-OFN-LS

Table 12.2
Loose Tube Cable Construction

Mechanical Specifications	Small Number of Fibers	Large Number of Fibers
Fiber count	5–36	37–96
Number of buffer tubes	6	6–8
Maximum fibers per tube	6	12
Nominal outer diameter	0.46 in	0.48–0.58 in
Minimum bend radius, installation	5.9 in	5.9 in
Minimum bend radius, long-term	7.9 in	7.9 in
Maximum tensile strength, installation	600 lb	600 lb
Maximum tensile strength, long-term	200 lb	200 lb

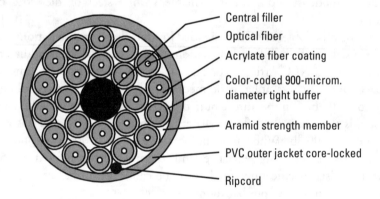

Central filler
Optical fiber
Acrylate fiber coating
Color-coded 900-microm. diameter tight buffer
Aramid strength member
PVC outer jacket core-locked
Ripcord

Figure 12.2 The indoor/outdoor tight buffered distribution cable.

(Underwriter Laboratories-Optical Fiber Network-Limited Smoke) specifications and do not require transition splices when entering a building. They can be used in outdoor installations but only feature a buffering tested to 50 kpsi. All cables feature 12 fiber strands per bundle and are available with a fiber strand count from two to 288. Figure 12.3 shows the makeup of the tight buffer cable construction.

Each bundle is held with binder threads to increase the strength of the cable. Filling compounds are used between bundles, and the bundles are combined in a large buffer tube. A dielectric center strength member is used in all cable types. The specifications, listed in Table 12.3, only cover the most common fiber counts in a campus network.

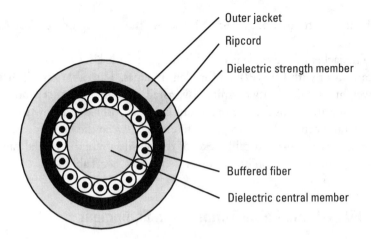

Figure 12.3 The tight buffer cable construction.

Table 12.3
Tight Buffer Cable Construction

Mechanical Specifications	Number of Fibers			
	under 30	36	72	96
Outside diameter	0.5 in	0.52 in	0.57 in	0.68 in
Fibers per bundle	12	12	12	12
Bundles per cable	12	12	12	12
Minimum bend radius, installation	5.1 in	5.4 in	6.1 in	6.9 in
Minimum bend radius, long-term	7.7 in	8.0 in	9.1 in	10.0 in
Maximum tensile strength, installation	600 lb	600 lb	600 lb	600 lb
Maximum tensile strength, long-term	135 lb	135 lb	135 lb	135 lb

12.1.3 Break-Out Cables

The application for break-out cables is horizontal or vertical riser spaces, requiring fiber counts below 24 strands. Both riser-rated and plenum-rated cables are available. The construction of the cable permits more flexing flexibility, but at a slightly higher attenuation. Figure 12.4 shows the makeup of a typical 12-fiber break-out cable.

The jacket of the cable is either compliant with UL OFNR/FT4 for riser applications or with OFNP/FT6 for plenum areas. The cable's small diameter, light weight, and flexibility permits easy installation, maintenance, and administration. Specifications for the break-out cable are provided below for 4-, 12-, and 24-strand cables. There are no differences between the handling properties of riser- and plenum-rated cables except the tensile strength. Special care must be taken in the installation of plenum-rated cables. See Table 12.4.

12.2 Fiber-Optic Cable Termination Equipment

Fiber termination equipment is required in all areas of premises distribution. At building entrance locations the termination equipment is usually referred to as the MCC. In riser areas, there are applications for IC and HC panels. While ICs and HCs can be wall-mounted fiber organizers, the MCC requires rack mountability and organization for cable management. The MCC is usually the entrance location for the backbone fiber network. It is also, in most cases, the location for fiber-optic transmission equipment, coaxial distribution equipment, CAT-5 termination and cross-connect panels, and other ancillary equipment. Sufficient rack space and orderly cable guides for patchcord management are important considerations.

There are numerous manufacturers of suitable fiber termination solutions. In a review of the offerings, it is important to obtain assurance that the equipment conforms to the EIA/TIA-568A commercial building cabling standard. This applies to connector module housings, patchcord management

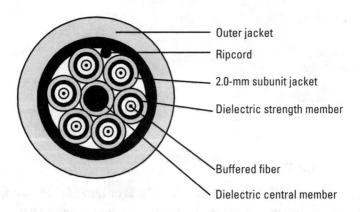

Figure 12.4 The break-out cable construction.

Table 12.4
Break-Out Cable Construction

Mechanical Specifications	Number of Fibers		
	4	12	24
Outside diameter	0.2 in	0.28 in	0.42 in
Minimum bending radius	2.1 in	2.8 in	4.6 in
Maximum tensile strength, riser (installation)	225 lb	405 lb	607 lb
Maximum tensile strength, riser (long-term)	67 lb	135 lb	225 lb
Maximum tensile strength, plenum (installation)	148 lb	297 lb	297 lb
Maximum tensile strength, plenum (long-term)	148 lb	74 lb	74 lb

guides, horizontal and vertical routing guides, as well as storage panels. A few examples of these items are provided in Sections 12.2.1–12.2.4.

12.2.1 Fiber Distribution Frames

The fiber distribution frame, sometimes referred to as a connector housing, provides interconnect and cross-connect functions between outside plant, riser, and distribution cables. The basic unit includes brackets for 19- or 23-in standard racks and contains mounting space for a number of connector modules. Connector modules are available for 6-, 8-, and 12-fiber connectors. Vertical positioning of these connector modules permits a space-saving arrangement for up to 144 fiber connections in a four-rack unit-high enclosure.

It is important to consider ordering of connector modules that are equipped with preloaded and factory-terminated cable assemblies. This will eliminate the time spent in the connectorization process and assure uniform performance. Other considerations are the existence of strain relief areas, the storage space for pigtails, grounding provisions, a removable top of the housing to accommodate patchcord routing, and a labeling area for each fiber to establish cable management and documentation. A removable or fold-down polycarbonate front door is also recommended for easy identification of circuits and rearrangement of interconnections.

Because connectorization is the most expensive installation cost, the user may opt to purchase the housing for ultimate capacity but only provide connector modules for the initial requirement. Blank connector panels can be used for unterminated fibers and exchanged with connector modules in a scalable manner.

12.2.2 Rack Mounting and Cable Management

Standard 19-in and 23-in open racks can be used. The 23-in version has more space to accommodate patchcord storage and routing. Many installations exemplify an oversight and lack of planning for cable management when it comes to the routing of patchcords. The patchcords are simply connected to the top housing, routed through vertical guides, folded at the lowest point, and fed upward to the destination panel. Typically, this type of installation can lead to patchcord fatigue, overbending, gravitational pull on the connectors, and an inability to determine which of the cords is used for what purpose. Properly equipped racks can eliminate these problems.

The rack layout should include single rack spaces equipped with horizontal troughs as well as half-shell guides, similar to splice organizers, on three-rack-unit panels to accommodate the individual routing of the patchcords. This method will provide strain relief on the patchcord connector caused by gravity and better identify the routing and purpose of each patchcord. To establish this functionality in the vertical plane, the rack should also be equipped with vertical guides and end caps, which is more easily accomplished in the 23-in rack configuration.

12.2.3 Single-Mode Fiber Connectors

Single-mode fiber strands can be connectorized using FC, ST, and SC connectors. The difference is mainly the construction properties of the ferrule and the hardware. All connectors are compliant with EIA/TIA specifications. The FC connector is available with a premium ceramic ferrule and metal hardware. The ST connector is available with a premium ceramic ferrule and either metal or composite hardware. The SC connector also features a premium ceramic ferrule and composite hardware but is available in a duplexible configuration compliant with ANSI/TIA/EIA-568 requirements.

Electrical performance of the single-mode connectors is equivalent. All of them feature a typical insertion loss of 0.35 dB. The reflectance is an important measure of connector performance. While all connectors feature a reflectance of better than −40 dB, the ST and SC connectors with ultra-PC polish can provide a reflectance of better than −55 dB. Reflectance is the amount of optical energy reflected back into the fiber strand. The higher the rejection of the returned optical energy, the better the performance of the system. Ultra-PC polish also reduces the insertion loss average by about 0.05 dB, which may be a consideration in dual-redundant rings with many cross-connect locations. The ST connector has a bayonette closure that may be less effective in certain environments than the push-pull closure of the SC connector. Vibrations in a dusty environment may reduce the performance of the ST connector over time and require periodic cleaning.

FC and SC connectors are also available with angled-PC. The angular polish will reduce the reflected optical energy by sending it into the cladding. An increase to a reflectance of −60 dB can be obtained with the angular polishing method.

While the SC connector is the most costly connector, the ability to combine transmit and receive directions in one assembly may offer many advantages for maintenance and cable management. For many applications the use of ST connectors with ultra-PC finish is the most cost-effective solution.

12.2.4 Preassembled Patchcords

Preassembled single and dual patchcords are available in every conceivable length directly from the manufacturer. Patchcords are available with FC, ST, and SC connectors with any of the polishing methods discussed in Section 12.2.3. Manufacturers also offer 100%-tested patchcords, but the surcharge for such individual handling is substantial.

There are new patchcord assemblies on the market that combine two patchcords in a dual configuration. The ability to route and to identify send-and-receive connections in cross-connect panels can simplify identification problems and rearrangements.

All the considerations discussed in Section 12.2.3 relative to the connectors in the fiber distribution frame also apply to the selection of patchcords. FC- and SC-equipped patchcords can be obtained with super-, ultra-, or angled-PC polish. ST is available with super- or ultra-PC only for optimized reflectance results.

12.3 Connectorization Considerations

The splicing of single-mode cables is not recommended, nor should there be a need to perform splicing in a properly designed system. Necessary operations include the termination of fiber cable in fan-out kits, connectorization, and polishing.

Buffer tube fan-out kits are available for the termination of 6- and 12-fiber buffer tubes. The kits provide the most compact and easy to install solution for field installation of pre-assembled connectors to loose tube cables. The fan-out assembly features the same color identification as the loose tube buffer and is available in 25- and 47-in lengths. Other fan-out kits are available for tight buffer cables in similar lengths.

In addition, the use of furcation kits is recommended to terminate the loose tube cable sheath and to water block the cable. Furcation kits are necessary for the termination of loose tube cables at outdoor termination locations, but they are also useful for indoor terminations.

A wide variety of connectorization tools are available. They consist of starter kits, complete connectorization kits, consumable kits, fan-out assembly kits, and polishing assemblies. The ease of connectorization of single-mode fiber has been improved over the years and can be completed using the proper tools and methods in less than 2 min per fiber strand.

12.4 Fiber-Optic Transmission Equipment

There are two major groups of fiber-optic transmission equipment, the baseband digital transmission system and the RF broadband transmission system. The baseband digital transmission system is commonly used by the data network equipment manufacturers. While the cost of the transmitters is lower, the baseband-oriented transmission is restricted to the speed and bandwidth requirement of the equipment. Typically, an OC-3 ATM transmitter will be built for the 155.2-Mbps speed and cannot be used for higher speed ranges. Higher speeds require a different optical transmission and a higher cost.

In an RF broadband network, the bandwidth of the laser transmission must be wide enough to transmit multiple channels at multiple RF frequencies. The RF broadband optical transmission system is AM-modulated and can provide a bandwidth of up to 10 GHz. Because this kind of bandwidth is not required, the typical laser operational bandwidth is limited from 20 to 750 or 860 MHz and equipped with impedance-matching circuitry at the conversion locations from electrical/optical and optical/electrical. The development of AM-based laser technology is only about seven years old. In this period of time, there have been spectacular improvements in the performance of the DFB laser. While the highest obtainable output level was about 2 mW in the 1988 time frame, present-day DFB laser configurations feature 10- and even 20-mW outputs. The DFB laser is the most cost-effective optical transmission engine for RF broadband transmissions over single-mode fiber. A brief description of the technology is provided in Section 12.4.1.

Fiber-optic transmission equipment is available from every major cable television manufacturer, such as General Instruments, Scientific Atlanta, Broadband Networks, Inc. (BNI), and Ortel, as well as through all major cable television distributors, such as Anixter and Toner.

When evaluating the many transmitter and receiver designs, it is important to consider the fault alarm capabilities. Local fault reporting is usually provided with LED indicators for loss of power, loss of signal, and optical output/input. Remote reporting of these performance parameters is seldom provided. Network management requires the reporting of equipment failures at a central location. Therefore, the ability to obtain performance and fault reporting external to the unit and in a format that can be forwarded through an alternative network is important.

12.4.1 High-Power DFB Laser Transmitters

The big advances that have been made during the past few years are related to the laser chip and fiber-coupling efficiency. Laser chip efficiency deals with the translation of electrical energy into photonic energy. Ideally, each electron produces one photon for a 100%-efficient transfer. The laser chip efficiency has been improved from an early 30% to 60%. Fiber coupling efficiency deals with the coupling of the laser output into the single-mode fiber core. The beam leaving the laser chip is divergent and slightly oval so the focus optimization must reshape the laser into the circular fiber strand. Coupling efficiencies have increased from an early 40% to about 80%. As a result, the output capability of the DFB laser transmitter has improved steadily by about 2 dB per year and the price has dropped about 20% per year.

RF broadband HFC systems can now be built with optical budgets of about 16 to 18 dB for the most demanding channel loading. This means that a forward RF broadband LAN could be built using an eight-way optical coupler connected to a single transmitter. This arrangement would require only a single fiber connection to each building and still have enough optical power to deal with cross-connect losses in a dual-ring architecture.

DFB lasers work with a low modulation depth. The reason for this is the trade-off between C/N and distortion. Because RF broadband uses a carrier frequency every 6 MHz to transport a different channel, there are beat products of second and third order. These distortion products increase by 2 dB for every 1-dB increase in the modulated signal. However, the lower the modulated signal, the better the C/N. Regardless of these observations, the DFB laser transmitter has matured to deliver up to 100 six-MHz channels side-by-side with good optical budgets, the best C/N, and reasonable costs.

12.4.1.1 Electrical/Optical Specifications

Table 12.5 lists the typical electrical specifications for a high-power DFB transmitter with an optical output power of between 25 to 30 dBm.

High-power DFB transmitter units are often packaged in a single rack space unit and easily connected with a standard fiber patchcord and a 75-ohm BNC connector.

The units must be ordered with the desired single-mode fiber connector but are available for FC, ST, and SC and all polish varieties.

12.4.2 Low-Power DFB Laser Transmitters

In cases where lower optical budgets can be tolerated, as it is likely in all direct campus wiring architectures, the low-power DFB laser transmitter offers an attractive low-cost solution. Assuming that single-mode fiber continuity exists between the control center and each building entry location, the low-power DFB laser can offer an optical budget of over 8 dB and easily cover all cross-connect losses in a ring architecture. The low-power DFB laser is ideally suited for direct return transmissions from the various buildings, whether in the 5–45-MHz, the 5–186-MHz, or the 5–750-MHz spectrum.

Table 12.5
Typical Electrical Specifications for a High-Power DFB Transmitter
With Optical Output Power of 25–30 dBm

Electrical Properties	Values
Channel capacity, AM-VSB (6-MHz)	40–100
RF input level	24–27dBmV
Composite triple beat (CTB)	65 dB
Composite second order (CSO)	62 dB
C/N	50 dB
Optical transmit power	25–30 dBm
Optical budget	18–16 dB
Power consumption	20W

12.4.2.1 Electrical/Optical Specifications

Table 12.6 lists the general electrical/optical specifications of the low-power DFB laser transmitter.

Low-power DFB laser transmitters are packaged in a 1-RU rack space unit and feature a 75-ohm BNC connector for the input of the electrical signals and an ST, SC, or FC connector for patchcord connectivity.

12.4.3 Receiving Equipment

Optical receivers for baseband digital applications generally use transimpedance type amplifiers and avalanche photodiodes (APDs), also known as PIN photodiodes. These devices exhibit good bandwidth properties but do not have adequate linearity to be used for RF broadband transmissions. The RF broadband receiver uses a PIN photodiode in conjunction with a carefully coupled wideband hybrid amplifier to keep the noise floor as low as possible. The resulting conversion of the optical power to electrical level is directly related to the number of channels or the utilized optical bandwidth.

Fiber-optic receivers are available with different gain blocks to accommodate different requirements. For instance, an optical receiver located at the headend only requires a nominal output level of 10 dBmV to deliver the signal to a small number of cable modems. In the forward direction, the electrical output level requires the highest possible output levels so that additional coaxial-based amplification equipment can be avoided.

Table 12.6
Electrical Specifications for Low-Power DFB Transmitters

Electrical Properties	Values
Channel capacity (6 MHz)	16–80
RF input level	34–27 dBmV
CTB	65 dB
CSO	62.5 dB
C/N	50 dB
Optical transmit power	18–20 dBm
Optical budget	8.5–6.5 dB
Power consumption	20W

12.4.3.1 Electrical/Optical Specifications

Table 12.7 provides the general specifications of typical receivers.

RF broadband receivers are usually packaged in a single RU unit. However, for incoming return traffic from different fiber nodes, they can be housed vertically in a multiple-receiver shelf. The optical input connector can, again, be ST, SC, or FC as desired. The electrical output is typically a 75-ohm BNC or F-connector.

12.5 Fiber-Optic Couplers

Fiber-optic couplers have a place in the design of an RF broadband distribution network. Similarly, as the splitter or the directional coupler in the coaxial distribution system, the optical coupler can be used to equalize the optical budgets of different fiber paths. The optical coupler can also reduce the number of fiber strands in common sections of the forward transmission system.

It is important to note that a single two-way coupler reduces the optical level by 50% or 3 dB. An eight-port coupler has an optical loss of about 12 dB, which can account for as much as a 6-km fiber transport. Optic couplers come in various configurations. Units with pigtails are quite common but are not usable in a networking environment. When considering the use of optical couplers, suitable packaging is important. Rack mountability and the ability to purchase the units with the right ST, SC, or FC connector and with the selected polish is an important consideration.

Table 12.7
Typical Electrical Specifications of Optical Receivers

Electrical Properties	Values	Values
Frequency range	5–200 MHz	5–860 MHz
Optical receive power	13 dBm	13 dBm
RF output level	10 dBmV	38 dBmV
CTB	67 dB	67 dB
CSO	67 dB	67 dB
C/N	52 dB	52 dB
Power consumption	16W	16W

13

Coaxial Transmission Equipment

Most of the equipment described in this chapter does not apply to the installation of an RF broadband LAN. In my opinion, the use of coaxial trunk cables and coaxial amplifiers is perpetuating the outdated technology of the conventional cable television distribution system. The existence of single-mode fiber reduces the use of coaxial cables, especially in campus-wide networking, even if a network is built only for the purpose of distributing cable television. HFC technology, using fiber for the interconnection of all buildings and coaxial service drops within the buildings, can eliminate the use of most amplifiers and deliver a higher quality product with better availability and the lowest maintenance requirements.

In the RF broadband LAN, the use of coaxial components can be limited to the passive devices and the service drop cables listed in this chapter. The use of amplification equipment can be avoided except in risers in high-rise buildings or whenever the outlet population exceeds 50. In these cases, the coaxial amplifier can be colocated with the fiber-optic receiver, and the higher output level can be used to feed a larger service area.

13.1 Coaxial Cables

Many years of use have produced a family of coaxial cables that feature low attenuation and extremely good mechanical properties. After all, coaxial cables have been exposed to the elements for decades. In the indoor environment, or even in conduit, the life expectancy of coaxial cable exceeds 20 years.

13.1.1 Outside Plant, Riser, and Plenum Cables

Cable manufacturers have designed an array of coaxial cables, in a variety of sizes and architectures, for the cable industry. The more air in the cellular structure of the dielectric, the lower the attenuation. The larger the diameter of the cable, the lower the attenuation. In the campus environment, there should be no need to use coaxial cables in the outside plant. Outside plant cables have a minimum diameter of 0.750 in and are, therefore, not discussed. Coaxial cable used in the riser of a tall building commonly features a diameter of 0.500 in. Figure 13.1 shows the makeup and cross-section of a typical 0.500-in diameter coaxial cable.

13.1.1.1 Specifications

The specifications discussed in this section cover only riser- and plenum-rated cables with a nominal diameter of 0.5 in. These cables have been specifically designed for in-building applications but can be used in the outside plant as well. For reference only, the model numbers described in Table 13.1 are CommScope P-3 500 JCAR and P-3 500 JCAP. Other manufacturers provide equivalent products. The riser-rated 0.500-in cable has a copper-clad aluminum center conductor, a dielectric of expanded polyethylene, a solid aluminum sheath, and a flame-retardant polyethylene jacket. The plenum rated 0.500-in cable features a copper-clad aluminum center conductor, a dielectric of foamed Teflon® fluorinated ethylene propane, a solid aluminum sheath, and a Kynar® PVDF jacket. Table 13.1 lists the physical and mechanical characteristics.

13.1.1.2 Attenuation

Table 13.1 shows differences of velocity and capacitance between the riser-rated and the plenum cable. These differences are caused by the use of foamed Teflon dielectric material in the plenum cable, which also produces substantial differences in the attenuation values over the frequency band. See Table 13.2. (All attenuation numbers in Table 13.2 are listed at a standard temperature of 68 degrees F.)

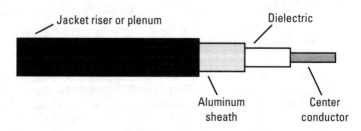

Figure 13.1 A typical coaxial cable.

Table 13.1
Properties of Riser and Plenum Cables

	0.500-in Riser	0.500-in Plenum
Physical Characteristics		
Nominal center conductor diameter	0.109 in	0.109 in
Nominal diameter over dielectric	0.450 in	0.450 in
Nominal diameter over outer conductor	0.500 in	0.500 in
Nominal outer conductor thickness	0.025 in	0.025 in
Nominal jacket thickness	0.033 in	0.012 in
Nominal diameter over jacket	0.566 in	0.524 in
Mechanical/Electrical Characteristics		
Minimum bending radius	6.0 in	5.0 in
Maximum pulling tension	300 lb	200 lb
Velocity of propagation	87%	86%
Capacitance	15.3 ± 1.0 pf/ft	16.4 ± 0.5 pf/ft
Impedance	75 ± 2 ohms	75 ± 3 ohms

Table 13.2
Attenuation Over Frequency for Riser and Plenum Cables

Frequency (MHz)	0.500 Riser (dB/100 ft)	0.500 Plenum (dB/100 ft)
5	0.16	0.17
54	0.54	0.61
186	1.01	1.25
220	1.11	1.38
300	1.31	1.67
400	1.53	2.02
550	1.82	2.46
750	2.16	3.43
1,000	2.52	4.31

Because these cables are mostly used within a vertical riser, the higher attenuation plenum cable will seldom be used. However, there may be a requirement for the plenum cable in large horizontal buildings requiring the installation of a horizontal riser.

The cables listed in Table 13.2 are designed to meet the requirements of the National Electric Code (NEC) and the safety guidelines of the National Fire Protection Association (NFPA). The cables must be marked by the manufacturer to indicate the type and classification and do not have to be installed in conduit.

13.1.2 Service Drop Cables—Riser and Plenum

The use of service drop cables can extend the bandwidth capabilities of the HFC-based RF broadband LAN directly to the workstation. The coaxial drop, in conjunction with a cable modem, can make a workstation or a group of workstations independent of authorization restrictions by using any of the supported RF frequencies. Service drop cables are used between multitaps in the riser room and the workstations. The length of the coaxial service drop is limited to about 250 ft when using an RG-6 type cable. The drop is terminated with standard F-connectors at both ends and has superior egress and ingress characteristics when compared with CAT-5 UTP wiring.

There are many types of drop cables available. The recommended type is an RG-6 cable with quad-shielding. The 6-series cable is available in both riser-rated and plenum configurations. These cables are manufactured by many companies and have comparable properties. For reference only, the CommScope type F6SSVR has been used for riser-rated, and type F6CATVP has been used for plenum applications. The riser-rated cable uses a 20-gauge steel-center, copper-covered center conductor, a flame-retardant polyethylene dielectric, an inner-shield aluminum polypropylene-aluminum laminated tape with overlap bonding to the dielectric, an outer shield of 34 AWG bare aluminum braid wire, and a jacket of flame-retardant black polyvinylchloride. The nominal outside diameter is 0.272 in. The plenum-rated 6-series cable uses an 18-gauge steel, copper-covered center conductor; a foamed Teflon dielectric; an inner-shield aluminum laminated tape with overlap bonding; an outer shield of 34 AWG bare aluminum braid wire; and either a Kynar or a plenum vinyl jacket. The nominal outside diameter is 0.244 in.

13.1.2.1 Specifications

Table 13.3 compares the physical and mechanical capabilities of the two service drop cable types.

Table 13.3
Properties of Service Drop Cables

Physical Characteristics	RG-6 Riser Supershield	RG-6 Plenum Supershield
Nominal center conductor diameter	0.035 in	0.040 in
Nominal diameter over dielectric	0.180 in	0.180 in
Nominal diameter over first shield	0.187 in	0.187 in
Nominal wall jacket thickness	0.030 in	0.020 in
Nominal diameter over jacket	0.300 in	0.284 in
Minimum bending radius	1.5 in	1.5 in
Minimum breaking strength	180 lb	180 lb
Nominal velocity of propagation	85%	83%

The differences exhibited by the dielectric explain the difference in propagation velocity and account for the attenuation difference.

13.1.2.2 Attenuation

Because the plenum-rated cable uses a larger gauge center conductor, the attenuation at lower frequencies is lower than of the riser-rated cable. At high frequencies, however, the Teflon dielectric increases the attenuation. The crossover frequency is around 350 MHz. See Table 13.4.

Table 13.4
Attenuation Over Frequency of Service Drop Cables

Frequency (MHz)	RG-6 Riser (dB/100 ft)	RG-6 Plenum (dB/100 ft)
5	0.67	0.53
54	1.86	1.58
186	3.18	2.88
220	3.35	3.15
450	4.60	4.86
550	5.00	5.45
750	5.80	6.60
1,000	6.75	7.90

In order to compensate for this attenuation difference at higher frequencies, it is advisable to pre-equalize the fiber receiver or amplifier output level (tilt) by an additional 1 or 2 dB.

13.2 Passive Devices

The existence of proven passive devices that are capable of splitting the spectrum in finite increments justifies the coaxial distribution area of the HFC system. Passive devices are also important in the functionality of the RF broadband LAN as they can be used to tap off the right amount of signal from the fiber-optic transmission equipment, split the signal into as many ports as required, and provide connectivity for the service drops or the workgroup cable modem. A number of passive units are always required next to the fiber-optic transmission equipment. These units can be required in the riser room, on each floor level, and at the MDF location at building entrance. Passive devices are ruggedized for outdoor use but are also available in indoor configurations. The most important feature is the quality of the shielding to prevent egress and ingress, a good reason to use outdoor housings that have proven water and electrical shielding protection.

13.2.1 Splitters and Directional Couplers

The term *splitter* refers to a device that splits a signal between two or three directions. The term *directional coupler* is used for devices that have a low insertion loss in the primary direction and a high insertion loss at the tap port. The higher the loss at the tap port, the smaller the insertion loss in the primary direction. Figure 13.2 indicates the physical layout of these devices.

The units are equipped with standard 5/8-in housing connector ports and contain a seizure screw module in each port to seize the center conductor. Important electrical specifications are the isolation and the return loss. Isolation refers to the signal loss between two output ports, while the return loss identifies the loss between the output and the input port. Good impedance matching assures high isolation and return loss figures.

13.2.1.1 Specifications

Table 13.5 lists the minimum specifications of splitters and directional couplers.

Splitters and directional couplers are subject to strict specifications covering operation in the outdoor environment as well as for ingress and egress. They feature special corrosion-protective coating over surface chromate, a

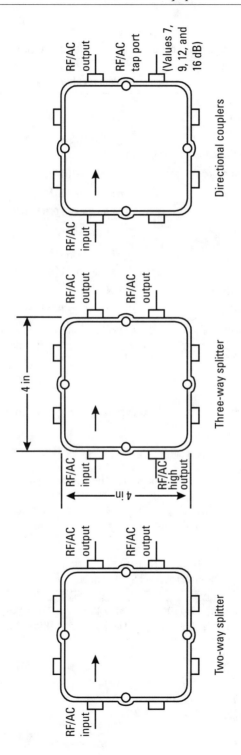

Figure 13.2 Splitters and directional couplers.

Table 13.5
Minimum Specifications of Splitters and Directional Couplers

Properties	Minimum Values
Passband	5–1,000 MHz
Passband flatness	±0.25 dB
Impedance across the band	75 ± 1 ohm
Return loss range	18 dB from 5–10 MHz
	20 dB from 10–450 MHz
	19 dB from 450–1,000 MHz
Isolation range	30 dB from 5–10 MHz
	27 dB from 10–450 MHz
	25 dB from 450–1,000 MHz
Power passing	10A, 60 volts, 60 Hz
Hum modulation	65 dB from 5–1,000 MHz
External dimensions	4-in W by 2.9-in H by 4-in L

molded neoprene weather seal, and a separate RF gasket to meet the FCC's radiation compliance.

13.2.1.2 Attenuation

Table 13.6 details the relationship between the insertion loss and the port loss of splitters.

Table 13.6
Insertion Losses of Splitters in Decibels

Frequency Band	Two-Way Splitter	Three-Way (Low) Splitter	Three-Way (High) Splitter
5–200 MHz	3.7	3.8	7.5
201–500 MHz	4.0	4.3	7.5
501–750 MHz	4.3	4.5	7.6
751–1,000 MHz	4.5	4.8	8.0

Because the insertion loss increases slightly with frequency, the design must compensate by adding pre-equalization at higher frequencies.

Directional couplers are available for 8-, 12-, 16-, and 20-dB attenuations at the tap port. Table 13.7 lists the insertion losses for these devices.

Because the insertion loss rises with frequency, it must be compensated by pre-equalization of the preceding amplification equipment.

13.2.2 Multitaps

Multitaps are available with two, four, and eight ports. Multitaps are used to provide exact levels to the subscriber. Equipped with F-connectors for easy installation of service drop cables, the multitap is the most important device in the HFC architecture of the cable industry. In the RF broadband LAN, the multitap can be compared to a coaxial cross-connect panel. The most frequently used device is the eight-port tap, which can either serve eight service drops connecting to workstations or feed eight cable modems. In this architecture, the cable modem outputs can be connected via a bridge circuit to the existing cross-connect panel and deliver a frequency assignment to a workgroup.

Assuming that the cable modem has a 40-Mbps Ethernet output, it could serve a number of workstations using the existing CAT-5 wiring system. In this manner, eight cable modems can be connected via a single multitap and deliver Ethernet service to eight different workgroups. The number of participants in a workgroup can be chosen as desired. Assuming an average of four workstations per workgroup, a total of 32 workstations can be served with eight cable modems, from a single multitap, each providing a 10-Mbps Ethernet transmission to every desktop. Eight-port multitaps are available with nine different attenuation values to provide correct levels to the service drop. They are the most economical device ever invented to distribute RF signals to the user

Table 13.7
Insertion Losses of Directional Couplers in Decibels

Frequency Band	Directional Couplers			
	DC-8	DC-12	DC-16	DC-20
5–200 MHz	1.4	0.9	0.7	0.5
201–500 MHz	1.6	1.1	1.0	0.8
501–750 MHz	2.1	1.6	1.5	1.0
751–1,000 MHz	2.3	1.8	1.6	1.1

in increments that could not be achieved with optical devices. In addition, multitaps are ruggedized for outdoor use, contain a weather seal, have corrosion protection, and feature a separate RF gasket to eliminate egress and ingress. Figure 13.3 illustrates the physical layout of the eight-port multitap.

The eight-port multitap features eight F-connector arrangements to provide for the connection of the service drops. The design of the tap assembly is modular and permits the change-out of the top cover to different values without removal of the housing from the main feeder. Multitaps incorporate self-sealing F-ports that will ensure the rejection of moisture even if the service drop is disconnected.

13.2.2.1 Specifications

Table 13.8 lists the minimum specifications of the eight-port multitap.

13.2.2.2 Attenuation

While there are nine different attenuation values for eight-port taps, the higher values of 33, 36, and 39 dB do not apply to most RF broadband LAN designs because they require coaxial amplifiers with high output levels. Table 13.9 lists the remaining tap values with their respective insertion loss values.

It can be seen that the insertion loss increases with frequency. This increase can affect the level at the outlet and adds to the higher attenuation of coaxial cables at higher frequencies. The designer can offset these higher

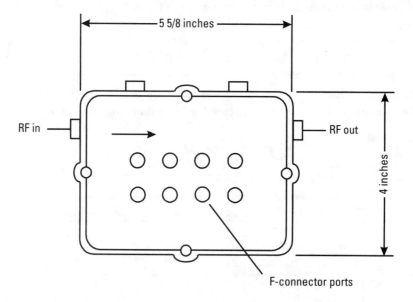

Figure 13.3 The eight-port multitap.

Table 13.8
Minimum Specifications of the Eight-Port Multitap

Properties	Values
Passband	5–1,000 MHz
Passband flatness	±0.25 dB
Impedance across the band	75 ± 1 ohm
Tap port to output isolation	30–40 dB across the band
Tap-to-tap isolation	20 dB from 5–600 MHz
	25 dB from 601–1,000 MHz
Return loss (in, out)	20 dB from 5–450 MHz
	18 dB from 451–1,000 MHz
Return loss (tap ports)	18 dB from 5–450 MHz
	16 dB from 451–1,000 MHz
Current capacity	60V at 7A, continuous
Hum modulation	−70 dB below signal
Physical dimension	4-in W by 2.9-in H by 5.6-in L

Table 13.9
Insertion Losses of Eight-Port Multitaps

Frequency (MHz)	Insertion Loss (dB)						
Tap Value	12	15	18	21	24	27	30
5	Term.	3.3	1.8	0.9	0.7	0.6	0.5
100		3.1	1.4	0.7	0.5	0.4	0.4
400		3.3	1.7	1.0	0.8	0.7	0.6
600		3.8	2.3	1.6	1.3	1.1	1.1
800		4.3	2.8	2.0	1.8	1.6	1.4
1,000		4.5	3.0	2.3	2.1	1.9	1.7

attenuations by adding pre-equalization to the output of the fiber receiver. It is also possible to insert passive cable equalizers into the circuit to compensate for these level differences.

13.2.3 Connectors

The selection of high-quality coaxial cable connectors is an important consideration. The connector, when not installed correctly, can become the main source of ingress and egress problems. Especially for RF broadband designs and when transmitting signals up to frequencies in the GHz range, the connector must feature accurate radiation sleeving, center conductor seizing, and outer sheath locking mechanisms.

13.2.3.1 Outside Plant and Riser Connectors

Outside plant and riser connectors are used to interconnect coaxial devices to the cable discussed in Section 13.1.1 above. They establish continuity between the cable and the standard 5/8-24 equipment entry port. There are three different connecting methods, described as follows:

- *Housing connectors:* The housing connector connects the cable with the standard 5/8-24 chassis port. It can be used with all housings whether they are used for active or passive devices. The connector must have an integral sleeve in the nut assembly to seize the outer conductor and fully shield the seized connection. The use of feed-through connectors that extend the center conductor of the coaxial cable into the housing are not recommended. The connector should feature a pin with a diameter of 0.067 in and a length of 1.6 or 2.31 in. The pin of the pin-type connector can then be properly inserted into the housing and tightened in the standard housing seizure screw assembly. The connector should feature three hex nut assemblies for accurate cable termination and connector assembly.

- *Housing-to-housing connectors:* These devices are used to interconnect the housings of different devices without the use of additional cable. In order to keep reflections between devices to a minimum, it is important that the interconnections of housings are made exclusively with housing-to-housing connectors. The housing-to-housing connector features a 1.6-in pin at both sides of the connector to permit the seizing of the center conductor in each housing. To keep multiple devices requiring interconnection with housing-to-housing connectors in the same orientation, the use of nonrotational connectors is recommended.

- *Housing terminations:* The last device in a string of multitaps or splitters has to be terminated. The connector that is used for this purpose is the housing terminator. It features a 1.6-in pin for center-conductor seizing and incorporates a 75-ohm termination in combination with a power-blocking capacitor. It is the purpose of the blocking capacitor to protect the terminating resistor from exposure to the 60V of line voltage often used to power remote amplifier stations. The use of cable-powered amplifiers in RF broadband LANs is not a very likely occurrence, and housing-to-F-connector adapters with an F-connector termination could be used. However, experience shows that a good 75-ohm termination is better assured by using the housing terminator.

13.2.3.2 Service Drop Connectors

The service drop connector, or F-connector, is used to terminate service drop cables at both ends. The F-connector was standardized in the early days of cable television and has undergone various transitions over the years to make it more user-friendly and to incorporate radiation protection. The new universal F-connector, type F-6-AHS/USA, permits the connectorization of different size service drop cables. When used with a 0.5-in crimped sleeve, the connector will provide a radiation-free connection. Proper preparation of the cable ends is required to assure a problem-free connection.

13.2.3.3 Coring Tools

The use of proper coring tools to prepare coaxial cables for connectorization is most important. The dimensional relationships of the jacket, the sheaths, and the center conductor cannot be maintained by using a knife. Every cable has its own coring recommendations that must be observed. Connector manufacturers such as Gilbert Engineering and Augat will provide coring tools specific to their products. RG-6 quad-shielded service drop cable requires the use of a 0.360-in hex crimp tool.

13.3 Amplification Equipment

The cable industry uses a wide variety of amplifiers to accommodate the need for flexibility. The architecture of the trunk amplifier addresses the optimization of performance in large amplification cascades. It features a low output level to keep the noise buildup to a minimum. Bridging amplifiers are used to produce a high-level output into multiple feeder lines. In addition, there are distribution and line-extension amplifiers that are used to reamplify the signals in the distribution network. Various gain block electronics are applied to optimize the amplification process. These methods include push-pull versions and feed forward models, and dual and quad output solutions with thermal and/or automatic level control.

The use of these devices in the RF broadband LAN is limited. The HFC technology extends the fiber to the building entrance areas and only utilizes coaxial wiring for short in-building extensions and for service distribution. As a result, the only amplification requirement exists in coaxial risers or to elevate the output level of the fiber receiver to be able to serve a larger user population. The most suitable amplifier for these purposes is the distribution amplifier described in more detail in Section 13.3.1.

13.3.1 Distribution Amplifiers

Distribution amplifiers are high-gain amplifiers that handle output levels of +46 dBmV at the highest passband frequency. Distribution amplifiers are not designed for large cascade numbers, as the noise floor of the amplifier is amplified as well. In the RF broadband LAN, there should never be a requirement to use more than two amplifiers in cascade. The normal installation location will be close to the fiber transmission equipment.

13.3.1.1 Spectral Performance

Present distribution amplifier designs have an upper frequency limit of 750 MHz and feature a narrow return band in the 5–45-MHz band. There are other models that can be used for operation in the high-split band. These amplifiers provide a forward transmission in the 220–750-MHz band and a 5–186-MHz return. Note that the introduction of coaxial amplifiers into the HFC-based RF broadband LAN only reduces the spectral capacity of the system. A single-mode fiber system connected directly to coaxial or copper service drops can provide continuous bandwidth between 20 to 750 MHz in both directions. While this range is typical for today's equipment, it is expected that both fiber-optic transmission equipment and coaxial amplifiers will be able to have spectral capacities beyond 1 GHz before the end of this century.

13.3.1.2 Specifications

Table 13.10 lists the minimum specifications of both subsplit and high-split distribution amplifiers.

Slope adjustment and gain control are usually provided by a combination of built-in variable control and external plug-in pads and equalizers. Plug-in devices can be ordered separately in accordance with system design data.

13.3.1.3 Powering Options

Most coaxial amplifiers are powered through the cable system and incorporate 60V power supply circuitry. For these amplifiers the use of an external 110/60V power supply is required. Some manufacturers, such as C-Cor, will

Table 13.10
Minimum Specifications of Subsplit and High-Split Distribution Amplifiers

Specifications	Subsplit		High-Split	
	Forward	Return	Forward	Return
Passband (MHz)	50–750	5–42	222–750	5–186
Input level (dBmV)	13	17	13	17
Output level (dBmV)	40–46	32	40–46	37
Noise figure (dB)	7.0	9.0	7.0	9.0
Cross modulation (dB)	65	76	67	78
Second order beat (dB)	67	68	68	73
Composite triple beat (dB)	68	91	84	81
Full gain (dB)	33	15	33	20
Flatness (dB)	0.5	0.5	0.5	0.5
Return loss, minimum (dB)	16	16	16	16
Slope adjustment (dB)	0–6	0	0–5	0–2
Gain control (dB)	0–20	0–20	0–20	0–20

provide 110V powered amplifiers that can be plugged directly into any electrical outlet. Because the RF broadband LAN only uses indoor-placed amplifiers, the use of 110V powered amplifiers offers an easier installation even when special ordering may be required. The power consumption of the circuitry in a distribution amplifier is typically in the range of 0.3–0.45A at 60 Vac and less than half for 110V powered units.

14

Implementation Considerations for the HFC-RF Broadband Network

14.1 Implementing the Planning Document

The planning document, developed in accordance with the guidelines presented in Chapter 8, is a summary of the major activities that must be implemented. This chapter addresses the scheduling and grouping of these activities into manageable segments.

There are many activities that can be conducted within the company organization. The facility department is accustomed to taking care of clean-up, small construction, and wiring tasks and, usually, has a cadre of readily available outside contractors to take care of larger room rearrangements, electrical wiring projects, and even in-building conduit runs. The first activity in a multifaceted implementation program is to separate in-house and outsourced activities. The dividing line between these two implementation methods is entirely dependent on in-house skills, experience, dedication, and available manpower. Every business establishment has different standards, organizational variations, and established work methods. It is helpful to employ the program management personnel that shared the experiences of the planning phase, augment this group with a few experienced implementation specialists, and map out the implementation program.

14.1.1 Implementation Categories

It is a good practice to analyze the planning document relative to the desired implementation schedule and to compare the required implementation speed with a detailed analysis of the various activities. Often, the total implementation time requirement is underestimated on the basis that the request for proposal (RFP) will determine the completion date and the contractor will comply. In many cases, incompatible work tasks are combined into a request for turnkey implementation, and the bid responses will vary substantially because task details are misunderstood and the vendor includes cost reserves to compensate for the lack of information.

To avoid some of these problems at the outset, it is desirable to categorize the implementation activities into manageable segments. This procedure has been called a work breakdown structure, but it is really only the logging of major and minor activities in a logical sequence. While it is logical that a conduit construction event must occur before cable pulling, detailing the work tasks can also include considerations relative to in-house completion and outsourcing. When it has been determined that outsourcing is the desired approach, it can then be determined whether the event of conduit construction and cable pulling should become a part of the equipment and in-service-taking or contracted separately. In this manner, it is possible to determine different implementation categories that have concise task descriptions and therefore can be bid more intelligently. Examples of implementation categories are:

1. Clearly identifiable small engineer, furnish, and install (EFI) activities that can be covered by a purchase order and by an informal bidding process from a group of predetermined suppliers—up to a maximum dollar amount;

2. Clearly identifiable larger EFI activities that require a formal RFP, bid evaluation, and award process;

3. Large multifaceted interdisciplinary EFI activities that require a formal request for information (RFI) and/or RFP bid evaluation and award process.

Good program management software can be an effective tool to separate sequential activities from those that can be performed in parallel and to categorize the work segments. It is often argued that an all-inclusive turnkey contract is easier to manage, requires less supervision, offers better control through a single contractor representative, and takes less time. There are, however, many examples of incurred overruns and schedule delays that are directly related to this type of contracting. The prime contractor will always attempt to adjust the contract value upward and use previously anticipated but changing conditions to excuse schedule delays and additional costs. A multiservice network consists of many multifaceted activities requiring a wide variety of skills that when grouped into compatible components can attract contractors with experience, be more easily supervised and managed, and eliminate the prime contractor's markup.

14.2 Outside Plant Installation Considerations

The work to be performed in the campus outside plant is most likely one of the first implementation events. The construction of new conduit sections and fiber cables can, of course, be conducted in parallel to in-building work, equipment room refurbishment, and the installation of a stand-by power network.

14.2.1 Underground Duct Installations

A detailed survey of both the existing and new conduit plant is important. Existing conduits require pullines to facilitate new construction. Installing pullines with distance marking establishes distances between manholes and reveals the condition of the duct system. Before starting the construction of new duct segments, it is important to locate and mark any existing utility services. In addition, in public areas, it is important to notify the owner, obtain the necessary permits, and inform the local before you dig (BUD) service. New conduit construction should be planned with 4-in diameter PVC, type-C duct, conforming to standards 8546, 8343, and TC-10. The standard type-C conduit has a wall thickness of 0.15 in and an outside diameter of 4.35 in. Each conduit section should feature a 6-in bell-coupling with an outside diameter of 4.7 in. Bends and sweeps should have features identical to the straight sections. The use of flexible conduit sections must be avoided. The following minimum specifications are provided as general guidelines for undergound installation projects.

- *Excavation* should be to a minimum depth of 30 in and all excavated material removed. The bottom 6 in of the trench should be filled with crushed stone or gravel with an aggregate size not in excess of 0.75-in diameter and not smaller than 0.25 in. The base material should be leveled and compacted before conduit placement. When using two

4-in conduit sections, side-by-side placement is recommended and more economical.

- *Manholes* or pull boxes should have a minimum size of 36 by 36 in and be manufactured from 28-day concrete with a compression strength of 6,000 psi and designed for HS-20-44 loading. Wall and floor thickness should be at least 6 in. The placement of the manholes requires a 6- by 6-ft excavation area and an identically compacted 6-in crushed stone and gravel base.

- *Innerducts* are used to separate different cable sizes and avoid problems with cable installations in different time periods. A four-channel, linear ribbed polyethelene product with a minimum inside diameter of 1.26 in and a wall thickness of 0.075 in can easily support two standard fiber-optic cables when pulled at the same time. The innerduct should meet the cell classification of ASTM D-1784 and feature tensile properties, external loading, and impact resistance compliant with ASTM D-638, D-2412, and NEMA TC-2 for electrical plastic tubing. The installation of the innertube system must be concurrent with the installation of the conduit and feature a flush-cut fit into the female ends for proper alignment with the conduit sections.

- *Bends and sweeps* should have features identical to any straight sections and be equipped with a 6-in bell at one end and with innerduct spigots for quick and weatherproof connections.

- *Backfill* material should be a flowable fill applied in layers to flood the excavation to the required compaction. Flooding must reach the pavement level and must be applied in a manner to prevent later settlements. Any restoration work should comply with any local requirements.

- *Junction boxes* may be placed at building entrance locations in cases where a conduit sweep cannot be accommodated through the foundation. Junction boxes should be of a galvanized steel construction with outside dimensions of a minimum of 24 by 24 in with a depth of 12 in minimum and type-C-to-steel-threaded or type-C-to-steel-to-epoxy adapters.

- *Continuity* of the innerduct in manholes and junction boxes should be assured with suitable termination kits that offer precast weather-tight termination and expandable plugs. Maximum expansion space should be provided through coiling using suitable support hardware.

14.2.2 Single-Mode Fiber-Optic Cable Pulling

Upon completion of the outside plant and the MDF refurbishing activities, the single-mode fiber cables can be installed and terminated. Familiarity with BICSI standards and an evaluation of the many fiber termination products is all that is required to translate the information of the planning document into an installation specification. The installation contract can be contracted to the outside-plant contractor or to an experienced local fiber contractor as long as the company's experiences include fiber terminating and testing expertise. It is also possible to combine the fiber-pulling activity with the conduit installation and separately contract the fiber termination activities. When specifying cable pulling activities, it is recommended that the pulling load never exceed 400 lb in a straight pull and be reduced to 200 lb when pulled through two 90-degree changes of direction. When there are more than two 90-degree changes of direction, backfeeding and center-pulling techniques must be applied. Tension should be constantly monitored during the pulling effort.

14.3 Inside Plant Considerations

A detailed effort to plan the inside plant implementation is essential to assure the quality and longevity of the network. Some of the essential considerations are discussed in Sections 14.3.1–14.3.3.

14.3.1 Conduit Installations

Conduit installations required inside buildings should follow the same specifications as those discussed for the outside plant. Generally, the loose tube cable used for outdoor installations should be brought to a termination location within 50 ft of the building entrance point. In cases where longer distances are required, the conduit and innerduct may be extended as required. For installations in the horizontal plane or when assuming difficult routing changes, the conduit should use sweeps and straight sections to reach its destination location. Inside construction of conduit does not require a type-C conduit material. A galvanized steel conduit with innerduct suffices and is easier to handle because it is bendable and offers a better grounding solution. The use of standard hangers with bottom and top support hardware is recommended at intervals of not more than 4–5 ft. TIA/EIA-569 covers commercial building standards for telecommunication pathways and spaces. Another useful document is BICSI's *Telecommunications Distribution Methods Manual* (TDMM), which addresses cable handling and support in more detail.

14.3.2 Equipment Rooms

Rack placement is required at all fiber termination locations. The rack can be a 19- or 23-in rack frame. The 23-in variety offers more space to integrate vertical and horizontal fiber routing guides. Figure 14.1 shows the fiber termination panels mounted in the top section of the rack. The empty 35-RU rack space can be used for the mounting of fiber-optic and coaxial transmission equipment. When multiple racks are to be placed, it is good practice to install fiber cable guides above the racks. Each fiber cable can be directly brought to a rack- or wall-mounted storage shelf and interconnected with the fiber termination panels using innerduct. The equipment rack requires the mounting of suitable AC strips for power distribution. Separate AC strips are required for standby power feeds. Every rack requires a solid copper grounding wire, minimum #6, which is connected in the most direct route to an established building ground with a resistance of less than 25 ohms. Single-point rack grounds often do not provide the required resistance value. Bonding between equipment shelves, wiretray sections, and covers is the recommended way to obtain the lowest ground resistance conditions. See Figure 14.1.

Fiber terminations and handling of single-mode fiber are not too different from the handling of multimode fiber except for connector preparation and polishing. To avoid mistakes in the area of connectorization, ordering factory assembled pigtails that reduce the work to standard splicing details is recommended. The use of dual SC connectors is recommended to simplify the routing and management of patchcords. Many manufacturers will incorporate the new AMP connector type MT-RJ double-density connector in their products. Using fiber termination panels with this duplex connector will reduce the rack space requirement but may not aid operations. When making the decision to utilize the new duplex connector, it may be good practice to check for BICSI recommendations and discuss the issue with a certified registered communications distribution designer (RCDD).

Incoming cables require the use of storage panels to provide for cable slack for future rerouting requirements. Testing of completed fiber segments is important. Simple optical time domain resolution (OTDR) testing, while it provides a good documentation of the installed quality, does not prove the integrity of the installation. Each segment requires loss testing and measurement of the return loss. In a campus zone, the segment loss is the accumulated loss of all sections between the buildings plus the patchcord losses. Typically, the segment between the fiber transmitter and the receiver at the last building in a zone may consist of a number of fiber sections and patchcord assemblies. Typical section loss should be better than 0.6 dB; typical patchcord loss should be in the area of 0.5 dB. The typical return loss in any section should be better

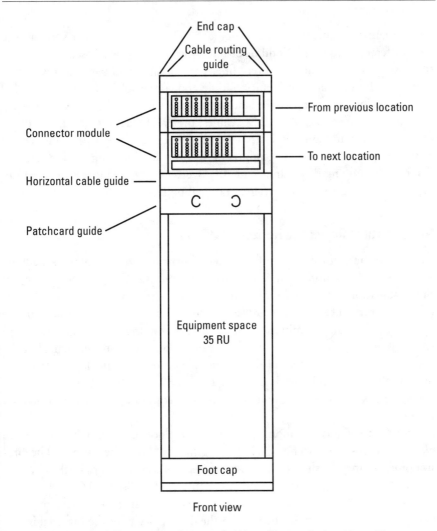

Figure 14.1 A typical fiber termination panel. (*Note:* The vertical cable guides are not shown.)

than −45 dB. Proper polishing methods must be applied to obtain these values. Single-mode fiber has two operating ranges, the 1310-nm and the 1550-nm range. Both ranges should be tested. The future may bring affordable DWDM equipment to the campus, or other equipment may become available for the 1550-nm band. When constructing dual-redundant or single-ring architectures, both sectional and end-to-end tests should be conducted. Even properly installed fiber strands vary in their performance because of their position in the bundle. This causes attenuation variations in each section, which may be more severe than some end-to-end measurements. Each individual fiber segment can have different test results. Good record keeping will provide a baseline for future fiber assignments. Assure that every fiber cable, strand, and patchcord is labeled at both ends. The same applies to module and connector housings. Assure that you can critique and approve the labeling method suggested by the contractor.

Patchcords should be dual SC cables, preferably factory tested, but delivered complete with connectors at each end. Use cable guides within the rack to route the patchcords to the equipment without leaving long slack segments. Patchcords hanging down are frequently overbent at the lowest point and can loosen the connector by its own gravitational pull.

14.3.3 Vertical Risers and Horizontal Wiring

Cabling in the vertical riser and additional horizontal wiring can be scheduled at any time after completion of riser room alterations. Copper wiring requires the installation of cross-connect panels. CAT-5E or CAT-6 installations require performance testing of at least the link segment. Fiber extensions in the vertical riser require termination panels and follow the same installation and testing requirements as outlined for fiber in the outside plant segment. For RF broadband systems the coaxial riser sections require level calculations to determine the value of multitap devices (see Chapter 15). Coaxial horizontal RG-6 wiring follows the same ground rules that are commonly used for horizontal wiring systems.

Fire-stopping is commonly used between floor levels. Conduit stubs are placed in each floor extending about 3–4 inches above the surface. The fire-stopping material should consist of elastomeric material that meets ASTME-814F and T ratings and conforms to UL listing 1479. The fire-stopping material should be inserted into the conduit stubs as soon as the floor opening has been made. At the time of the cable placement the material has to be removed and reapplied after completion of the cable installation. To separate the vertical cabling in an orderly manner, a series of conduit penetrations with fire-stopping material can be made at the time of the riser room refurbishing event.

Fiber and coaxial risers should be routed separately through the fire stubs. To identify the fiber from other cables, the routing may be protected by innerduct. Innerduct can be spliced for service feeds at each floor level. Vertical installations in a multistory building should use standard cable grips to support the cables at every fourth level. The fiber cable is brought to the termination panels, terminated, and installed in an identical manner to that previously described in Section 14.3.2. The connectorization of coaxial cables requires special proficiency and skills. The contractor should use the specified coring and cutting tools to properly mate the cable with the connector assembly. A riser cable connector may appear to be correctly installed, but the center conductor may not make contact. Electrical testing is recommended. When installing hard-line coaxial cable in a riser, it is a good practice to leave a 6-ft cable loop at each floor level. This will allow for proper splicing and connectorization and equipment installation at a later date. Some installations in large horizontal buildings require the installation of cabling in horizontal spaces. A horizontal riser can use wiretrays or be ceiling mounted with support members every 3–4 ft. Unless laid in a wiretray, the use of special mounting hardware is required. Coaxial horizontal risers sometimes require passive components to be mounted in ceiling spaces. This equipment requires individual support hardware for devices such as splitters and multitaps. Tie-wraps should not be used.

Coaxial equipment such as amplifiers, taps, and splitters can be placed in riser rooms. The conventional installation practice has been to place a 3/4-in plywood wallboard on one of the walls and to mount the equipment using stand-off brackets. This type of installation is permissible in the absence of a rack frame. A better looking installation can be achieved by mounting the devices in the back of standard blank panels and interconnecting the cables neatly within the rack. Figure 14.2 indicates the mounting arrangement in a rack rearview.

CAT-5 plus wiring such as CAT-5E or CAT-6 originates at a cross-connect panel in the riser room. Because the performance of the horizontal wiring system is a direct result of the components used and the workmanship applied, it is recommended that BICSI standards are followed and that compatible components are used throughout. Jacks, patchcords, patch panels, and cabling must be compatible. CAT-5E and CAT-6 require matching components to meet the increased performance specifications. If the contractor is responsible for the installation of the link only, it is important that all performance parameters are tested and meet specification values.

Coaxial service wiring begins at the multitap installation in the riser room. The RG-6 cable is terminated by a standard F-connector at both ends. The contractor should use coring tools approved by the cable and connector manufacturer and properly mate the cable with the connector assembly. An easy way to check the workmanship is to inspect for leftover braid material and check the tightness of the crimped connector sleeve. Electrical testing of attenuation at the highest transmitted frequency is the final indicator of acceptance. Figure 14.3 shows the F-connector continuity at the outlet and shows the connection with a cable modem and a multimedia computer with videoconferencing equipment.

Figure 14.3 shows a cable modem serving the workstation with 25.6-Mbps ATM and feeding a videoconferencing codec for JPEG video/audio

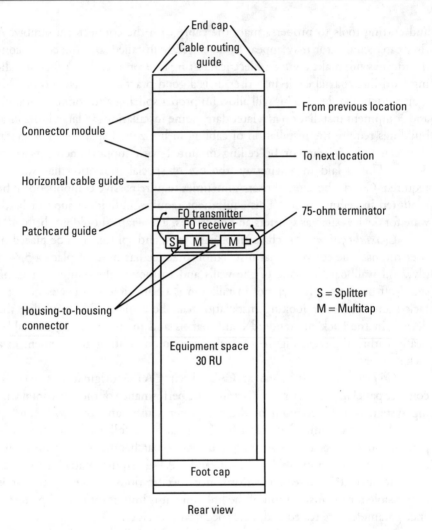

Figure 14.2 Mounting of coaxial riser equipment. (*Note:* The vertical cable guides are not shown.)

at 4.96 Mbps translated to NTSC video for presentation on a standard television set.

14.4 Vendor Capability Assessments

With a quite comprehensive list on hand relative to the work activities to be completed, it is possible to develop manageable segments for the bidding

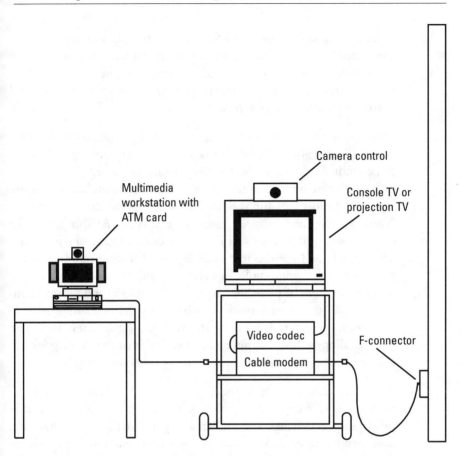

Figure 14.3 A multimedia workstation and videoconferencing equipment.

process. Using the bidding categories discussed in Section 14.1.1, the project can be scheduled and subdivided. Using standard project management software will help to establish realistic timelines and determine any conflicts that may exist between the activity segments. The work activities to be completed are listed as follows:

- *Preliminary scope of work:* It is good practice to develop scope-of-work descriptions for every work detail and to prepare sketches or simple line drawings for specific requirements. A number of campus locations may have to be visited again to determine the exact location of a rack or a cable route. Developing the guidelines for the bidder and good detailing of the expected scope of work will save money and eliminate misunderstandings and aggravation later. To determine the desired location of a rack accurately is better than finding out later that the contractor has put it in the wrong place and requires a change notice to move it. A few notations and sketches included in the bid documentation, illustrating how the work is to be done, can sustain peace of mind during the implementation phase.

- *Preliminary scheduling* includes a review of the timelines of the proposed RFP categories. Activities that can be conducted in parallel need to be studied relative to their activity locations. When there are two independent activities required at the same time at the same location, both contractors will use the conflict as an excuse that they cannot "work around" and must be given an extension. At this point, the development of the various RFPs and purchase orders can commence. The cleanup and refurbishing tasks as well as underground construction, conduit laying, cable installation, and termination tasks are activities that can be performed by a number of local and regional contractors. Most of these activities can be clearly defined and usually do not require a pre-evaluation of the vendors' capabilities. However, RFPs dealing with system integration and ultimate network performance are important and require a vendor-evaluation effort.

- *Capability assessments* of vendors have become a necessary event in these times of technological changes. Often, the available marketing information is deficient and outdated. A good first step is to invite the data network equipment vendor to make a presentation and to discuss the pending project on an informal basis and to learn about the vendor's technical and managerial strengths and weaknesses.

- *Technical concerns* include the ability to handle overall performance requirements and system integration issues. ATM and VOIP are familiar topics, but CTI, CBXs, RF broadband, cable telephony, and coaxial service drops may not be. Previous experience in functional and performance testing is a good subject for discussion. What test equipment and test sequences has the contractor used to determine reconvergence time of switching equipment, and has the performance of a cable modem been checked?

- *Scheduling concerns* can also be discussed informally to determine the realism of the project plan that has just been completed and approved. The sequencing of cut-over activities is another important item that may change the assumptions that have previously been made. The verification of expected performance time lines is paramount.

- *Managerial concerns* relative to contractual issues and project supervision require a discussion about project responsibilities, personnel assignment, accessibility, storage, working hours, and progress reporting. The vendor's past performances and references will help to provide a better insight.

The completed round of vendor evaluations may either confirm or reject the approach taken in the planning document. A second close look at the project can now be taken. The planned work activities, RFPs, and purchase orders can be reviewed and changed to optimize the project timeline, to distribute cash flow and capital expenditure requirements in a more viable manner, and to map the final and agreed-upon approach to project implementation.

14.5 Preparation of RFPs and Bid Specifications

Preparing purchase orders, RFPs, and bid specifications requires attention to detail. Some important guidelines are provided in Sections 14.5.1–14.5.3.

14.5.1 RFP Preparation

A comprehensive RFP should contain a detailed scope of work, clearly define the responsibilities of the contractor, provide a desired project time table, express the bid response requirements in detail, state the project management requirements, and define the owner-provided activities. The RFP contains the framework to permit competitive bidding but requires organization to promote concise responses that can be used for a valid comparison of the bidders. A few reminders follow:

- *Scope of work* provides an overview of the project with an emphasis on goals, expectations, and quality requirements. The system description should be detailed enough to convey all work activities in the desired sequence. It is important to include routing diagrams or sketches to avoid misinterpretations by the bidders and to format the statement of work in a manner that will permit concise point-to-point responses and pricing categories.
- *Contractor's responsibilities* require extensive detailing. A clear set of rules is required to hold the contractor responsible for detailed engineering, procurement and delivery of all equipment, installation of all system components, and the provision of an operational system

performing to the provided performance specifications. It should be clearly stated that any provided information, such as owner-provided designs and bill-of-materials, are offered as examples only and that the contractor is fully responsible for detailed design, design verification, and compliance to performance specifications.

- *Contract data requirements* are a part of the scope of work but require special emphasis. This listing should include all information that requires approval before a final acceptance is granted. This entails weekly progress reports; reporting of delays; the periodic delivery of documentation identifying completed installation work; project meeting scheduling and reports; functional subsystem test data; test data of sectional testing; end-to-end, operational, and acceptance tests; and the delivery of the final documentation package. It is important to emphasize that any requested remedial action activities have to be completed within a limited time frame and to the satisfaction of the owner.

- *Scheduling requirements* require itemization. A listing of milestone events in the RFP provides guidance. In addition, the bidder should be requested to provide both best and worst timeline milestones. Categories of critical milestones such as completion dates for construction details, cable installation, splicing and terminating, equipment installation, activation, operational and acceptance testing, as well as final documentation delivery require emphasis. If the project is a large campus requiring activities by campus zones, these details are required for each zone.

- *Project management* assurances by the bidder will provide owners with a good insight into the organizational capabilities of the bidder. Information should be requested for a detailed description of the proposed project organization, safety issues, scheduling of activities, manpower requirements, scheduling management, progress reporting, the delivery of contract data items, and quality control measures.

- *Bid response requirements* are the basis of an expedient bid evaluation. A properly organized RFP can provide responses that can easily be compared. A request for point-to-point responses alone does not provide sufficient insight. The bidder should be asked to address specific information requirements. Subject items, such as a bill-of-material, routing and design, compliance to technical specifications, system performance data, project scheduling, project management, staffing plan, test methods, and a list of test equipment to be made available for the project, are important indicators. Because the implementation activities for multiservice LANs consisting of single-mode fiber, coaxial cables, and UTP are more demanding, it is interesting to obtain an indication of the experiences and capabilities of the personnel that will be assigned to the project.

- *Owner-provided activities* require a detailed statement. The ground rules for providing power service, storage space for material, office space for the contractor's personnel, or rules related to accessibility and working hours are important inputs to the bidder. This section of the RFP can also include explanations of work activities that have been covered by other contracts such as conduit construction, fiber cable installation, provision of emergency power, or other third-party activities.

14.5.2 Technical Specifications

Hardware and equipment specifications and any data developed in the planning document and from vendor evaluations can now be listed as desirable preferences. Functional requirements, electrical performance, mounting, and environmental requirements should be included. Preliminary bill-of-materials can be attached to convey the detail of the work already performed, as long as a disclaimer states the preliminary nature of the listing and leaves the responsibility for completeness with the bidder. It may be desirable to procure major hardware items directly and modify the turnkey requirement accordingly. While this method can save money, it also reduces the bidder's markup and transfers the responsibilities of purchase and delivery to the owner. Before making a decision in favor of the monetary benefits, assure that the additional contractor/owner interfaces do not lead to technical interface problems, questions of responsibility, and opportunities for unanticipated delays.

Software components required for network management, system performance management, and traffic management require a concise description of the expected requirements. There are no minimum standards, and the offered services are vendor-proprietary solutions. The RFP should require a separate section requesting detailed information on all desired operational software capabilities. Standard network management software (NMS) must configure, monitor, and control all hardware platforms. When using a mixture of equipment such as cable modems, ATM and IP hardware, single-mode fiber, and coaxial electronics, the integration of QoS levels, traffic, and network management into a single NMS is desirable.

Installation requirements include the more important notes that have been compiled during the planning phase, some sketches, and the requirement to follow BICSI structural wiring standards explicitly. Requesting sketches of any proposed equipment placement in racks is a good way to determine the quality consciousness of the bidders. In cases where exact locations have not been determined, a request for submission of installation particulars for approval before commencement of the work can be helpful and assure expectations.

Testing consists of functional tests, operational tests, and acceptance testing. Functional tests include fiber-optic cable tests such as OTDR; section, segment, and end-to-end attenuation; reflection measurements; and transmitter/receiver performance. In coaxial segments, the functional tests include continuity testing and attenuation measurements of service drops, activation and alignment of coaxial amplifiers, the RF sweep-testing of forward and return HFC segments, outlet level verification, and radiation testing. Compliance with the CLI rules of the FCC is a requirement for all coaxial cables but may also be required for CAT-6 cables with a bandwidth performance up to 200 MHz. Operational testing consists of performance testing of all ATM- or IP-based segments, assembly and tear-down intervals of switching equipment, reconvergence testing for out-of-service conditions such as cable cuts, equipment and power outages, cable modem and RF broadband end-to-end performance, and NMS performance.

Acceptance testing sequences must include the verification of major performance requirements of the various subsystems and randomly selected overall network end-to-end performance. In RF broadband segments, acceptance test sequences should include C/N measurements, cable modem and RF network loop testing, and a confirmation of CLI results.

Delivery of documentation and acceptance are the last project activities. It is important that the RFP conveys these requirements in detail. A complete record of both installation data and test results is an absolute necessity for in-service taking, operation, and maintenance. All too often, the delivery of documentation items is viewed as a miscellaneous requirement. CAD drawings have to clearly identify installation particulars, rack-up details, cabling and wiring methods, and powering and marking details. Final project documentation must collect all installation and test data in a chronological manner and identify interfaces with existing equipment as well as contain conceivable methods to perform troubleshooting and repair.

14.5.3 Terms and Conditions

Corporations usually have a set of general terms and conditions that can be readily applied to the RFP document. There are, however, specific terms that need to be compiled and approved before a contract can be executed. To reduce or even eliminate the contract negotiation period with the successful bidder, incorporating these specific terms into the RFP is recommended. This process can save time and reduce contract negotiations to the issuance of a purchase order with the RFP and the bidder's proposal becoming the legal contract documents.

Specific terms identify the purpose of the RFP; determine the bidding events; identify the bidders' conference, site survey, and walk-through; and deal with the proposal format, the required bid responses, the desired pricing structure, price validity, project schedule requirements, optional proposals and pricing, confidentiality, work-around requirements, changes and alterations, bidders' questions and responses, and bid bond and performance bond requirements, if any.

Warranty requirements are also specific to the project. The procurement of equipment warranted for a period of time by the manufacturer may have expired at the time of installation. The extension of equipment warranties and the inclusion of a service warranty for a minimum of one year from the date of acceptance is important. In addition, there are subjects such as good quality, fit-for-purpose, good title, EMI and RFI affect on warranties, voidance of warranties, and repair requirements during the warranty period that require attention.

Vendor qualifications are required as part of the bid evaluation process. In most cases, the requirement addresses the desire for technical and financial capability. It is believed that a contractor must have sufficient assets and working capital to undertake a project. In many cases, the financial capabilities are rated higher than the technical ability, and the most dedicated contractor is rejected. A good balance is desirable. References and a good evaluation of the technical know-how of key personnel is an important input.

Bid evaluation can be based on the evaluation of four different ratings: (1) the technical matrix that compares the technical offerings with the RFP requirement; (2) the financial matrix, consisting of pricing levels, unit prices, and system pricing; (3) the operational matrix, consisting of the evaluation of project management, scheduling, staffing, quality assurance, and organizational matters; and (4) the bidders' qualifications such as references, past experiences, and understanding of the project requirements. The evaluation process should be conducted with all bidders but the selected highest rated bidders may also be requested to submit a best and final pricing.

General terms are usually prepared and can easily be added to the RFP. Subject items usually covered by general terms are insurance requirements, payment responsibility, liens, liquidated damages, fire prevention, code compliance, work plans and approval, changes and alterations, extra scope, permits and records, and the tax-exempt status, if any.

14.6 Implementation Supervision

Any large implementation contract requires the attention of a dedicated group of people to assure that the quality and scheduling promises made by the various contractors are met. This project management group can consist of delegates from a number of departments that have knowledge in areas of facility management, electrical distribution, and digital and voice communications. Periodic visits to all activity areas are essential to supervise the progress of the work, evaluate the workmanship, and monitor the activities of the contractors. Routinely made, short-duration, daily contacts with crew leaders and contractor-provided project management personnel are a good practice and will establish better control and appreciation of problems encountered and progress made. Acceptance of any contracted activity requires a physical walk-through of all work locations and attention to detail. Contractors will be more than happy to address any "punch list" rework items to speed the payment process. All electrical and performance testing sequences that have been outlined in the bid specifications require the witnessing of the owner's project management staff. Test records offered by the contractor are a poor substitute to actually witnessing the individual acceptance testing activities.

15

The Design of the RF Broadband Network

15.1 Overview of the Design

For the purpose of this discussion, it is assumed that the planning document has been completed, that all outside and inside plant construction requirements are in progress, and that conduit continuity exists between all campus buildings. It is also assumed that riser spaces and riser rooms are suitable for equipment installation and that standard or even emergency power will be made available, as required.

15.1.1 Fiber Design Criteria

The design of the single-mode fiber infrastructure is simple, assuming a few basic ground rules are established and the evaluation of the various routing options is complete. The following is a partial listing of requirements and solutions to be considered by the designer before starting the design process. While this list does not cover all the particular conditions readers may face, it is offered as a guide toward optimizing the functionality of the network and adapting the functionality to any future eventualities.

1. Single-mode fiber will be used for connectivity between all buildings.

   Reasoning: While the use of some coaxial cables may be more economical, the installation of passive devices in manholes decreases the reliability of the network.

2. All information centers will be interconnected.

    Reasoning: An interconnection between the MPOP, the PBX, the data center, the video production center, and the server locations will permit a scalable transition to multiservice networking.

3. The interconnection between the information centers will use a redundant ring architecture and fiber throughout.

    Reasoning: While a coaxial interconnection may be more economical, it cannot provide route redundancy. Using a dual-redundant fiber ring architecture for the interconnection of all information centers, all traffic origination points are protected from a catastrophic failure. The development of an APOP or alternative WAN service location is recommended to secure uninterrupted enterprise communications in cases of emergencies.

4. A new network center will be the main fiber cross-connect location and the origination location for the campus network.

    Reasoning: The dual-redundant ring, interconnecting all information centers, requires a gateway for campus connectivity. It is now possible to use either star or ring architecture for campus interconnectivity and consider more economical solutions. The MCC location permits flexible rearrangement of the multiservice network. Figure 15.1 illustrates the ring/ring and the ring/star multiservice campus network architecture. The MCC acts as the source location for all information transmitted throughout the campus and the WAN and gives the multiservice LAN a new control and switching center.

While there are numerous advantages to this networking concept, it is only intended as an example of a flexible networking technology. Network managers are encouraged to use this model as a baseline in their efforts to arrive at more economical solutions.

15.1.2 Coaxial Design Criteria

The use of coaxial distribution service is not required in every RF broadband LAN. Interconnection with existing CAT-5 wiring systems may offer satisfactory solutions for RF services. However, in cases where the entire frequency band has to be brought to a group of workstations or when an economical riser design is required in a high-rise building, the use of coaxial cables and equipment is desirable. Once the decision has been made to use coaxial cables to

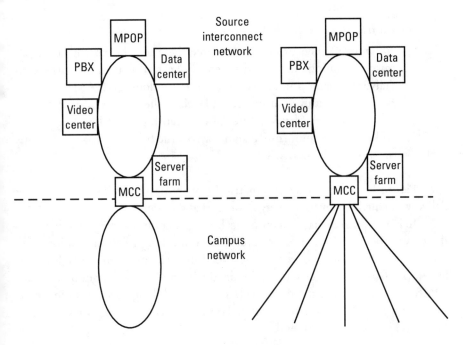

Figure 15.1 The double-ring and ring/star architectures.

extend the reach of the fiber-based RF broadband network, consideration should be given to a few ground rules, alternative solutions, and recommendations.

1. The length of a coaxial service drop should be limited to 250 ft.

    Reasoning: It is desirable to deliver equal levels to an outlet at both higher and lower frequencies. Because the signal is attenuated more at higher frequencies, the level difference between high and low frequencies needs to be controlled. At a length of 250 ft, the outlet level at 750 MHz is about 12 dB higher than the level at 50 MHz.

2. All service drops should have a minimum length of 150 ft.

    Reasoning: To be able to compensate for the service drop attenuation difference over the frequency band, it is necessary to pre-equalize the levels into the service drop in a standard manner. Providing all service drops to a length of 150 ft and using longer drops for distances between 150 and 250 ft will keep the outlet level values uniform. Excessive drop wires for service drops under 150 ft are coiled, taped, and stored in the ceiling area.

3. All riser designs should be symmetrical.

Reasoning: When using coaxial cables between the fiber-optic transmission equipment and different floor levels, it is important to deliver the same amount of signal to every floor level. A symmetrical riser design assures not just equal levels at the various floor levels but also controls the level difference between high and low frequencies and delivers a uniform RF signal across the band to the service drops or to the cable modems.

Figure 15.2 illustrates a symmetrical distribution segment design for the delivery of equal RF levels to every floor level and outlet.

It is recognized that a symmetrical design requires more cable between floor levels, but provides the same levels to all outlets regardless of location. The symmetrical design will assure that the level differences between the highest and lowest passband frequencies are maintained within identical ranges of less than 12 dBmV, regardless of the floor level. Symmetrical design also aids the return transmission of RF frequencies, as it controls level variations for the cable modems located at the headend.

15.2 The Design of the Fiber Backbone

The availability of the latest DFB laser technology for the transmission of the RF broadband spectrum over single-mode fiber-optic cables simplifies the design task of the fiber backbone. A variety of optical transmission budgets are available to assist the designer in devising flexible networking solutions.

One of the more important considerations is the network architecture. In a campus environment, the danger of physical damage to the outside plant is limited. Therefore, a star architecture requiring a direct connection between every building and the MCC is a logical approach. In cases where the conduit

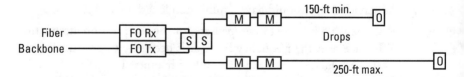

Figure 15.2 The symmetrical distribution segment design. (*Note:* S = splitter; M = 8-port multitap; and O = outlet.)

plant provides a physical ring interconnectivity between all buildings, the designer can use the pathway to place the star topology into the ring. The added advantage is the ability to serve every building from both directions and to add additional routing diversity into the network.

The ring architecture is good protection when interconnecting the information source locations, such as the PBX and the data center or the satellite farm. The application of a star topology for service between the MCC and the campus extremities is, however, recommended for both forward and return direction. In this manner, a direct connectivity is established for each building.

The optical design task is then limited to the determination of cross-connect and patchcord losses for each of the paths and the selection of suitable transmission equipment. The other important consideration is the assessment of the number of 6-MHz channels that will be used, initially and in the future. Optical budgets are related to the number of channels being transmitted. The more channels that are being transmitted, the lower the optical budget. These tradeoffs are discussed in more detail in Section 15.2.1.

15.2.1 Forward and Return Considerations

The spectral capacity of the single-mode fiber is unlimited. Bandwidth limitations are directly related to the presently available DFB laser transmission equipment. The frequency passband of this equipment can transport any information between 20 and 860 MHz. Because fiber is unidirectional, this bandwidth is automatically available in both directions. In a multibuilding campus, the selection of a dedicated fiber pair for each building will transfer this 20–860-MHz spectrum to the building entrance locations. System design consists of the determination of the optical losses that are produced by cross-connect panels. To keep the number of cross-connect locations to a minimum, it is advisable to break a large campus area into multiple zones. Using zoning as a design tool provides for a better control of cross-connect losses, as shown in Figure 15.3.

The example in Figure 15.3 uses a 24-strand fiber cable feeding a group of six buildings, while the cable is fed through five cross-connect locations on its way to the last building. The total distance is less than 1 km with an assumed optical attenuation of 0.35 dB. The five cross-connect losses are estimated at 0.6 dB each to account for connectorization and patchcord. The total optical attenuation, and the worst case estimate for the last building will be 6.95 dB. Zoning has several advantages. It simplifies the design, permits the use of low-power fiber transmission equipment, and offers the opportunity to use optical

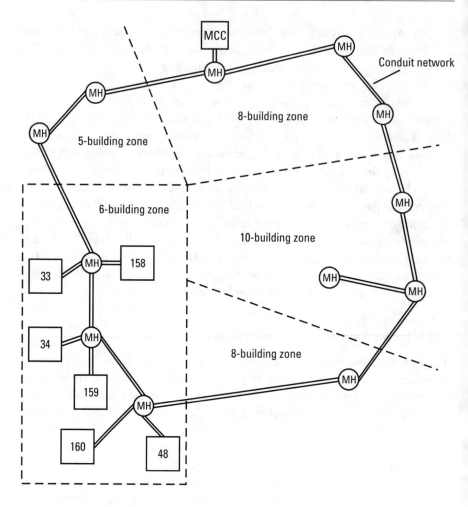

Figure 15.3 Campus zoning.

coupler functions in the forward transmission direction. In addition, it is possible, during the construction phase, to pull a number of 24-strand cables at the same time and to use innerduct branching for each zone. The only disadvantage of the zoning approach is the requirement for more cabling. It is left to the ingenuity of the designer to determine offsetting equipment savings. When conduit space occupancy is not at a premium, it may be possible to assign one innerduct for each zone of buildings. The number 24 was selected only to provide a spare fiber pair for each building. When using high-power DFB laser equipment, it is conceivable that the number of buildings in a zone can be increased and that cables with larger strand count numbers can be used. The designer has the flexibility to choose after carefully evaluating the economical impact of each alternative.

15.2.2 Optical Design Budgets

Since RF broadband fiber transmission equipment is available with optical budgets between 6.5 and 18 dB, the design of the single-mode fiber backbone consists of the addition of all optical losses in a section. Because optical budgets are related to channel loading, the designer is well advised to base the calculations on the future demand of the network and to allow a margin of 2–3 dB for networks expected to carry over sixty 6-MHz channels. In the above example of six buildings in one zone, it is only necessary to determine the longest distance and the highest number of cross-connect locations.

Single-mode fiber working at 1,310 nm has an optical attenuation of 0.35 dB per kilometer. Because most plant records are measured in footage, a conversion is appropriate. A kilometer corresponds to 3,048 ft, resulting in an optical attenuation of about 0.01 dB for every 100 ft. The sectional optical loss is the addition of all individual distance segments. In addition to the optical loss in the fiber section, there are patchcord and connector losses. These losses vary from one fiber strand to the next because of the variation in the connectorization process and the position of the fiber in the bundle. The losses in the connection of the fiber strand to the fiber distribution panel and through the patchcord to the second fiber distribution panel and the fiber strand termination can vary in a range of 0.3–1.3 dB. These values are based on ST connectors with ultra PC polish. While it is possible to order 100%-tested patchcords, the cost of the testing exceeds the cost of the patchcord and, therefore, is not considered a viable option. However, it is not necessary to use the maximum loss value of the patchcord assembly because a string of cross-connect assemblies will follow random statistics. Using an average cross-connect loss of 0.5 dB through fiber terminations and patchcord is a good assimilation.

Using the above hypothetical six-building zone, and assuming a total distance of 5,000 ft, the total optical section loss would be the addition of six cross-connects plus the distance that results in a total section loss of 3.5 dB. The geographic distribution of the campus buildings and the conduit routing govern the sizes of the zones. The forming of larger zones and the use of fiber couplers in the forward direction may be considered to reduce the number of fiber transmitters. The return direction requires a transmitter for each building location. The low 3.5-dB section loss qualifies for the use of low-power fiber transmitters for each of the buildings in the zone. It is also possible to incorporate a two-way coupler into the design and use the second output to provide an alternative routing of the return fiber.

15.2.3 Fiber Couplers

Fiber couplers are tools for limiting the number of fiber transmitters in the forward direction and enhancing the flexibility of the design. In the campus environment, fiber couplers can be used to combine multiple zones and to provide alternative routing solutions.

When using a fiber coupler for redundant routing, the destination building will have two active fibers arriving from two different directions. New switching designs exist that will perform automatic switching when the optical energy in one of the fibers is lost.

Other new switching designs will become available to replace couplers and cross-connect panels that can be used for transparent switching or for the multiplexing of different speed ranges and drop-and-add functions. For the time being, the fiber coupler is the most economical solution.

Fiber couplers at the fiber transmitter can be used to equalize the optical losses between the different campus zones. In cases where geography dictates that a particular zone includes 10 buildings and is located in the most distant part of the campus, the fiber coupler can be used to adjust the optical energy to the required level. Figure 15.4 depicts a simple hypothetical campus with zones and optical budgets.

Couplers are available with different coupling ratios. Figure 15.4 shows a high-power transmitter feeding a two-way coupler with a 30/70 coupling ratio in the forward direction. The primary port has a 2.0-dB attenuation while the secondary port has a 6.0-dB attenuation. The coupler's primary port is connected to a 20/80 ratio coupler with 1.3 dB in the primary port and 7.8 dB in the secondary port. The resulting three-way split of the transmitted optical signal is 3.3, 6.0, and 9.8 dB to correspond to the larger and smaller campus zones.

The design of the single-mode fiber backbone is not a complicated task. The designer can apply flexible solutions and ingenuity to the design process and optimize performance and economy. With ever-increasing optical budgets, the fiber segment of the RF broadband LAN is fully transparent to the electrical input and output signals and a solid foundation for the multiservice LAN of the future. The fiber coupler also has a place in high-rise buildings. It can be used to extend the single-mode fiber to the various floor levels for the forward distribution. The extension of the single-mode fiber will reduce or eliminate the use of coaxial riser distribution equipment. When using a fiber coupler arrangement for the forward transmission in a high-rise building, it should be remembered that the return path cannot be combined using this technology. Either separate fiber return routes are required from each of the floor levels or a coaxial riser must be used for the combining of all return traffic to the MDF location where the coaxial transmission is connected to a fiber-optic transmitter.

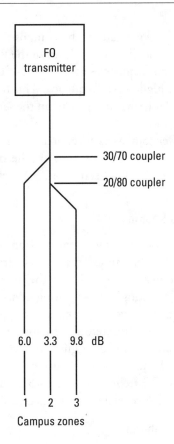

Figure 15.4 Campus zoning and the use of fiber couplers.

15.3 The Design of Coaxial Segments

The RF broadband LAN contains coaxial sections at both ends of the fiber backbone. In the forward direction, the output of the cable modems is combined using coaxial hardware to produce a single RF level. This input level is in the range of +24 to +32 dBmV, depending upon the number of channels. A larger number of channels require lower input levels. At the receive end, the fiber-optic receiver produces a combined RF level in the range of +34 to +38 dBmV into a coaxial output port. The fiber path is essentially transparent except for a small overall gain.

In the return direction, the coaxial input level range into a low-powered DFB transmitter can be in the range of +18 to +24 dBmV. The receiver output level can be either high-gain within a range of +34 to +38 dBmV or low-gain with only about +10 dBmV to feed a number of cable modems. The fiber segment will transmit optically whatever the composition of the electrical signals present at the input of the fiber transmitter. In the RF broadband LAN it is desirable to pre-equalize the frequencies at the front of the system by assigning a 6-dB higher level to the highest RF frequency. This 6-dB slope will counteract the higher attenuation at higher frequencies in the service drop segment and minimize outlet level variations.

The coaxial design encompasses three major segments: the service drop segment, the riser distribution segment, and the headend interconnect segment. Design details are provided in Sections 15.3.1–15.3.3.

15.3.1 The Service Drop Segment

The service drop segment extends between the horizontal cross-connect and the user population. In smaller buildings, the service drops can be connected directly to the fiber equipment in the MDF at the building entrance location. The designer can employ existing CAT-5 wiring or use coaxial service drops. The choice is directly related to workstation service requirements. Because the RF broadband technology uses discrete RF frequency assignments, workstations or groups of workstations can be assigned for service from one or more of the available frequencies. In these cases the use of CAT-5 wiring has advantages. However, when all RF frequencies have to be available at a group of workstations, the use of RG-6 coaxial cabling is necessary.

15.3.1.1 Using CAT-5 Distribution

In order to illustrate the application of CAT-5 wiring in an RF broadband network, it is helpful to use a hypothetical example. For this purpose, it is assumed that the campus building is a three-level structure with a workstation population of 320. Ethernet service has been used and CAT-5 wiring exists between the MDF installation in the basement and all individual workstations. A cross-connect panel is available. In addition, there is a telephone cross-connect for 100 telephones on each floor using CAT-3 wiring. The RF broadband fiber-optic equipment has been installed in available rack space. The coaxial portion of this installation requires the interconnection of the fiber-optic equipment with cable modems using passive devices. Figure 15.5 illustrates the equipment arrangement.

In Figure 15.5, a connection is made between the fiber-optic transmitter and receiver to a coaxial splitter. The splitter is connected to a second splitter that feeds two eight-port multitaps in series. The multitap ports are connected to cable modems. Figure 15.5 assumes a total of eight cable modems, each carrying Ethernet traffic. The baseband ports of the cable modems are interconnected with the cross-connect panel through a bridge, each providing service to

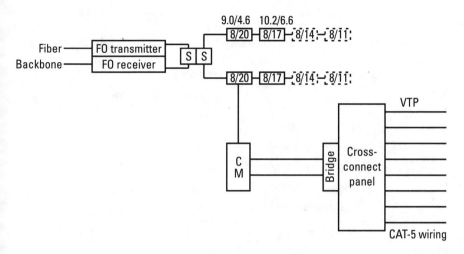

Figure 15.5 The CAT-5 distribution service. (*Note:* S = splitter; 8/20 = eight-port tap value 20; and CM = cable modem.)

20 workstations. The number 20 has no significance; the user population can be broken up into any desired combination of workgroups. As a result, the building is now served with 16 separate Ethernet loops and could handle VOIP as well.

The second multitap in each branch is reserved for future expansion. These taps can also be used with the existing CAT-3 wiring system in conjunction with phone hubs. Phone hubs are available for 24 telephones and may operate on a 25.6-Mbps ATM circuit using a cable modem as the transmission device.

The design of the coaxial subnetwork is simple. In the forward direction, the level at the fiber receiver output has been measured to be +38 dBmV at 860 MHz and +32 dBmV at 50 MHz. The calculation for each branch is outlined in Table 15.1.

The example contains four eight-port taps with a total of 32 ports. In order to use additional ports for coaxial distribution, the tap values would have to change to values 14 and 11. Service levels for cable modems can be in the +3.0 to +7.0 dB range. In the above example, additional taps with values of 14 and 11 could be added to the two branches. The result is a passive system with 64 tap ports for the connection of cable modems.

15.3.1.2 Using RG-6 Coaxial Distribution

Using RG-6 service drop cabling enables the entire RF frequency band to reach the outlet. This installation permits the workstation to be connected through a

Table 15.1
Design Calculations of a Coaxial Subnetwork for Connections up to 64 Cable Modems

Calculation	Frequency	
First Tap	860 MHz	50 MHz
Receiver level (dBmV)	+38	+32
Splitter loss (dB)	−4.5	−3.7
Second splitter loss (dB)	−4.5	−3.7
Multitap loss (dB), value 20	−20	−20
First tap output level (dBmV)	+9.0	+4.6
Second Tap		
Receiver level (dBmV)	+38	+32
Splitter loss (dB)	−4.5	−3.7
Second splitter loss (dB)	−4.5	−3.7
First tap insertion loss (dB)	−1.8	−1.0
Multitap loss (dB), value 17	−17	−17
Second tap level (dBmV)	+10.2	+6.6

frequency-agile cable modem that can select any one of the transmitted frequencies. Whenever workstations require connectivity to different data streams, or if a workstation is authorized to receive all RF frequencies, the use of a coaxial service drop is the best solution. RG-6 coaxial wiring is also required for cable television applications to students in the dormitories of educational campuses. The ability to tune in any spectral segment or channel is the big advantage.

Again, the drop cable length should not be shorter than 150 ft and should be limited to a maximum of 250 ft in order to assure outlet levels for all frequencies to fall into a close range and to assure level control of return transmissions. Return transmissions are required for access to the Internet via Ethernet; for VOIP, voice over ATM, videoconferencing, and VOD; and for any high-speed access to the corporate network that may require different RF frequencies. The design of the coaxial distribution segment is similar to the coaxial segment discussed for CAT-5 service, except that higher levels are required for signal delivery over RG-6 coaxial wiring. Figure 15.6 shows the calculations for a passive coaxial connection to the fiber receiver.

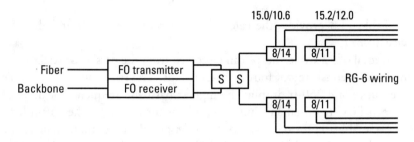

Figure 15.6 The RG-6 coaxial distribution segment.

The output level of the fiber receiver is +38 dBmV in the forward direction at 860 MHz and +32 dBmV at 50 MHz. To produce equal levels at all tap ports, symmetrical design of all components is paramount. The receiver output and the transmitter input are combined using a splitter. A second splitter follows to permit the symmetrical placement of the multitaps. The multitap values are chosen to provide levels of about +15 dBmV into the 250-ft service drop. Table 15.2 outlines the calculations.

Table 15.2
Design Calculations for a Coaxial Subnetwork for 150–250-ft Coaxial Service Drops

Calculation	Frequency	
First Tap	860 MHz	50 MHz
Receiver output level (dBmV)	+38	+32
First splitter loss (dB)	−4.5	−3.7
Second splitter loss (dB)	−4.5	−3.7
First multitap loss (dB), value 14	−14	−14
First tap out level (dBmV)	+15.0	+10.6
Second Tap		
Receiver output level (dBmV)	+38	+32
First splitter loss (dB)	−4.5	−3.7
Second splitter loss (dB)	−4.5	−3.7
First tap insertion loss (dB)	−2.8	−1.6
Second multitap loss (dB), value 11	−11	−11
Second tap out level (dBmV)	+15.2	+12.0

Table 15.3 provides the calculation for both the 150-ft and the 250-ft service drop.

A total of 32 service drops can be served from the passive coaxial distribution segment consisting of four eight-port taps. The resulting 0.0-dBmV level at the outlet of a 250-ft drop at the highest passband frequency of 860 MHz still meets FCC requirements but also explains the reason for the 250-ft limitation. To be able to provide service to an outlet population of more than 32 users, the addition of a coaxial amplifier is indicated. Distribution amplifiers have a +46-dBmV output level and can provide sufficient levels to 64 service drops.

15.3.2 The Riser Distribution Segment

A coaxial riser distribution is one of the options available to the designer. It is the appropriate and economical choice for multistory buildings. In high-rise buildings with more than six floor levels, it may be more advantageous to extend the fiber-optic backbone to selected floor levels and use a coaxial riser only for return transmissions. Another alternative is to treat the high-rise building as a multibuilding zone and provide fiber-optic equipment at selected floor levels. In this scenario, each zone has its own fiber pair assigned and is connected directly to the MCC.

The designer can devise numerous coaxial riser architectures as long as the ground rules of symmetry and level control are observed. When using amplifiers, it is wise to limit the number of amplifiers in cascade to no more than two. Some of these preferences and good design practices are discussed in Sections 15.3.2.1 and 15.3.2.2.

15.3.2.1 Passive Riser Distribution

The passive riser distribution system looks very similar to the service distribution segment discussed in Section 15.3.2. The only difference is the coaxial cabling between the devices. Figure 15.7 illustrates the passive riser design.

Table 15.3
Service Drop Attenuations and Outlet Levels

Calculation	150-ft Service Drop		250-ft Service Drop	
Frequency	860 MHz	50 MHz	860 MHz	50 MHz
Tap level (dBmV)	+15.0	+10.6	+15.0	+10.6
Attenuation (dB)	−9.3	−2.79	−15.0	−4.65
Outlet level (dBmV)	+5.7	+7.81	+0.0	+5.95

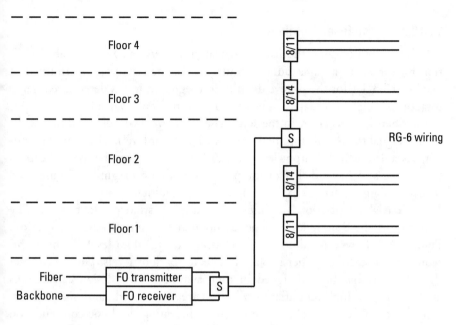

Figure 15.7 The passive coaxial riser.

In Figure 15.7, it is assumed, again, that the fiber-optic transmission equipment has been placed at the MDF location in the basement of the building. The first splitter, again, is used to combine the transmit and receive directions. A 0.500-in cable is used to the second floor level, where the second splitter is placed to feed an eight-port tap in the second floor and a second eight-port tap in the first floor. The other leg of the splitter feeds an eight-port tap on the third floor and a second tap at the fourth floor. The added cabling consists of the length of the riser between the basement and the second floor as well as (worst case) the length of the coaxial cable between the second, third, and fourth floors. Assuming a distance of 12 ft between floors, the total added cable length to interconnect the devices from the above distribution segment is about 60 ft. The riser-rated 0.5-in diameter cable has an attenuation of 2.25 dB/100 ft at 860 MHz and an attenuation of 0.54 dB/100 ft at 50 MHz. The 60 ft of cable required in Figure 15.6 modifies the calculations of the distribution segment by −1.35 dB at 860 MHz and by −0.324 dB at 50 MHz. The passive riser design covers four floor levels with eight tap ports at each floor level, follows the same calculations as the distribution segment design, and meets all symmetrical requirements for equal outlet levels. The only exception is that the outlet level at 860 MHz for the 250-ft service drop has become marginal. A somewhat shorter drop length is desirable.

15.3.2.2 Active Riser Distribution

In a high-rise building, the use of coaxial amplifiers can be a valuable addition for service to a larger outlet population. In the calculation performed in Section 15.3.2.1 for the passive distribution segment, it was determined that a total of four eight-port taps can be served from the level available at the output of the fiber-optic receiver. In the active riser this level is simply used to distribute the signal to coaxial amplifiers at the various floor levels. The coaxial amplifier has a +46-dBmV output level at 860 MHz, which can be used to extend the service area beyond that of the passive distribution segment. Figure 15.8 depicts a high-rise active coaxial riser and its capabilities.

Coaxial distribution amplifiers are placed at strategic floor levels. The selection of these floor levels depends upon the required outlet population. Figure 15.8 shows a requirement for 16 drops at each floor level. To meet this requirement, each amplifier is placed to serve six floor levels. Due to the fact that the riser has been designed in total symmetry, the design calculations are limited to a single amplifier distribution area. This modular design concept provides equal levels to all service drops by repeating the basic configuration and by symmetrical feeding of each amplifier. Except for minor tilt and gain adjustments at the amplifiers, the installation can be considered plug-and-play.

Table 15.4 outlines the calculations for the forward levels of the active service distribution module.

Table 15.4 shows that equal levels are available for two multitaps in series at every floor level. The cable attenuation of the 0.500 cable between floors has been purposely disregarded. The maximum length of this cabling is the height of three floors, or about 40 ft, and only contributes to the losses by less than 1 dB. Table 15.5 presents the calculation of the worst-case distribution segment.

While the use of additional splitters and directional couplers has increased the low frequency level by a little, the high frequency levels are identical to the above calculations for the passive distribution segment. Equal levels will be provided to 16 outlets in each of six floors. The symmetrical arrangement of the components following the distribution amplifier maximizes the number of coaxial drops. A total of 96 outlets can be served by a single amplifier.

In the previous calculation under Section 15.3.1.1, for service into a CAT-5 wiring system, it was determined that the tap levels could be substantially lower. The same conclusion can be reached for riser distributions using amplifiers. An additional 16 outlets can be made available for the interconnection of cable modems. This raises the number of cable modems for CAT-5 cross-connect installations to 36 per floor. As a result, the RF broadband LAN could provide for 36 different workgroups per floor.

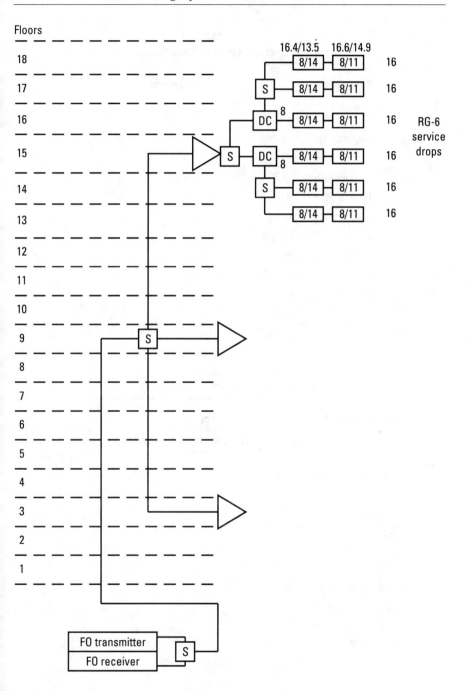

Figure 15.8 The active riser distribution system.

Table 15.4
Active Riser Level Calculations of Figure 15.8 Diagram

Calculation	Frequency	
	860 MHz	50 MHz
Amplifier out level (dBmV)	+46	+40
First splitter loss (dB)	−4.5	−3.7
DC-8 insertion loss (dB)	−2.1	−1.4
Level to other floors (dBmV)	+39.4	+34.9
Second splitter loss (dB)	−4.5	−3.7
Level to next floor (dBmV)	+34.9	+31.2
Third splitter loss (dB)	−4.5	−3.7
Level to multitaps, each floor (dBmV)	+30.4	+27.5
DC-8 tap level (dBmV)	+31.4	+26.4

Table 15.5
Design Calculation for 16 Coaxial Service Drops at Each of Six Floors

Output Level	Frequency	
First Tap	860 MHz	50 MHz
Splitter level (dBmV)	+30.4	+27.5
8-port tap loss (dB), value 14	−14	−14
First tap out level (dBmV)	+16.4	+13.5
Second Tap		
Splitter level (dBmV)	+30.4	+27.5
First tap insertion loss (dB)	−2.8	−1.6
Second tap loss (dB), value 11	−11	−11
Second tap out level (dBmV)	+16.6	+14.9

Return transmission calculations have been made for the above coaxial distribution examples. While they meet the return level requirements for subsplit and high-split spectral architectures, they have not been itemized in this treatment of the network design details for fear of information overflow. The return design is simply the assessment of the return transmission at the highest frequency of the selected passband. These frequencies can be either 45 or 186 MHz in subsplit or high-split systems when single coaxial service segments are used. The return can also encompass the entire band in a dual coaxial configuration. The network designer is at liberty to follow the path of the signal through the coaxial components and in the opposite direction. There are only three common rules in return design: (1) the coaxial return amplifier requires an input level of +17.0 dBmV, (2) the fiber-optic transmitter requires an input of between 27 to 34 dBmV depending on channel loading, and (3) the cable modem output level shall be around +55.0 dBmV and should be equipped with level adjustment.

15.3.3 The Headend Interconnect Segment

The headend interconnect segment is the coaxial area between the headend equipment or the cable modem bank and the location of the fiber-optic transmission equipment. Often, these locations are colocated and may be a short distance from each other and require coaxial interconnection. The equipment complement required to interconnect all RF broadband sources for transmission on the single-mode fiber backbone depends solely on the number of sources. These sources can be multichannel satellite-receiving equipment, cable modem banks for data and voice, or any other RF source. It is not the intent of this section to design the interconnect wiring for baseband interconnectivity between telephone, data, and video equipment. Only interconnections between RF sources/destinations and the interface with the optical transmission equipment are described herein.

15.3.3.1 Combining the Forward Direction

The headend interconnect segment has to be a dual coaxial interface. Cable modems have input and output ports for both transmission directions. Unlike the coaxial distribution tail of the RF broadband network, which allows the combining of the two directions in a single coaxial cable, connections in the headend must be made for unidirectional transmission.

Combining several sources into a single transmission for transfer to the fiber-optic transmitter requires the use of passive components. Because it is also important to deliver equal levels of each RF frequency to the transmitter, it is required that the passive component configuration be designed in a symmetrical manner. Figure 15.9 shows the combining network of a cable modem bank using passive and active components to assure the delivery of required levels of each of the RF carriers to the fiber transmitter input port.

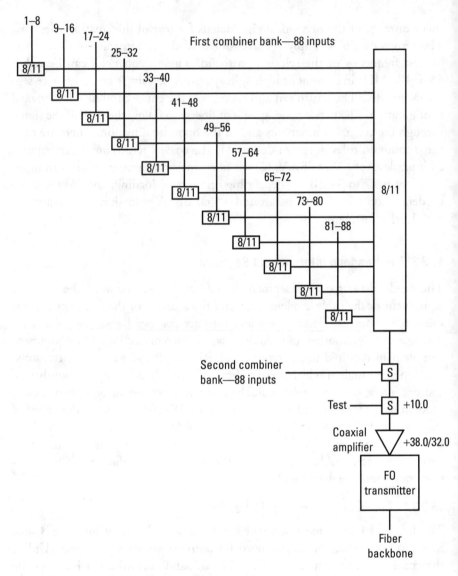

Figure 15.9 The forward headend combining network.

Cable modems, whether they are used for video, data, or telephony transmission, typically have an output level of +55.0 to +58.0 dBmV. The input level to the fiber transmitter typically is +38 dBmV for the high channel and +32 dBmV for low frequencies. Since it is not a good practice to set the output levels for different values dependent on the frequency assignment, the flat output level is first combined using eight-port taps followed by splitters and a combiner amplifier. The use of eight-port taps for combining signals will limit return loss problems and increase the isolation between the ports. The outputs of the eight-port taps are then combined into a second eight-port tap. Because value 11 taps are used, the total attenuation of the combining network is 22 dB.

Combining the signal of only 11 devices can still be accomplished within the +55.0 minus +38.0-dB or 17-dB differential. However, combining more than 11 devices requires amplification. In Figure 15.9, there are two levels of eight-port taps followed by two splitters designed to accept signals from two alike arrays. The total losses of the passive network are 22 dB from the two eight-port taps plus two 4.5-dB splitters or 48 dB. The amplifier input level of about 10.0 dBmV is amplified and provides the required pre-equalization for the +38.0/32.0-dBmV input to the fiber transmitter. A total number of 176 RF devices plus test inputs can be combined in this manner.

15.3.3.2 Splitting the Return Direction

The return direction of the RF broadband network may utilize separate fiber-optic receivers for single, or groups of, buildings. Due to this architecture, the splitting of the incoming RF signals can be arranged using only passive components. Figure 15.10 shows a return splitting arrangement to feed the input of 22 cable modems from each of four fiber receivers.

The return splitting network can serve a combined total of 64 different RF frequencies. The output level of the receivers are at +38.0 dBmV. The losses in the passive combining network are 11 dB for the eight-port tap plus 4.0 dB for the splitter. The resulting level of +23 dBmV requires external attenuation to approximate the desirable cable modem input level of +10 dB. Other combinations of passive components can be used. Again, symmetrical layout of all components will assure equal levels to all units.

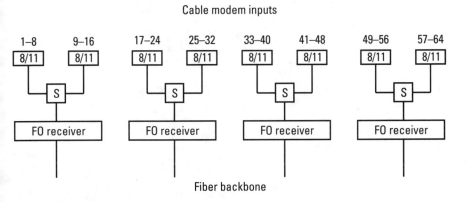

Figure 15.10 The return headend splitting network.

16

Cost Considerations

16.1 General Assumptions

The transition from the standard triple wiring concept used today in the corporate LAN to the single multiservice facility is a gradual process. Because existing infrastructures are complex and have depended upon evolutionary developments, it is not possible to propose a detailed conversion plan or to recommend a finite plan of action. It is certain, however, that the convergence toward multiservice networking is proceeding in the long-distance segment and in the local loop and that it will influence the future of every corporate network to some degree. The introduction of competitive local and long-distance services and the offerings of a variety of QoS levels will transfer the need for ATM-based multiservice traffic into the LAN backbone. Cost considerations provided herein have been restricted to the use of RF broadband technology in this environment and do not attempt to cover the many and diverse cost elements that are an integral part of the transition process.

16.1.1 Standard Network Components

The multiservice LAN of the future requires an access control that combines voice, data, and video for FR, ATM, and Internet connectivity. Various offered QoS levels and rates will offer an incentive to incorporate many of these service levels into the corporate LAN. The cost comparisons discussed in this chapter assume that an ATM-based backbone exists or that the existing legacy LAN is not going to be upgraded and that RF broadband forms an overlay to permit multimedia services. For the comparison of costs it is therefore assumed that

333

such existing components are standard network components that are exempt from the pricing models that have been identified in Section 16.1.2.

16.1.2 RF Broadband Applications

RF broadband cannot replace an ATM-based backbone operating at 2.4 Gbps. Cable modems cannot handle speeds of more than 40 Mbps at this time. RF broadband can become an extension to an ATM-based or a legacy LAN and carry multiple services over a single fiber to remote campus locations. RF broadband can also help to reduce the fiber construction requirement and offer an economical expansion of videoconferencing and training for VOD in an established legacy LAN. Integrated RF broadband and cable modem technology can be used to do the following:

1. Reduce the construction of new fiber facilities by using a single fiber pair to a new location;

2. Eliminate the requirement for workgroup switches at outlying buildings or floor levels by consolidation of the workgroup equipment at the backbone;

3. Consolidate maintenance at backbone locations;

4. Provide single-fiber continuity for massive new telephone requirements and elimination of new copper cable construction;

5. Provide a single-mode fiber overlay to an existing LAN to establish new telephone, videoconferencing, and VOD services.

The use of RF broadband in new multiservice LAN architectures may tax the ingenuity of the network manager, but it can become a useful tool in the step-by-step process toward convergence and permit more economical solutions. A few examples are discussed in Section 16.2.

16.2   A Comparison of IP and RF Broadband Costs

The example of Figure 16.1 assumes the existence of an ATM-based high-speed backbone and an edge switch at the MCC location. As defined earlier, the MCC is the location that serves the outlying campus, and it is usually equipped with legacy switch/router equipment that translates the ATM backbone traffic to 100-Mbps fast Ethernet segments to feed workgroup switches in the buildings of the various campus zones. It has been determined that a total of four

100-Mbps services are required to feed the 72-workstation population in a particular building. Twenty-four workstations are earmarked for 100-Mbps service, and 48 additional workstations are considered standard 10-Mbps clients. Figure 16.1 shows the intended equipment configuration and a possible RF broadband alternative.

Figure 16.1 assumes that 12 multimode fiber pairs are required for the service and that other existing services have used up any available fibers. Since there are two unused single-mode fibers in the bundle, the question arises how to interconnect the building. One option is to install an additional multimode fiber cable; another is to utilize the spare single-mode dark fibers. To provide the same service to the workstations, a comparison study was conducted. The plan entailed using four 100-Mbps services for the 48 workstations operating at 10 Mbps. This means that each workstation will operate at a minimum speed of 8.33 Mbps. It was also planned to use eight fibers for the 24 workstations equipped for 100-Mbps service. The minimum speed of these workstations would be 8×100 divided by 24 or 33.33 Mbps. This means that a new 24-pair multimode fiber has to be installed.

Using RF broadband and cable modems, only one single-mode fiber pair and a total of 36 cable modems are required. Each cable modem provides for a 40-Mbps transfer in the downstream direction. Twenty-four cable modems are used for the direct connection to the 24 fast Ethernet workstations, and a minimum speed of 40 Mbps can be maintained. RG-6 service drops are used for these clients, and the cable modem is installed next to each workstation. The remaining 12 cable modems are connected via UTP to the 48 low-speed

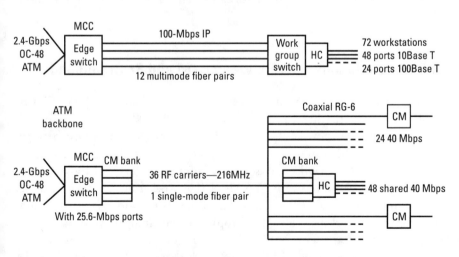

Figure 16.1 An IP-based LAN and RF broadband comparison.

clients. Four stations are served by one cable modem through the existing cross-connect, and a minimum speed of 10 Mbps can be maintained.

A comparison of the costs of the two approaches consists of the installation and termination costs of the 24-strand, multimode fiber cable against the costs of 36 cable modems and the installation of 24 RG-6 service drops. While the total costs for cable modems and service drops are estimated to be under $25,000, the fiber construction is dependent upon distance, availability of conduit space, and rack space for fiber termination equipment. In addition, there are other benefits in using RF broadband technology, including the following:

1. The setting of QoS parameters right at the desktop;

2. The installation of workgroup switching equipment at the MCC ATM backbone location, which reduces the number and costs of the required modules;

3. Additional cost savings from the simpler UTP port connectivity with the cable modems;

4. The flexibility of expanding the services to additional workstations by simply adding cable modems.

In addition, there are cost savings in data networking equipment. Typically, a 100Base-T module commands a price of about $7,000 for a four-port module, while a 24-port 10Base-T module can be obtained for about the same amount. By moving the location of the workgroup switch to the backbone, the more expensive fast Ethernet modules can be eliminated at a savings of about $7,000.

16.3   A Cost Comparison Between ATM and RF Broadband

The ATM-based LAN extension versus RF broadband is quite similar. Figure 16.2 compares the two.

The example shows 155-Mbps and 622-Mbps ATM service between the MCC and the workgroup switch in the new building. It has been planned to serve 24 workstations with a minimum of 155 Mbps and the remaining 48 workstations with 25.6 Mbps. The workgroup switch requires six 155.2-Mbps modules that feed the 155.2-Mbps workstations. The minimum speed for the 24 workstations is 38.8 Mbps. The remaining 48 workstations are equipped with 25.6 ATM-PCIs and are fed from six 155.2-Mbps circuits. The minimum speed reserved for the 48 workstations is 19.8 Mbps. In order

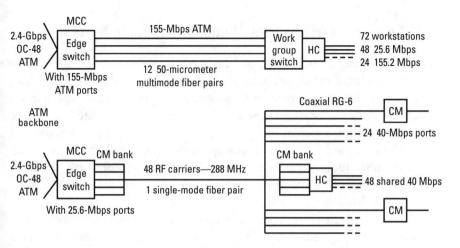

Figure 16.2 An ATM-based LAN and RF broadband comparison.

to cover the distance, it was planned to use 50-μm multimode fiber with 24 strands.

The RF broadband alternative can provide 40-Mbps service to the 24 high-speed and 20-Mbps to the 48 low-speed workstations. The total number of required cable modems is 48 on the basis that 24 are served with RG-6 coaxial service drops and that the 48 remaining workstations are served from 24 cable modems in the MDF using existing UTP wiring for 2 workstations per cable modem. The comparison of costs can be based again on the need for a new fiber cable installation versus the costs for 48 cable modems and 24 RG-6 coaxial drops. These costs are estimated to be below $30,000. The benefits of using RF broadband are identical to those described for the IP comparison and provide QoS control from the desktop.

A comparison of the costs of various ATM modules indicates an even greater difference between a two-port 155-Mbps module and an 18-port 25-Mbps module. The 25-Mbps module can be obtained for about $7,500 while the same price level applies to the 155-Mbps module. Our example uses either twelve 155-Mbps modules or forty-eight 25-Mbps modules at the edge switch. The cost savings can be on the order of $50,000. While these savings are partially offset by the higher costs of the ATM computer NIC cards ($150 for the 25-Mbps card versus $60 for an Ethernet card), Figure 16.2 shows that a relocation of workgroup switches to backbone locations, and using RF broadband cable modems for the extensions to the workstations, can be an economical alternative.

16.4   Costing an ATM-Based Videoconferencing System

Videoconferencing in an ATM-based LAN can be an economical addition to limit the amount of time spent in meetings. Using JPEG technology, the bandwidth can be chosen to provide full-motion, high-quality video presentations between workstation locations. The cost for the equipment required at the core switch location for a system interconnecting four users is about $4,500. An ATM 25-Mbps card in the participating workstation is all that is required. Especially suited for multimedia communications between corporate staff, such a system can easily be justified and form the basis of future high-quality videoconferencing through the WAN.

Many corporate video systems feature board rooms and conference rooms that have been equipped with expensive projection sets and automatic camera control systems. The existence of an ATM-based backbone with ATM-based or RF broadband-based tributaries can replace the existing coaxial plant. A codec unit at each of these locations can be obtained at about $5,000 and interconnect the digital video with existing analog NTSC-based video equipment.

16.5   Costing an HFC-Based VOD Network

In case the existing IP-based campus LAN fills all the present requirements, the decision toward establishing a multiservice facility can be delayed. An intermediate step can be taken to overlay a single-mode fiber infrastructure to provide a transmission path for future integration. This example assumes that the delivery of VOD is a desirable first step. VOD for employee training purposes can be established without requiring any rebuilding of the data network, by adding a DVD video library, an MPEG-2 statistical multiplexer, and a cable modem system. Figure 16.3 shows such an example.

The MPEG-2 statistical multiplexer can combine 10 videostreams into a 20-Mbps data stream that can be accessed by any computer terminal and/or by a standard digital set-top converter for presentations in lecture rooms on existing analog television or projection sets. Figure 16.3 shows the interconnection with the ATM backbone and a single cable modem at the ATM edge switch location. Provided that the single-mode fiber facility has been extended to selected locations in the campus, the VOD can be made available at any desired computer location and through coaxial RG-6 drops at selected presentation rooms. The cost elements consist of DVD recording expenses, the video library, the MPEG-2 statistical multiplexer, the single-mode fiber facility, the coaxial service drops, the ATM switch, and the cable modem facility. The installation cost of an RG-6 drop is similar to the cost of UTP. The costs of

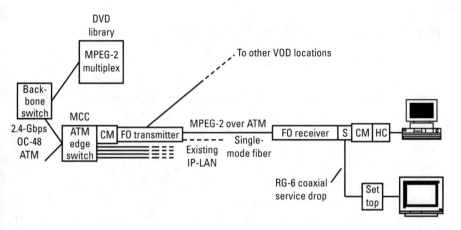

Figure 16.3 An HFC-based VOD overlay.

set-top converters and cable modems are in the range of $250–300. Fiber-optic transmission equipment consists of a single DFB laser transmitter in the range of $3,000 and multiple receivers at about $500 each. The central cable modem unit with network management software is estimated at about $20,000. Not considering the pricing of the single-mode fiber construction or the video library equipment complement, a 10-location VOD network is estimated in a range of $50,000. DVD authoring and recording is presently an expensive proposition, but since DVD will soon replace the VCR, the costs will steadily decline. Multiple MPEG-2 encoding equipment can cost $40,000 because it is marketed mainly to broadcast and satellite entities. This pricing will also drop when the cable industry converts to digital video transmission.

While it is difficult to estimate equipment costs in a period of introduction and change, the costs of establishing a single-mode fiber facility can be readily established by an addition of the cost of the fiber cable, the laying of conduit, the pulling of the fiber cable, and the terminating and connectorization of the fiber. A 15 to 20% increase over multimode fiber construction should be taken into consideration.

Once the single-mode fiber network has been established, it is easy to find immediate and future uses. The VOD network is just one example of the many multiservice LAN and RF broadband applications that can be accommodated. A single-mode fiber overlay network forms the foundation of the convergence toward the multiservice LAN.

It can first be used for VOD or cable television, later to bypass the massive copper wiring of the telephone network, and finally to integrate all multimedia services into a single wiring system.

16.6   Costing an ATM-Based Telephone Overlay

Once an ATM-based LAN backbone has been established, the addition of voice networking can be added quite economically. At the PBX location, a computer-based unit can be added to transform eight, 16, or 24 outside or extension lines into 25.6- or 155.2-Mbps ATM cell streams for a cost of about $3,000. The ATM-based output is then connected to the ATM backbone switch for connection-oriented transmission to workstations or stand-alone telephones. A three-port NIC card for 25.6-Mbps ATM service is valued at about $350 and translates the digital voice back to analog for use with standard telephone sets. The same arrangement for 155.2-Mbps is about $100 more. Stand-alone telephone sets require interconnection through a phone hub that translates the ATM-based voice back to the analog format and, at the same time, provides conventional power to the telephone sets. The prices range from $1,800 for an eight-telephone connectivity to $2,800 for 16 phones.

While it may not be immediately necessary to integrate telephones at the workstation level, the utilization of the ATM backbone for telephony can eliminate the need for additional copper cable construction. ATM-based telephony and cable modem-based tributaries can also be used to provide an alternate route for telephony and, for the first time in the history of telecommunications, facilitate the removal of copper cables from the duct system. As explained in Chapter 10, the use of RF broadband-based cable modems can be used to reduce the conduit occupancy by removing copper cables in groups of 240 pairs. The economic impact of such a transition in the campus LAN can be measured by comparing the planned multiyear construction cost budgets with the required initial capital outlay.

16.7   Price Forwarding

The costing information provided in this chapter reflects 1998 prices. New products are introduced almost on a monthly basis and will change the assumptions made herein. While it is the user's impression that ATM-based products carry a surcharge because of their complexity, and while cable modems are still awaiting public evaluation, it is evident that add-on services such as telephony, videoconferencing, and VOD can be economically implemented and added to an existing single-mode fiber infrastructure. The user is invited not only to assess initial capital outlay but to consider the many ancillary benefits. A continuous review of pricing levels of the various equipment and technologies will indicate a downward trend over the next decade, as it has for IP-based products in the past.

17

Outlook

17.1 The Changing Telecommunications Environment

In planning a path toward integration of voice, data, and video for the corporate LAN, it is important to consider the explosive changes taking place in the telecommunications environment. Taking a bird's eye view of global developments is essential to avoid pitfalls and short-lived solutions. After two decades, digital technologies are just now beginning to affect the structure of the telecommunications infrastructure. It may well take another two decades to implement the new range of services and to reap the benefits within every enterprise and corporation.

One of the goals of this book is to convey the number of important technical developments toward an integrated single-wiring system for the corporate network. While it is impossible to cover all of the technological developments under way, it is important to consider that any development within the corporate network has a direct relationship to the global changes in every area of telecommunications. This chapter summarizes some of the trends in the public network arena. Readers are reminded that planning the multiservice LAN is directly related to the service offerings available in the local loop and may continue to be different in many geographical areas of the country for several years. Every CLEC and every ISP looks first toward service in high revenue-producing areas and profits from these lucrative areas. Because we are entering an era of diverse and confusing service offerings, it is wise to stay abreast of the conditions in other areas of the nation and compare before signing any long-term service contract. This chapter discusses some of the ongoing developments in the telecommunications industry.

17.2 The Multiservice Public Network

The transition of the public network to an interoperable multiservice facility is well under way. Some highlights are presented in Sections 17.2.1 and 17.2.2.

17.2.1 The Long-Distance Segment

Since the establishment of a competitive environment by the Telecommunications Act of 1996, a frantic building program has begun. AT&T, MCI, Sprint, GTE, PSI, Worldcom, Qwest, and a host of others have used railroad and pipeline right-of-ways to construct single-mode fiber connectivity throughout the nation. Simultaneously, there are multiple submarine cable projects under way to interlink every continent. Also in the implementation stage are LEO satellite projects and sun-powered aircraft with the ability to fly at 70,000 ft for an unlimited period of time and act as relay stations for wireless global communications. These tremendous efforts are directly related to the phenomenal growth of the Internet, which is also undergoing total renovation.

The industry is now in the second phase, the merger phase, so important to establish dominance and versatility. Typical merger examples are Worldcom and MCI, Bell Atlantic and GTE, AT&T's acquisition of TCI and its working arrangement with British Telecom, and SBC's buyout of SNET and merger with Ameritech.

While this is not the place and time to analyze the many benefits of these mergers, it is interesting to note that the AT&T/TCI merger provides service continuity to AT&T from long distance to the local loop. Inclusive of TCI affiliates, the RF broadband network covers 30% of all households. In other areas of the country, AT&T can levy local access through its Teleport Communications Group (TCG) and offer its own DSL services by collocating the equipment in existing LEC central offices. This new regulatory regime also opens the doors to accommodate other long-distance carriers in offering DSL services in the local loop.

The transmission technology in the long-distance backbone is merging as well. The days of circuit-switched services are numbered, and the new SONET/ATM/DWDM platform is merging with the IP platform of the Internet. All long-distance carriers will soon be able to route their voice traffic in the IP format with the aid of ATM cell transport tributaries. Sprint has built its network solely based on SONET/ATM technology. The concept of others may be IP-oriented and may rely on multiprotocol label switching (MPLS) to assure some QoS levels. MPLS lets a router or switch assign a tag to each of its routing table entries and identify the next hop or hops without looking up the full address. MPLS can also flow over ATM-based networks by employing the 53-byte cell as the transport. MPLS can be used over pure IP networks such as packet over SONET by inserting an additional header. The result is transparency for IP transmission in the long-distance segment, whether it is IP, ATM, SONET, or frame relay. The long-distance segment will become versatile and flexible. Multiple QoS levels may be the only indication that an ATM-based network is being used by a particular service provider. Most likely, the network core may employ DWDM and transport multiple OC-192 transmissions in a point-to-point architecture. The outer regions of the core will employ SONET/ATM technology to control loading and provide easy multiplexing of tributaries arriving at different speeds using cell switching to assure the shortest time intervals for circuit assembly, tear down, and reconvergence.

17.2.2 The ISPs and the Internet

To accommodate telephony services over the Internet, the ISPs must expand their switching technology to provide intelligent network (IN) capabilities. The switching fabric must support SS7 to compete directly with the carriers. VOIP lives in the environment of an Internet telephone service switch point (IT-SSP), which provides the interface between the PSTN and the IP network. An IT-SSP unit can serve as a direct interface to the local CO and the IN using SS7 trunking and enables the ISP to use SS7 messaging to transport VOIP traffic directly over the Internet. The application of the SS7 switch can provide the ISP with fully compliant SS7 intelligent switching solutions for any type of VOIP call. SS7 is also required to support future portable number transportability, calling card services, e-commerce activities, and 800/888/877 services. Before deploying VOIP, the user is well advised to determine the IN capabilities of the ISP.

17.3 The Multiservice Local Loop

Changes in the local loop will dominate the next decade. The beginning of these changes can already be seen by the DSL, ATM-T-1, and FR offerings of the established LECs. The present selection of competitive services, however, remains limited, a condition that has been caused by the copper-cabling monopoly of the serving LEC and the inability of the RF broadband cable TV providers to expand into data and telephony services. The end of this post-Telecommunications Act inactivity is near. The development of competitive services is depending upon the demographics of each local loop area and the status of the technological development of a particular service provider. Since the regulatory environment permits the collocation of a CLEC in the CO of an LEC to offer DSL services, it is obvious that lucrative markets will be chosen first. This regulatory ruling came about because of Section 706 of the Telecommunications Act, which required the FCC to ease the regulations affecting data services. A separate subsidiary must be set up by the LEC to accommodate the CLEC. A final ruling in the matter is expected during the year of 1999.

The status of present plant upgrading by the cable industry is another factor in this transition. Only a fully HFC-upgraded cable company can offer voice and data services. Many cable companies have not even started to rebuild their plants and, therefore, can only offer high-speed Internet access with a low-speed telephone-based return. AT&T, as another example, can provide local services in New York, Massachusetts, and Florida, and because of its ownership in TCG, can implement Internet high-speed access in TCI's upgraded networks. However, it has to market DSL services in other geographic areas in competition with other providers.

While the local loop is considered a formidable revenue source for new CLEC entries, it can also become a battleground between aggressive marketing organizations as numerous companies are trying to develop their territories at the same time. As much as competitive services and rates are appreciated by the user, unsolicited telemarketing efforts can produce a significant annoyance component.

17.3.1 CLECs

A CLEC can be copper wire-based, RF broadband cable-based, or RF broadband wireless. Wire-based CLECs will concentrate on the many forms of DSL services and offer IP- and ATM-based voice and data services at symmetrical and asymmetrical speed ranges. RF broadband wireless services will enter the local loop of the major metropolitan areas within a few years. The technology is based on microwave frequencies in the millimeter band that can provide a 750-MHz spectrum for delivery of multiple television channels, videoconferencing, massive telephony, and high-speed data. The frequency band in the 28–30-GHz region was auctioned in 1998 and is identified as LMDS.

Often called cellular television, LMDS could become another competitor to the cable industry, even though it appears that most of the winning bidders are interested in serving the business community. Initial offerings of DSL and LMDS services will not be uniform throughout the nation and may never be realized in lower density local loop areas. The competitive impact, however, in lucrative demographic areas will offer a range of new service offerings at considerably lower rates.

Vendors engaged in the development of LMDS technology include Nortel, Lucent Technologies, Alcatel Telecom, Siemens, Bosch Telecom, Ericsson, and a number of smaller companies. The interest of these companies is not only in the U.S. market, but in the global needs of both developed and developing countries.

A good example is the recent buyout of Bay Networks by Nortel. Prior to this transaction, Bay Networks had purchased LanCity, a symmetrical 10-Mbps Ethernet cable modem manufacturer. The merger permits Nortel to provide IP- and ATM-based switching equipment and RF-based cable modems to both the cable television industry and the new cellular RF broadband LMDS service providers.

LMDS service, which has been called "the race to the final mile," features a number of interesting winners in the FCC auctioning process. The top ten winners in alphabetical order are ALTA Wireless Inc. of Denver, a group backed by EchoStar with 4 basic trade area (BTA) licenses; ARNet Inc. of Amarillo, TX, an ISP with 16 licenses; Baker Communications LP, based in Coudersport, Pennsylvania, also known as Adelphia Cable TV, with 232 licenses; Cortelyou Communications Corp., based in New York, with 15 licenses and cellular and paging services in Puerto Rico and the Virgin Islands; BTA Associates, a group of Colorado-based electric cooperatives with seven licenses; Eclipse Communication Corp. of Issaquah, Washington, a paging and cellular service provider in the western United States with 51 licenses; CoServ LLC, based in Corinth, Texas; the Denton County Electric Company with six licenses; NEXTBAND Communications LLC, based in Bellevue, Washington, an affiliate of Nextel Communications, formed by Craig McCaw and Wayne Perry of the former McCaw Cellular Communications, Inc. with 42 licenses; Winstar LMDS LLC of New York City, providers of local, long-distance, and Internet services, with 15 major licenses; and WNP Communications, based in Earlysville, Virginia, a group composed of seven venture capital funds with 80 licenses. The cross-section of these companies is as interesting as the services that will become available first in large business centers and later in suburban areas and MDU establishments. They can include IP- and ATM-based transmissions of any kind with direct interfaces with long-distance and ISPs.

17.3.2 RF Broadband Cable Networks

Close to 90% of all households are wired for cable television. Except for business segments, every local loop contains a second infrastructure that can compete with the traditional copper plant of the serving LEC and the facility-sharing CLECs. The percentage of HFC-upgraded cable networks is presently only around the 30% mark. However, upgrading to HFC is well under way and will reach the 90% completion mark by 2003. The first data service, already offered in many systems, is the delivery of high-speed Internet access to the home. @Home, a TCI company, holds affiliate agreements with MSOs representing 50% of homes passed by cable. Road Runner, a Time Warner/Media One affiliate, has obtained major funding from Microsoft and Compaq to expand the Internet access business. As discussed earlier, the interoperability specifications for the MCNS-compliant cable modem have been finalized, and the stage is set for an intense marketing program to promote Internet access at megabit speeds. Voice over the cable modem is sure to follow, as are several MPEG-2-based, real-time VOD movie-selection methods.

The headend is undergoing a transition to the digital domain with a consolidation of diverse services. Using SONET fiber transport rings between serving cable properties, the super headend combines analog and digital video services, ad insertion equipment, IP- and ATM-based network access control equipment, cable modem control equipment, VOD servers, and automated maintenance equipment. The days when a particular channel audio had an annoying level and had viewers running to adjust the volume, will soon be forgotten. New computer-based performance monitoring of all video and audio carriers as well as status monitoring of every node and branch line will automatically adjust all levels to specification performance. A tight integration between the cable's backbone network and regional data centers, long-distance carriers, CLECs, and ISPs is the ultimate goal, and it will transform the cable industry into a formidable multiservice provider in the local loop.

17.4 The Multiservice WAN

Enterprise communications includes direct LAN-to-LAN connectivity between branch offices and affiliates, telecommuters, and personnel traveling in remote parts of the globe. It is important that voice, data, and video communications can be provided in a fully transparent manner—seeming as if the users are in the same building. Such transparency requires that access to the corporate LAN, to voice mail, and to servers be identical and conductible at comparable speed ranges and in complete privacy. Due to the evolutionary development of the telecommunications business, this premise remains a work-in-progress. The leasing of private lines and T-1 connectivity has proven a costly proposition especially in areas where the interLATA connectivity requires employment of multiple long-distance company POPs. A digital integration of voice and data is the basis to reduce the costs of communications and to apply alternative routing solutions. The explosive growth of the Internet has been the driving force in the transition of the long-distance carriers from standard circuit-based to packet-based technology.

For the most part, the transition to IP- or IP-over-ATM-based packet switching networks has been completed and competes directly with the IP-based Internet. The bottleneck is no longer the long-distance segment but the access arrangements in the local loop.

17.4.1 ATM and IP Choices

Current enterprise connectivity is based on the relative location to the LEC's central office and the nearest ISP. The connectivity consists of copper wires that were laid half a century ago and that are good carriers for 3-kHz analog voice and about 34-Kbps data speeds. The ISDN concept was promoted as the best solution to increase speeds by employing copper wires in parallel, but it has not progressed beyond the T-1 speed of 1.544 Mbps and the ability to dial up when needed. When compared with speeds within the corporate LAN, ISDN appears as an evolutionary dead-end branch. High-speed accessibility to the long-distance segment or to the Internet still remains the problem in many parts of the country. However, there are hopeful signs of DSL service, LMDS wireless, and the transition of the cable television provider to an RF broadband-based multiservice communications company. At present, FR and low-speed ATM at 1.544 Mbps offer the opportunity for businesses to integrate voice and data into an ATM-based digital stream. While these services use the public long-distance network, the demand for IP and VOIP remains strong. An end-to-end prioritization across the WAN that would reduce congestion across the network is required. The IETF has suggested type of service (ToS) options, such as low latency, high throughput, high reliability, low cost, and normal, to be inserted in the header of the packet. Recently, however, the IETF working group Differential Services or "diffserv" began defining classes of service to provide customers with "gold" and "silver" service levels that may replace the ToS end-to-end control required for WAN services. Essentially, this change would transfer the packet identification function to the ISPs that would overwrite the ToS byte and undo the end-to-end packet identifier that was selected for the WAN locations. This development requires monitoring and may lead to a loss of control on the part of the WAN network manager. As the alternative, the WAN manager has the option to employ ATM-based equipment at each of the WAN locations and use cell-based transmissions of IP packets for end-to-end connectivity.

One-way traffic such as voice messaging and fax can be routed safely over today's Internet. Gallup/Pitney Bowes recently reported that Fortune 500 companies will spend $15–20 million annually on telecommunications, 36% of which will be used to support fax traffic. Using fax-over-IP (FOIP) can reduce these expenditures drastically and should be a high priority.

17.4.2 VPNs

Many companies are offering solutions for VPN. The emphasis of these hardware and software offerings is on private and secure traffic in the WAN, often referred to as the extranet. There are two kinds of VPNs, virtual private trunking (VPT) and virtual private remote network (VPRN). VPTs establish the equivalent of a leased line and use FR or ATM connectivity through the public network. VPRNs deliver end-to-end connectivity over the Internet. IP traffic can be transmitted over both and can internetwork the LANs of the various branch offices of an enterprise. VPNs can provide fully secure communications with QoS assurance and establish various degrees of network management to provide an enterprise with a unified view of the traffic, the access equipment, and even third-party routers. While tunneling protocols are used by many vendors to establish a direct data connectivity between end locations, some manufacturers offer features that assure critical traffic and bandwidth guarantees that can also be used for telephone communications.

There are a number of IETF standards that apply to VPN traffic. RFC 1701 defines the generic routing encapsulation (GRE) of the IP-in-IP tunneling or bridge tunneling process. RF 1702 specifically addresses the GRE over IPv4 networks. RFC 1701 also addresses IPX tunneling and Apple Talk tunneling. The security measures are defined in RFC 1825, 1826, 1827, and 1828. The equipment complement frequently consists of Internet access routers that are installed at ISP locations and at branch offices as well as tunneling/WAN routers, which are required at the central location. The equipment is available with or without firewall protection and provides for authentication, encapsulation, encryption, and, in some cases, policy-filtered access control to network resources that can be based on the authenticated identity of the person or the network address at the other end of the tunnel. The problem with the technology is, again, the local loop which, at the present time, does not provide any faster access than T-1 speeds. VPNs can drastically reduce the costs of long-distance communications and, through voice compression, reserve most of the bandwidth for data exchange. It is also interesting to note that the encryption process is CPU-intensive and reduces throughput and firewall performance.

The IPSec standard is flexible and offers multiple implementation options. The tradeoff between speed, encryption, and firewall performance requires analysis before arriving at a purchase decision.

17.5 The Multiservice Corporate LAN

While the subject of this book is the changing environment in the corporate LAN, the transformation of the triple wiring system to a multiservice single infrastructure, and the application of the RF broadband technology, it is important to note that these activities cannot be undertaken without a good understanding of the converging global developments. Every corporation is well advised to first assess the knowledge level within its own organization. Many corporations have separate telephone and data administrations with highly specialized knowledge in their fields and a dedication to their cause. As global telecommunication changes, an integration of the knowledge levels within the organization is an important requirement. IP packet technology, ATM cell technology, compression of digital voice, and WAN interfaces are of primary importance to reduce corporate telecommunications expenses. The establishment of a dedicated group of experts can bridge the gap and put the corporation on the fast track to first tackle WAN and Internet connectivity problems and develop a step-by-step plan with alternative solutions. Fiscal responsibility to appropriate savings in the long-distance and WAN segments toward a progressive LAN upgrade plan is the basis for a step-by-step multimedia integration. In this phase, it is hoped that the guidance provided in this book can help readers to develop a constructive path and avoid some of the pitfalls that are common and many in this age of rapid technological changes.

17.5.1 ATM and IP Choices

Opinions are based on experience. An IT-trained individual will insist that IP is the only way to the future. It is human nature to select the path of least resistance to reach the goal of higher speeds without having to learn a new technology. Whenever the end goal is to just move data faster, the gigabit Ethernet technology will be highly cost-effective. However, when the long-range plan incorporates the integration of telephony, data, and video, an ATM-based backbone is the winning architecture. Telecommunications industry forecasts predict that ATM-based switching equipment installations will increase by about 50% per year through 2002 and that almost 30% of all corporate LANs will utilize an ATM backbone. This estimate does not include the expected increased use of ATM equipment when upgrading PBXs and integrating a multimedia single wiring system.

Every data network vendor is in the process of adding ATM interfaces to its equipment offerings. With LANE and MPOA specifications finalized, it is not very difficult to provide a transition path between Ethernet-based tributaries and an ATM backbone. ATM switch fabrics are already available to support 2.4-Gbps speeds and can provide multilevel QoS and true voice and video integration, features that the gigabit technology cannot provide. The development and release of new IP-ATM-compatible products requires the continuous attention of network managers. Recent product introductions feature switching fabrics that combine high-performance routing for IP packets over SONET as well as ATM-based cell switching. The walls between the packet- and cell-switching camps are being removed. The switches can be configured to both gigabit Ethernet and ATM backbone services and permit the installation of gigabit ports as well as 2.4-Gbps and 622- and 155-Mbps ATM ports. Called multiservice switches, they widen the choices for corporations and offer a path toward a homogenized multiservice LAN. DWDM technology can also be used in the corporate campus. A good example is Microsoft's campus in Redmond, which uses a four-fiber SONET backbone in a bidirectional ring configuration. DWDM trials are under way to handle 16 wavelengths on each fiber with a mix of 2.4 Gbps and 622 and 155 Mbps to expand the capacity of the fiber transport and to avoid the construction of additional SONET fiber rings.

17.5.2 The Integration of RF Broadband

One of the priority activities, besides expanding the knowledge level in the fast-changing field of telecommunications, is to analyze the existing wiring infrastructure and use single-mode fiber in the backbone and for any distant campus location. As discussed in this book, the integration of RF broadband into this planning process can be an alternative consideration. When taken through a cost-effectiveness study, RF broadband may prove to be the more economical solution. Consideration for spare fiber should be made in the backbone. An RF broadband overlay can be used whenever required to establish a range of new services that may not be contemplated or even considered desirable. This may include cable telephony, a separate network for wireless services, VOD for employee continuous education, interactive video services, cable television, special executive services, or any other new service requiring separate transmission paths.

17.6 Electronic Commerce

Commerce on the Internet or e-commerce has been compared with the industrial revolution. It is forecasted that the conduct of business as we know it will soon be replaced by global electronic transactions. There are already many examples of increased production, lower overhead costs, and more economical services. The airline ticketing business is a good example. Airlines pay about $8 when a ticket is issued through normal channels or travel agencies. These costs are reduced to $1 when the ticket is electronically purchased. Another good example is the online stock trade fee. To buy or sell up to 5,000 shares for $8 is more convenient than trying to locate a broker by telephone, getting his or her voice-mail, and having to pay $20–40 for the same transaction. Any corporation that uses its extranet for material procurement, supplier coordination, vendor evaluation, and delivery confirmation can point to large cost reductions. E-commerce can reduce project costs and completion time periods and streamline established business practices. An RFP placed online for a product or service will encourage global responses and can lead to acceptable bids at greatly reduced price levels. Electronic processing can speed the design cycle of a new product and bring the product to market earlier and at lower product costs. Customer service and order taking no longer rely on structured business hours but can be conducted 24 hours a day and seven days a week. The ability to reduce overhead expenditures and at the same time increase productivity will be the driving force in the transition toward establishment of fully automated business transactions over the Internet. E-commerce promotes global competition and removes time and space limitations. With the advent of e-commerce and multimedia communications in the WAN, the Internet, and the extranet, the infrastructure and functionality in the corporate LAN must accommodate seamless interconnectivity for every employee in the conduct of the daily business.

17.7 Thoughts in Closing

This book aims to provide a brief overview of the ongoing explosive development in the telecommunications industry and to bring the various trends into context. Since new developments are taking place at a rapid pace, the book was written with the goal of taking readers away from their day-to-day stressful duties to overview the fabric and makeup of technological advances and interrelationships in the telecommunications industry.

Such an overview shows that the transition to digital transmission formats is well under way and that it will influence our lives in many ways. The finalization of technical standards in conjunction with the enormous investments already made assures competitive services in the long-distance and local loop segments in the near future. Every business establishment and household will eventually benefit from this multimedia convergence. Existing enterprise and corporate communications networks have been established in accordance with concepts, equipment, and cabling available at the time of their inception and have been upgraded only to satisfy immediate needs. Telecommunications personnel in many organizations have achieved a high degree of expertise in their respective fields. Moreover, decisions are made on the basis of a collective set of experiences. The transformation of the corporate infrastructure to a multiservice facility requires a knowledge of many diverse technologies as well as an appreciation of the speed of change. Cross-fertilization of this specialized knowledge combined with continuous research into new developments and visionary thinking are essential ingredients of the corporate communications network's successful transition to an integrated multiservice facility.

If this book has conveyed that telecommunications is entering an exciting time period and has increased its readers' understanding of developing trends; if it has communicated that this changing telecommunications environment can be leveraged to improve the productivity, economy, and conduct of business; or if it has circumvented a single pitfall or wrong decision, it has served its purpose.

List of Acronyms and Abbreviations

AAL ATM adaptation layer

ABR available bit rate

ACR actual cell rate (ATM)

ACR attenuation to crosstalk ratio (CAT-5)

ADSL asymmetric digital subscriber line

AMS audio/visual multimedia service

ANI access network interface

ANSI American National Standards Institute

APD avalanche photodiodes

API application program interface

APOP alternative point of presence

ARP address resolution protocol

ASIC application-specific integrated circuit

ATM   asynchronous transfer mode

ATMP   ATM protocol

ATP   acceptance test plan

ATV   advanced television

BDF   building distribution frame

BICI   broadband intercarrier interface

BICSI   Building Industry Consulting Services, International

B-ISDN   broadband integrated services digital network

BNI   Broadband Networks, Inc.

BPI   baseline privacy

B-picture   bidirectional picture

BT   burst tolerance

BTA   basic trade area

BUD   before you dig

BUS   broadcast and unknown server

CAC   connection admission control

CAD   customer access device

CAT-3   category 3 wire

CAT-4   category 4 wire

CAT-5   category 5 wire

CAT-5E   category 5 enhanced wire

CBR constant bit rate

CBX computer branch exchange

CCITT International Telegraph and Telephone Consultative Committee

CCTV closed-circuit television

CDPD cellular digital packet system specifications

CDV cell delay variation

CES circuit emulation services

CI commercial insertion

CLEC competitive local exchange carrier

CLI cumulative leakage indicator

CLP cell loss priority

CLR cell loss ratio

CM cable modem

CMR cell misinsertion rate

CMTS-DRFI cable modem termination system-downstream RF interface

CMTS-NSI cable modem termination system-network side interface

CMTS-URFI cable modem termination system-upstream RF interface

C/N carrier-to-noise ratio

CO central office

CoS class of service

CPCS common part convergence sublayer

CPE   client/customer payroll envelope

CPI   common part indicator

CRC   cyclic redundancy check

CS   convergence sublayer

CSO   composite second order

CSB   Cable Service Bureau

CTB   composite triple beat

CTI   computer telephony integration

DARPA   Defense Advanced Research Program Agency

DAVIC   Digital Audiovisual Council

DCO   digital CO

DCS   digital cross-connect system

DCT   discrete cosine transformation

DES   digital encryption standard

DFB   direct feedback

DOCSIS   data over cable service interface specifications

DOCSS   data over cable security system

DSL   digital subscriber line

DTMF   dual tone multifrequency

DVB   digital video broadcast

DVD   digital video disk

DWDM   dense wavelength division multiplexing

DWMT   discrete wavelet multitone

DXI   data exchange interface

ECTF   Enterprise Computer Telephony Forum

EFI   engineer, furnish, and install

EIA   Electronic Industry Association

EMAC   European Market Awareness Committee

EMI   electromagnetic interference?

ENG   electronic news gathering

FAA   Federal Aviation Administration

FAR   Federal Acquisition Regulations

FC   fiber cross-connect

FCC   Federal Communications Commission

FDDI   fiber distributed data interface

FEBE   far-end block error

FEXT   far-end crosstalk

FOIP   fax-over-IP

FR   frame relay

FRAD   FR access device

FUNI   frame user-network interface

GFC   generic flow control

GOP   group of pictures

GRE   generic routing encapsulation

HALE   high-altitude long-endurance

HC   horizontal cross-connect

HDLC   high-level data link control

HDT   host digital terminal

HDTV   high-definition television

HEC   header error check

HF   high frequency

HFC   hybrid fiber-optic/coaxial

IC   intermediate cross-connect

ICR   initial cell rate

IDF   intermediate distribution frame

IEC   International Electrotechnical Committee

IETF   Internet Engineering Task Force

ILMI   integrated local management interface

IMTC   International Multimedia Teleconferencing Consortium

IN   intelligent network

IP   Internet protocol

I-picture   intraframe picture

IPSec   Internet protocol security

IPX   internetwork packet exchange

ISDN   integrated services digital network

ISO   International Organization for Standardization

ISP   Internet service provider

ITR   interoperability test report

IT-SSP   Internet telephone service switch point

ITU   International Telecommunications Union

ITV   instructional television

JPEG   Joint Picture Experts Group

JTAPI   Java telephony application programming interface

L2TP   layer 2 tunneling protocol

LAN   local access network

LANE   LAN emulation

LATA   local access transport area

LDAP   lightweight directory access protocol

LEC   LANE client (ATM)

LEC   local exchange carrier

LECS   LANE configuration server

LED   light emitting diode

LEO   low Earth orbit

LES   LANE server (ATM)

LMDS local multipoint distribution service

MAC media access control

MAP manufacturing automation protocol

MATV Master Antenna Television

max CTD maximum cell transfer delay

MBS maximum burst size

MCC main cross-connect

MCNS Multimedia Cable Network Systems

MCR minimum cell rate

MDF main distribution frame

MDU multiple dwelling unit

MEO medium Earth orbit

MMDS multichannel multipoint distribution service

MPC MPOA client (ATM)

MPEG Motion Picture Experts Group

MPLS multiprotocol label switching

MPOA multiprotocol over ATM

MPOP main point of presence

MPS MPOA server

MSO multiple system operator

MTA major trading area

MTBF   meantime between failure

MTTR   meantime to repair

MTU   maximum transmission unit

NAMAC   North America Market Awareness Committee

NCTA   National Cable Television Association

NEC   National Electric Code

NEXT   near-end crosstalk

NFPA   National Fire Protection Association

NHRP   next hop resolution protocol

NIC   network interface card

NMS   network management software

nrt-VBR   non-real time VBR

NSAP   network service access point

NT   network termination

NTSC   National Television Standards Committee

NVOD   near video-on-demand

OC   optical carrier

OC-1   optical carrier level 1

OFDM   orthogonal frequency division multiplexing

OSI   Open Systems Interconnection

OSS   operation support system

OTDR   optical time domain resolution

PAD   packet assembler/disassembler

PBX   private branch exchange

PCR   peer cell rate

PCS   personal communication services

PGL   peer group leader

PHY   physical layer

PICS   protocol implementation conformance statement

PIP   picture-in-picture

PKI   public key infrastructure

PMD   physical medium dependent

PNNI   private network-to-network interface

POP   point of presence

POS   packet over SONET

POTS   plain old telephone service

P-picture   predictive picture

PPP   point-to-point protocol

PPTP   point-to-point tunneling protocol

PRI   primary ISDN

PS-ACR   power sum-attenuation-to-crosstalk ratio

PS-ELFEXT   power sum equal-level far-end crosstalk

PS-NEXT power sum near-end crosstalk

PSTN public switched telephone network

PT payload type

QAM quadrature amplitude modulation

QoS quality of service

QPSK quadrature phase shift keying

RBB Residential Broadband

RBOC Regional Bell Operating Company

RCDD registered communications distribution designer

RFC request for comments

RFI radio-frequency interference; also request for information

RFP request for proposal

RIP routing information protocol

rt-VBR real-time VBR

SAC Strategic Air Command

SAN server area network

SAR segmentation and reassembly sublayer

SCR sustainable cell rate

SCTE Society of Cable Telecommunications Engineers

SDH synchronous digital hierarchy

SDLC synchronous data link control

SEAL   simple efficient adaptation layer

SECBR   severely errored cell block rate

SKIP   simple key-exchange Internet protocol

SMATV   master antenna television system

SNA   System Network Architecture

SNET   Southern New England Telephone

SNMP   simple network management protocol

SONET   synchronous optical network

SPE   synchronous payload envelope

SPI   serial program interface

SS7   signaling system 7

SSAA   shelf-space allocation algorithm

SSCS   service-specific convergence sublayer

STM   synchronous transport method

STP   shielded twisted pair

STS   synchronous transport signal

STS-1   synchronous transport signal level 1

STU   set-top unit

SVC   switched virtual circuit

TC   transmission convergence

TCG   Teleport Communications Group

TCP transmission control protocol

TDM time division multiplexing

TDMM Telecommunications Distribution Methods Manual

TIA Telecommunications Industry Association

TII technology independent interface

TOP technical and office protocol

ToS type of service

TR task recommendation

TVOD true video-on-demand

UBR unspecified bit rate

UDP user datagram protocol

UHF ultra high frequency

UL-OFN-LS Underwriter Laboratories-Optical Fiber Network-Limited Smoke

UNI user network interface

USB universal serial bus

UTP unshielded twisted pair

VBR variable bit rate

VCI virtual channel identifier

VHF very high frequency

VLAN virtual LAN

VOD video-on-demand

VOIP   voice over IP

VPI   virtual path identifier

VPN   virtual private network

VPRN   virtual private remote network

VPT   virtual private trunking

VT   virtual tributary

WAN   wide area network

WDM   wavelength division multiplexing

WLL   wireless local loop

Bibliography

Books

Azzam, Albert, *High-Speed Cable Modems, IEEE 802.14 Standards,* New York: McGraw-Hill, Inc. 1997.

Bartlett, Eugene R., *Cable Communications,* New York: McGraw-Hill, Inc., 1995.

Black, Ulysses, *Emerging Telecommunications Technologies,* Upper Saddle River, New Jersey: Prentice Hall, Inc. (a Simon & Schuster Company), 1997.

Chiong, John, *Internetworking ATM for the Internet and Enterprise Networks,* New York: McGraw Hill, Inc., Inc., 1997.

Minoli, D., and A. Schmidt, *Network Layer Switched Services,* New York: John Wiley & Sons, Inc., 1998.

Minoli, D., and T. Golway, *Designing and Managing ATM Networks,* Greenwich, CT: Manning/Prentice Hall, 1997.

Tunmann, Ernest, *Hybrid Fiber-Optic Coaxial Networks,* New York: Flatiron Publishing, Inc., 1995.

Standards (Samples)

ANSI T1.629, B-ISDN ATM Adaption Layer 3/4 Common Part Functions & Specifications

ANSI T1.662, B-ISDN ATM End System Address for Calling and Called Party

ATM Forum, ATM Traffic Management Specifications 4.0

ATM Forum, LAN Emulation Over ATM (LANE) Specifications, Vs 1.0

ATM Forum, Multi-Protocol Over ATM (MPOA), Vs 1.0

ATM Forum, PNNI Specification 1.0

ATM Forum, User-Network Interface (UNI) Specification 4.0

IEEE 802.14, RF Working Group Documents, 1997

ITU-T Recommendation I.361, B-ISDN ATM Layer Specification

ITU-T Recommendation I.363, B-ISDN ATM Adaptation Layer (AAL) Specification

ITU-T Recommendation Q.2130, B-ISDN ATM Adaptation Layer, Signaling at the UNI

ITU-T Recommendation J.83, Digital Video Standard (SCTE-DVS-068)

RFC 1972, IPv6 Stateless Address Autoconfiguration

RFC 1932, IP Over ATM: A Framework Document

RFC 2022, Support for Multicast over UNI 3.0/3.1-Based ATM Networks

SCTE, IPS-SP-001, Flexible RF Coaxial Drop

SCTE, IPS-SP-200, On-Premises Bonding and Safety

SCTE-IPS-SP-205, Active Network Interface Devices (NID)

SCTE, IPS-TP-103, Air Leak Test Method for Trunk, Feeder, and Distribution Cable

SCTE, IPS-TP-201 to 204, Insertion Loss, Return Loss, Isolation, Hum Modulation

SCTE, DSS-97-2, MCNS Data-Over-Cable RF Interface Specification

SCTE, DVS-018, ATSC Digital Video Standard

SCTE, DVS-031, Specifications for Digital Transmission Technology

Standards Sources

ATM Forum Worldwide Headquarters
303 Vintage Park Drive
Foster City, CA 94404
Telephone: (415) 578-6860

American National Standards Institute (ANSI)
1430 Broadway
New York, NY 10018
Telephone: (212) 642-4900

Electronics Industries Association (EIA)
2001 I Street, NW
Washington, D.C., 20036
Telephone: (202) 457-4966

Institute of Electrical and Electronics Engineers (IEEE)
445 Hoes Lane
Piscataway, NJ 08855
Telephone: (908) 564-3834

International Organization for Standardization (OSI)
1 Rue de Varembe, Case Postale 56
CH-1211 Geneva 20, Switzerland
Telephone: (41) 22 734-1240

International Telecommunications Union (ITU)
CH 1211 Geneva 20, Switzerland
Telephone: (41) 22 730-5853

National Institute of Standards and Technology (NIST)
Technology Building 225
Gaithersburg, MD 20899
Telephone: (301) 948-1784

Fiber Channel Association
12407 MoPac Expressway North 100-357
Austin, TX 78758
Telephone: (800) 272-4618

Cable Television Laboratories, Inc.
400 Centennial Parkway
Louisville, CO 80027
Telephone: (303) 661-9100

Society of Cable Television Engineers (SCTE)
1 Philips Road
Exton, PA 19341
Telephone: (610) 363-6888

MCNS Multimedia Cable Network System
c/o Arthur D. Little Inc.
Acorn Park
Cambridge, MA 02140
Telephone: (617) 498-5000

About the Author

Ernest O. Tunmann graduated as an electrical engineer in 1952, with the degree of Diploma Engineer, from the Technical University of Berlin, Germany. Thereafter, he worked for the Siemens Corporation, where he developed vehicle mobile systems for the Allied forces in Berlin and established interconnectivity with command centers.

In 1955, Mr. Tunmann joined Bell Telephone Company of Toronto. There, he originated the concept of Frontier Radio Service and designed the application components for DID from fixed mobile stations located in remote areas. The service was later renamed Improved Mobile Telephone Service (IMTS) and grew at an exponential rate. In 1958, he joined Entron Inc., a Washington, D.C.-based cable TV provider, to design and implement the first CATV systems in the nation. While becoming the chief network designer for the company, he also developed pole attachment standards jointly with Bell Telephone Company of New York and influenced the make-ready and construction standards for outside plant implementations. He directed a number of cable TV installation projects in five different states from inception through make-ready engineering, network design, construction supervision, and project management. In 1961, Mr. Tunmann was employed by Raytheon to design and implement 6-GHz, long-haul microwave systems for large commercial customers. The work included propagation analyses and frequency coordination of every network. He developed program management tools and systems to assure on-time completion schedules and to supervise multivendor implementation teams.

Mr. Tunmann formed Tele-Engineering Corporation in 1973, to partake in the development of cable TV systems throughout the nation. The company

371

designed and implemented over 2,000 miles of broadband coaxial systems for major franchise holders in the United States and Puerto Rico. In 1976, he formed the Community Cablevision of Framingham, obtained the municipal franchise for the area, and built the first 36-channel cable TV network in the region with the help of Tele-Engineering Corporation. In 1989, he formed TE Consulting, Inc., to provide expertise to enterprises and corporations in the areas of high-speed data networking. The company was organized to provide a complete range of consulting services, from needs assessment through detailed network design and RFP preparation to acceptance testing and maintenance. In 1997, TE Consulting, Inc., expanded its expertise in determining economical network solutions to the wireless local loop. Other current applications of expertise are the fast-growing areas of Internet telephony (IP telephony), VOIP, voice over IP/ATM in corporate LANs, the integration of computer telephony (CTI), and the establishment of secure VPNs for WAN services over the Internet.

In addition to this book, *Practical Multiservice LANs: ATM and RF Broadband* (Norwood: Artech House, 1999), Mr. Tunmann wrote and published *Hybrid Fiber/Coaxial Networks* (New York: Flatiron Publishing, 1993). Mr. Tunmann can be reached at the web site of TE Consulting, Inc., www.teconsulting.com, or by e-mail at etunmann@bigfoot.com.

Index

Acceptance testing, 308
Active riser distribution, 326–29
  design calculations, 328
  distribution amplifiers, 326
  frequency level, 326
  level calculations, 328
  return design, 329
  system illustration, 327
  See also Vertical risers
Addressability, 21
Address resolution protocol (ARP), 115
Amplification equipment, 289–91
Analog video cable modulators/
    demodulators, 216
Analog videoconferencing, 250–51
  automation equipment, 250–51
  ITV, 250
  See also Videoconferencing
Apartment complexes, 172–75
  alternatives illustration, 174
  conventional wiring methods, 172–73
  multiservice infrastructure
      alternatives, 173–75
  "private cable" systems, 172
  tenants' obligations, 175
  See also ATM-based network applications
Application-specific integrated circuits
    (ASICs), 44–45
Architectures, 77–81

ATM, 95, 131–35
  building network, 147–55
  campus network, 136–47
  double-ring, 313
  dual, 80–81
  enterprise network, 132–35
  fiber backbone dual-ring, 145–46
  fiber backbone single-ring, 144–45
  fiber backbone star, 142–44
  global multiservice network, 132
  high-split, 79–80
  preferences, 203–7
  RF broadband star, 146–47
  ring, 315
  ring/star, 313
  subsplit, 77–78
  VOD, 260–62
Asymmetric data communications, 35
Asymmetric digital subscriber line
      (ADSL), 53
Asynchronous transfer mode. *See* ATM
AT&T, 342
ATM, 6–7, 37, 95–113
  applications for RF broadband
      LANs, 129–30
  architecture, 95
  cable modem standards
      and, 125–29, 233
  cable telephony, 42

ATM (continued)
  cell illustration, 97
  cell loss priority (CLP), 97
  cell structure, 45, 96–97
  cost comparison, 336–37
  defined, 7, 95
  enterprise network, 34
  Ethernet and, 118–19
  fast Ethernet and, 118–19
  FDDI and, 118–19
  frame relay and, 114
  importance of, 83–130
  interoperability, 34, 113–21
  IPv6 and, 116
  JPEG over, 254–55
  in LANs, 45–47, 129–30, 336–37
  low-speed, in WAN, 40
  MPEG-2 over, 255–56
  multiprotocol over (MPOA), 7, 112–13
  multiservice technology, 113–21
  QoS, 105–9
  reference model, 98
  RF broadband and, 50–52
  service classes, 105–9
  signaling, 101
  SNA in, 116–18
  switches, 10, 11
  switching, 101–2
  switching equipment, 46
  TCP/IP and, 114–16
  technology, 7
  telephony and, 119–21
  traffic management, 105
  as unifying technology, 84
  UNI signaling, 100–101
  versatility, 96
  video and, 119–21
  virtual channel identifier (VCI), 97
  virtual path identifier (VPI), 97
  for VOD, 261–62
  voice channels, 40
ATM adaptation layers (AALs), 97, 98–100
  AAL1, 99
  AAL2, 99
  AAL3/4, 99
  AAL5, 99, 100
  in ATM reference model, 98

  functions, 98
  standards, 100
ATM-based cable telephony, 244–46
ATM-based network applications, 165–85
  ancillary, using RF broadband, 181–87
  apartment complexes, 172–75
  campus network, 175–81
  intelligent high-rise building, 169–71
  SOHO and small office LAN, 166–69
ATM-based network architectures, 131–55
  building network, 147–55
  campus network, 136–47
  enterprise network, 132–35
  global multiservice network, 132
ATM Forum, 100
  activity coordination, 124–25
  broadband intercarrier interface working
    group, 123
  defined, 121
  LANE working group, 123
  membership, 121
  MPOA working group, 124
  network management working
    group, 122
  physical layer working group, 121–22
  PNNI working group, 122
  reference model for HFC, 126
  residential and small business working
    group, 124
  Residential Broadband (RBB) working
    group, 125
  service aspects and applications working
    group, 123
  signaling working group, 122
  specifications, 121–25
  testing working group, 124
  traffic management working
    group, 122–23
  voice and telephony working group, 123
ATM-RF broadband campus, 179–81
  benefits of, 179, 181
  illustrated, 180
  See also Campus network
Attenuation
  coaxial cable, 26, 71, 203, 277, 278–80
  directional coupler, 284–85
  multitap, 286–87

service drop coaxial cable, 281–82
single-mode fiber cable, 66
splitter, 284–85
UTP, 70
Available bit rate (ABR) service, 101, 108–9
   cell rate traffic adjustment, 109
   flow control, 109
   in HFC system, 225

Bandwidth, 62–76
   coaxial cable, 70–73
   fiber-optic cable, 62–67
   infrastructure and, 74–75
   limitations, 73–76
   planning and management, 205–6
   transmission equipment and, 75–76
   UTP cable, 67–70
   See also Spectral capacity
BICSI standards, 195–203
   defined, 195–96
   presentation summaries, 197
Bid
   evaluation, 309
   specifications, 307–8
Break-out cables, 267–68
   defined, 267–68
   illustrated, 268
   jacket, 268
   mechanical specifications, 269
   See also Single-mode fiber
Buffer tube fan-out kits, 272
Building Industry Consulting Services,
      International. *See* BICSI standards
Building network architecture, 147–55
   high-rise building, 148–52
   horizontal wiring, 153–55
   small building, 147–48
   See also ATM-based network
      architecture
Business telephone systems, 240–42
Byte-interleaving multiplexing, 94

Cable data, 183
   defined, 3, 183
   offerings, 183
Cable Labs, 128–29
Cable modems, 11, 215–35

access control parameters, 225
applications, 233–35
Broadcom Corporation design, 230, 231
defined, 215
development status, 215–35
digital, 216–17
evolution of, 215–17
functioning of, 215
in HFC network, 224
for Internet access, 234
management, 242
market, 217–19
output, 155–57, 330
providers, 219–20
in RF broadband LANs, 235
standards, 125–29
technological developments, 218–19
Cable modem technology, 220–28
   MAC layer, 222–24
   MAC management, 224–25
   physical layer, 221–22
   QoS considerations, 225–26
   security considerations, 226–28
Cable Service Bureau (CSB), 22
Cable telephony, 237–47
   ATM-based, 244–46
   business telephone systems, 240–42
   in cable industry, 237–43
   in corporate LANs, 243–47
   customer access device (CAD), 241
   defined, 3, 237
   features, 240
   host digital terminal (HDT), 241, 246
   network management, 242–43
   powering problems, 239
   RF broadband, 183, 246–47
   status of, 237–47
   synchronous transfer method (STM)
      and, 238
   VOIP, 238–40
   See also Telephony
Cable television, 14–23
   cascades of noise, 17–18
   composite triple beat, 19
   customer service, 21–22
   deficiencies, 16–22
   distribution network, 18

Cable television (continued)
  early systems, 15–16
  headend, 32–34
  HFC, 29
  ingress and egress, 19–20
  multipath problems, 20–21
  regulatory environment, 22–23
  RF broadband and, 181
  set-top converters, 20
  Telecommunications Act of 1996
    and, 24–25
Campus network, 175–81
  computer telephony integration, 178
  conventional wiring systems, 176
  data networking alternatives, 177–78
  multiservice ATM and RF
    broadband, 179–81
  See also ATM-based network applications
Campus network architecture, 136–47
  fiber backbone dual-ring
    architecture, 145–46
  fiber backbone single-ring
    architecture, 144–45
  fiber backbone star architecture, 142–44
  migration toward multiservice
    network, 137–39
  RF-ATM corporate, 139–42
  RF broadband star architecture in dual
    ring, 146–47
  versatility of, 142–47
  work-in-progress, 138
  See also ATM-based network
    architectures
Campus zoning, 316
  fiber couplers and, 319
  requirements, 208
Carrier-to-noise ratio, 159–60
  decibel definition, 159
  of fiber-optic transmitter, 159
  thermal noise, 160
Category-3 (CAT-3) UTP, 39
Category-4 (CAT-4) UTP, 43
Category-5 (CAT-5) UTP, 43, 52, 68,
    69, 147, 177
  distribution, 320–21
  standards, 198, 199–200
  wiring, 301

  See also Unshielded twisted pair (UTP)
Category-6 (CAT-6) UTP, 69, 200
Category-7 (CAT-7) UTP/STP, 69, 201
CBR service, 107
Cellular digital packet system specifications
    (CDPD), 184
Cellular television, 54
CMTS-NSI, 229, 230
Coaxial cable, 62, 277–82
  attenuation, 26, 71, 277, 278–80
  backbone, 72–78
  bandwidth, 70–73
  design criteria, 312–14
  illustrated, 278
  marking, 280
  outside plant, 278–80
  plenum, 278–80
  reasons for using, 26
  RG-6, 71, 321–24
  riser, 278–80
  service drops, 70, 71–72, 280–82
  sizes/compositions, 70
  specifications, 278
  spectral capacity of, 25
  standards, 202–3
  testing, 202
  transmission equipment, 76
  types of, 62
  wiring, 301
Coaxial segment design, 319–31
  calculations for connections, 322
  headend interconnect, 329–31
  riser distribution, 324–29
  service drop, 320–24
  See also Coaxial cable
Coaxial transmission equipment, 277–91
  amplification equipment, 289–91
  connectors, 288–89
  directional couplers, 282–85
  multitaps, 285–87
  passive devices, 282–89
  placing, 301
  splitters, 282–85
  See also Coaxial cable
Commercial insertion (CI) business, 34
Competitive LECs (CLECs), 344–45
Composite triple beat, 19, 160–61

buildup, 161
  defined, 160
Computer
  branch exchanges (CBXs), 41
  center, 193–95
  telephony integration, 11, 178
Concatenation, 95
Conduit installations, 297
Congestion control, 104–5
Connectivity, 53–55, 271–72
Connectors, 288–89
  coring tools, 289
  housing, 288
  housing-to-housing, 288
  outside plant, 288–89
  riser, 288–89
  server drop, 289
  single-node fiber, 270–71
Copper cables, 62
Cornerstone, 241
Corporate communications, 37–60
Cost, 333–40
  assumptions, 333–34
  ATM-based telephone overlay, 340
  ATM-based videoconferencing
    system, 338
  ATM vs. RF broadband
    comparison, 336–37
  HFC-based VOD network, 338–39
  IP vs. RF broadband
    comparison, 334–36
  price forwarding and, 340
Crystal Exchange, 241
Cumulative leakage index (CLI), 23
Customer access devices (CADs), 241
Cyclical redundancy checks (CRCs), 99

Data networks
  center, 195
  early, 47–49
  evolution of, 42–52
  routers, 44–45
  video, 49–50
  wiring systems, 43–44
Data over cable service interface
  specifications (DOCIS), 129

Dense wavelength division multiplexing
    (DWDM), 11, 257
Design, 311–31
  coaxial, criteria, 312–14
  coaxial segment, 319–31
  fiber, criteria, 311–12
  fiber backbone, 314–19
  headend interconnect segment, 329–31
  optical, budgets, 317
  overview, 311–14
  riser distribution segment, 324–29
  service drop segment, 320–24
  See also RF broadband networking
DFB laser transmitters, 273–75
  electrical/optical specifications, 274, 275
  high-power, 273–74
  low-power, 274–75
  modulation depth, 273
Digital Audiovisual Council
    (DAVIC), 126–28
Digital cable modems, 216–17
Digital cross-connect system (DCS), 87
Digital videoconferencing, 251–57
  JPEG over ATM, 254–55
  JPEG vs. MPEG-2, 253–54
  MPEG-2 over ATM, 255–56
  RF broadband and, 256–57
  standards, 251–53
  in the WAN, 257
  See also Videoconferencing
Digital video disk (DVD), 258, 259
Digital video modems, 216
Direct asynchronous multiplexing, 87
Directional couplers, 282–85
  attenuation, 284–85
  defined, 282
  illustrated, 283
  insertion losses, 284
  specifications, 282–84
  See also Coaxial transmission equipment
Distribution amplifiers, 290–91
  powering options, 290–91
  specifications, 290
  spectral performance, 290
DiviTrack, 260
Documentation, 308

Dual system architecture, 80–81
  cable industry and, 80–81
  defined, 80
  illustrated, 81
  See also Architectures
Dual tone multifrequency (DTMF) dial
  tones, 240

Electronic commerce, 351
Encryption/decryption, 227–28
Enterprise communications, 55–60
Enterprise network, 132–35
  illustrated, 135
  shared ATM service, 134
  See also ATM-based network
    architectures
Equipment rooms, 298–300
  fiber termination panel, 299
  fiber terminations, 298
  patchcords, 300
  rack placement, 298
  storage panels, 298
Ethernet
  ATM and, 118–19
  fast, 44, 118–19
European Market Awareness Committee
  (EMAC), 124–25
Excavation, 295–96

Fast Ethernet
  ATM and, 118–19
  MTUs, 44
Fiber backbone design, 314–19
  fiber couplers, 318–19
  forward and return
    considerations, 315–17
  optical budgets, 317
  See also Design
Fiber backbone dual-ring
    architecture, 145–46
  illustrated, 146
  use of, 145
Fiber backbone single-ring
    architecture, 144–45
  illustrated, 145
  use of, 144
Fiber backbone star architecture, 142–44
  connectivity, 143

  defined, 142
  illustrated, 143
  negative factors, 144
Fiber distributed data interface (FDDI), 7
  ATM and, 118–19
  backbones, 118
  MTUs, 44
Fiber distribution frames, 269–70
Fiber node, sizing, 28–31
Fiber-optic cable
  bandwidth, 62–67
  spectral capacity, 25
  termination equipment, 268–71
  transmission equipment, 75–76
  See also Multimode fiber; Single-mode
    fiber
Fiber-optic couplers, 276, 318–19
  campus zoning and, 319
  coupling ratios, 318
  defined, 318
  at fiber transmitter, 318
  for redundant routing, 318
  See also Fiber backbone design
Fiber-optic receivers, 275–76
  electrical/optical specifications, 276
  gain blocks, 275
Fiber-optic transmission equipment, 272–76
  cable termination, 268–71
  connectorization and, 271–72
  couplers, 276
  high-power DFB laser
    transmitters, 273–74
  low-power DFB laser
    transmitters, 274–75
  receiving equipment, 275–76
  See also Fiber-optic cable
Fiber termination panel, 299
Fire-stopping, 300
Frame relay, 114
Frame user-network interface (FUNI), 134
Furcation kits, 272

Generic flow control (GFC), 97
Generic routing encapsulation (GRE), 348
Gigabits per second (Gbps), 2
Global multiservice network
    architecture, 132

Group addressing, 101

HDTV transmissions, 216
Headend, 32–34
   defined, 32
   SONET/ATM, 33
Headend interconnect segment, 329–31
   combining forward direction
      and, 329–31
   defined, 329
   splitting return direction and, 331
   See also Coaxial segment design
HFC-RF broadband
   bandwidth, 4
   cable television, 29
   campus-wide, 28
   capacity, 8
   components, 5, 8, 27
   defined, 1
   implementation considerations, 293–310
   infrastructure, 6
   network management, 242
   nonobsolescence of, 8–9
   peer relationship, 221
   reasons for obscurity, 9–10
   services, 34–36
   technology use, 5
   upgrading to, 9, 25–32
   vendor capability assessments, 302–5
   vendors, 10
High-altitude long-endurance (HALE)
      platform, 55
High frequency (HF) band, 13
High-power DFB laser transmitters, 273–74
High-rise building, 169–71
   cabling, 171
   design illustration, 170
   innovations, 169
   riser rooms, 171
   See also ATM-based network applications
High-rise building network
      architecture, 148–52
   dual ring, 150, 151
   RG-6 wiring to desktop, 152
   service options, 151
   single ring, 150
   star, 150

   UTP wiring to desktop, 152
High-split architecture, 79–80
   defined, 79
   forward spectral capacity, 80
   illustrated, 79
   See also Architectures
Horizontal wiring, 153–55, 300–302
   records, 192
   requirements, 211
   See also Vertical risers
Host digital terminals (HDTs), 241
Hybrid fiber-optic/coaxial. *See* HFC-RF
      broadband

IDF equipment, 192
   requirements, 210
   uses, 192
IEEE 802.14, 125–26, 127
Implementation supervision, 310
Infrastructure
   architecture comparison, 74
   bandwidth and, 74–75
   limitations, 75
Innerducts, 296
Inside plant, 297–302
   conduit installations, 297
   equipment room, 298–300
   horizontal wiring, 300–302
   vertical risers, 300–302
Inside plant records, 190–93
   horizontal wiring systems, 192
   IDF equipment, 192
   MDF equipment, 191
   outlet installations, 192–93
   vertical risers, 192
   See also Plant records
Installation
   conduit, 297
   outlet, 192–93
   outside plant, 295–97
   PBX, 39
   PCS campus, 185
   requirements, 307–8
   UG duct, 295–96
Instructional television (ITV), 250
Integrated network management, 242–43

Integrated services digital network
    (ISDN), 6, 252
International Multimedia Teleconferencing
    Consortium (IMTC), 252
Internet
    ATM and, 114–16
    ISPs and, 343
Internet protocol (IP)
    IPv6, 116
    LAN costs, 334–36
    LANs, 46
    networks, 6
    tunneling, 40–41
    See also TCP/IP
Intranets, 55
ISO/IEC 11801
    CAT-6, 200
    CAT-7, 201
    coaxial wiring, 202–3
    enhanced CAT-5, 199–200
    equipment interconnection, 198
    fiber termination and cable
        management, 197
    implementation approaches, 196
    single-mode fiber, 197–98
    task groups, 196
    UTP wiring, 198–202

Joint Picture Experts Group. *See* JPEG
JPEG
    advantages, 253
    defined, 253
    MPEG-2 vs., 253–54
    over ATM, 254–55
Junction boxes, 296

Kilobits per second (Kbps), 2

LAN Emulation (LANE), 110–11
    components, 110
    operation, 111
    standard, 110
LANs
    ATM-based, 45–47, 48, 336–37
    connectivity, 57
    future, 81–82
    IP-based, 46
    multiservice broadband, 187–213

multiservice corporate, 349–50
    RF broadband, 47–52, 81–82, 129–30
    small office, 166–69
Layer 2 tunneling protocol (L2TP), 57
Link access procedure-D, 114
LMDS
    band, 13, 14
    defined, 344
    development, 54, 345
    services, 344, 345
Local exchange carriers (LECs), 6
    competitive, 344–45
    LMDS wireless, 14
Local multipoint distribution services, 54
Long-distance segment, 342–43
Loose tube cables, 264–65
    defined, 264
    illustrated, 265
    mechanical specifications, 266
    recommended use of, 265
    See also Single-mode fiber
Low-Earth orbitals (LEOs), 54–55
Low-power DFB laser transmitters, 274–75

MAC layer, 222–24
    areas, 223
    defined, 222
    management, 224–25
    objectives, 223
    standardization issues, 224
    See also Cable modem technology
Manholes, 296
Manufacturing automation protocol
    (MAP), 47
Maximum transmission units (MTUs), 44
MCNS cable modems, 228–33
    compliance confirmation, 233
    customer premises interface, 229–33
    reference model, 228
    universal network interface, 229
    See also Cable modems
MDF equipment, 191
    concerns, 191
    space and powering
        requirements, 209–10
MDU service, 241
Meantime between failure (MTBF), 161

Meantime to repair (MTTR), 161
Medium Earth orbitals (MEOs), 54
Megabits per second (Mbps), 2
Megahertz (MHz), 2
MPEG-1, 35–36
MPEG-2, 36
    advantages, 254
    defined, 253
    DVD and, 258
    JPEG vs., 253–54
    over ATM, 255–56
    picture type definitions, 256
    transmission, 256
Multimedia Cable Network System
        (MCNS), 217
    See also MCNS cable modems
Multimedia workstations, 303
Multimode fibers, 61
    bandwidth, 62–65
    cladding diameter, 63
    core diameter, 62–63
    limitations, 64, 65
    transmission equipment for, 65
    types, 62–63
    uses, 65
    See also Fiber-optic
Multiprotocol over ATM
        (MPOA), 7, 112–13
    advantages, 113
    framework, 113
    server support, 112
    support services, 112
Multiservice
    corporate LANs, 349–50
    global network architecture, 132
    infrastructure alternatives, 173–75
    migration toward, 137–39
    SONET/ATM networking, 132
    switches, 350
    wiring, 202
Multiservice LANs, 187–213
    conformance to BICSI
            standards, 195–203
    existing plant records and, 188–95
    facility construction requirements, 212
    implementation of, 213
    planning, 187–213

planning document, 207–11
    project plan, 211–13
Multiservice local loop, 343–44
    CLECs, 344–45
    RF broadband cable networks, 345–46
Multiservice public network, 342–43
    Internet and, 343
    long-distance segment, 342–43
Multiservice WANs, 346–49
    ATM and IP choices, 347–48
    VPNs, 348–49
Multitaps, 285–86
    attenuation, 286–87
    eight-port, 286
    insertion losses, 287
    ports, 285
    specifications, 286, 287
    uses, 285

Near-Earth satellite communications
        systems, 54–55
Network architecture
    bandwidth planning/management, 205–6
    powering issues, 207
    preferences, 203–7
    preplanning considerations, 204–7
    RF spectrum planning/management, 205
    routing considerations, 204
    traffic management and QoS
            considerations, 206–7
    VOD, 260–62
    zoning considerations, 204–5
    See also Architectures
North America Market Awareness
        Committee (NAMAC), 125
Nrt-VBR service, 108
NTSC analog video, 35

OC-3, 10, 34, 76, 120
OC-12, 10, 34, 93, 120
OC-48, 93, 120
Open shortest path first (OSPF), 103
Operation support systems (OSSs), 88
Optical couplers, 276
Outlet
    installations, 192–93
    requirements, 210–11

Outlook, 341–52
    electronic commerce, 351
    multiservice corporate LAN, 349–50
    multiservice local loop, 343–46
    multiservice public network, 342–43
    multiservice WAN, 346–49
    telecommunications environment
        and, 341
Outside plant
    construction requirements, 212
    installation considerations, 295–97
    requirements, 208–9
Outside plant records, 188–90
    building entrance locations, 190
    conduit occupancy, 189–90
    routing, manholes, distances, 188–89
    See also Plant records

Packet cable, 240
Packet fragmentation, 229
Passive riser distribution, 324–25
    defined, 324
    floors/taps, 325
    illustrated, 325
    See also Vertical risers
Patchcords, 271, 300
PBXs
    analog and digital technology, 38–39
    CAT-3 cable, 39
    defined, 38
    development of, 38–42
    equipment records, 195
    installation illustration, 39
    transformation, 41–42
    See also Telephony
PCS wireless services, 241
Peer group leaders (PGLs), 103
Personal communication services
        (PCSs), 184–85
    campus installation, 185
    frequencies, 184
PHY layer, 221–22
Planning document, 207–11
    horizontal wiring requirements, 211
    IDF and vertical riser requirements, 210
    implementation categories, 294–95
    implementing, 293–95

outlet and user requirements, 210–11
    outside plant requirements, 208–9
    power and space requirements, 209–10
    single-mode fiber interconnect
        facility, 207–8
    See also Multiservice LANs
Plant records, 188–95
    computer center and server
        farms, 193–95
    data network center, 195
    inside plant, 190–93
    outside plant, 188–90
    PBX equipment, 195
    See also Multiservice LANs
Point-to-point protocol (PPP), 115
Point-to-point tunneling protocol
        (PPTP), 57
Power
    construction requirements, 212
    network architecture issues, 207
    requirements, 209
Price forwarding, 340
Private network-to-network interface
        (PNNI), 102–4
    addressing, 103
    defined, 102
    integrated (I-PNNI), 115
    network interface and, 103
    peer group leaders (PGLs), 103
    routing, 104
    signaling, 104

Quality of Service (QoS), 105–9, 161–62
    assurances, 106
    classes, 106
    parameters, 106
    See also ATM

Racks
    mounting, 270
    placement, 298
Registered communications distribution
        designer (RCDD), 298
Regulations
    cable, 22–23
    federal, 22–23
    state/local, 23

Request for comments (RFCs), 114–15
  bid response requirements, 306–7
  contents of, 305
  contract data requirements, 306
  contractor's responsibilities, 305–6
  owner-provided activities, 307
  preparation, 305–7
  project management, 306
  scheduling requirements, 306
  scope of work, 305
Return splitting network, 331
RF-ATM corporate campus
        network, 139–42
RF broadband
  architecture limitations, 77–81
  for asymmetric data communications, 35
  ATM and, 50–52
  cable data, 183
  cable networks, 345–46
  cable telephony, 183, 246–47
  cable television, 181
  in college campus, 30
  cost comparisons, 334–37
  digital and analog services, 6
  evolution of, 13–36
  frequency spectrum, 7
  for high-speed Internet access, 35
  in horizontal wiring, 153–55
  integrated voice/data and video facility, 5
  in the local loop, 53
  for MPEG-1, 35–36
  for MPEG-2, 36
  as multivendor network, 10
  for NTSC analog video, 35
  PCSs and, 184–85
  scalability, 9, 52
  services, 34–36
  spectral capacity, 31–32, 61–82
  spectrum management, 8
  for symmetric data communications, 35
  technology, 1
  for telephony, 35
  upgrading to HFC, 25–32
  versatility, 165–66
  videoconferencing and, 182–83
  for VOD, 181–82, 262
  See also HFC-RF broadband

RF broadband LANs, 47–52, 81–82
  ATM applications for, 129–30
  cable modem applications in, 235
RF broadband networking, 4–5
  architecture selection, 162–64
  availability, optimizing, 164
  carrier-to-noise ratio, 159–60
  composite triple beat (CTB), 160–61
  design, 311–31
  early, 158
  economical considerations, 164
  flexibility, 155–58
  functions, 155
  performance, 158–62
  reliability of service, 161–62
  schematic, 156
  spectral capacity, optimizing, 162–63
RG-6 coaxial distribution, 321–24
  calculations, 324
  illustrated, 323
  service drop attenuations and outlet
        levels, 324
Riser distribution segment, 324–29
  active riser distribution, 326–29
  passive riser distribution, 324–25
  See also Coaxial segment design
Risers. *See* Vertical risers
Routers, 44–45
Routing information protocol (RIP), 115
Rt-VBR service, 107–8

SAW filters, 6
Scalability, RF broadband network, 9
Security
  cable modem, 226–28
  encryption/decryption, 227–28
Serial line Internet protocol, 115
Server area networks (SANs), 193
Service classes, 105–9
  ABR service, 108–9
  attenuation, 203
  CBR service, 107
  list of, 106
  nrt-VBR service, 108
  rt-VBR service, 107–8
  UBR service, 108
  See also ATM

Service drop coaxial cable, 280–82
    attenuation, 281–82
    length, 322
    properties, 281
    specifications, 280–81
    types of, 280
    See also Coaxial cable
Service drop segment, 320–24
    attenuations and outlet levels, 324
    CAT-5 distribution, 320–21
    design calculations, 323
    RG-6 distribution, 321–24
    See also Coaxial segment design
Set-top converters, 20, 216
Shelf-space allocation algorithm (SSAA), 260
Shielded twisted pairs (STPs), 68
Signaling messages, 101
Signaling system 7 (SS7), 56
Simple key-exchange Internet protocol
    (SKIP), 57
Simple network management protocol
    (SNMP), 115
Single-mode fiber, 61–62, 263–68
    attenuation, 66
    bandwidth, 65–67
    break-out cables, 267–68
    broadband capabilities, 67
    cladding diameter, 65
    coating around the cladding, 264
    connectors, 270–71
    core diameter, 65
    design criteria, 311–12
    electrical specifications, 264
    interconnect facility, 207–8
    limitations, 67
    loose tube cables, 264–65
    optic cable pulling, 297
    spectral capacity, 315
    standards, 197–98
    tight buffer cables, 265–67
    transmission forms, 67
    trunk, 27–28
    types of, 66
    See also Fiber-optic
Small building network architecture, 147–48
Small office LANs, 166–69
    RF-ATM, 168–69

typical system, 167
Society of Cable Telecommunications
    Engineers (SCTEs), 128
SONET, 84–95
    advantages, 86–88
    applications, 83
    frame structure, 88–92
    hierarchies, 84–86
    importance of, 83–130
    mesh configuration, 93
    network illustration, 87
    networking, 93
    OSSs, 88
    overhead bytes, 91
    overhead functions, 90
    performance control, 91
    SDH comparison, 86
    switches, 102
    synchronous payload envelope (SPE), 88
    transport system, 86, 92–95
    virtual tributary frame structure, 92
SONET/ATM, 83
    headend interconnection, 33
    multiservice networking, 132
    standards, 46
    See also ATM
Spectral capacity, 31–32, 61–82
    coaxial cable, 25
    downstream, 31–32
    fiber-optic, 25
    high-split architecture, 80
    optimizing, 162–63
    single-mode fiber, 315
    upstream, 31–32
    See also Bandwidth
Splitters, 282–85
    attenuation, 284–85
    defined, 282
    illustrated, 283
    insertion losses, 284
    specifications, 282–84
    See also Coaxial transmission equipment
Standards
    AAL, 100
    BICSI, 195–203
    cable modem, 125–29, 228–33
    CAT-5, 198, 199–200

coaxial cable, 202–3
single-mode fiber, 197–98
SONET/ATM, 46
UTP, 198–202
Structured cabling system, 68
STS
  defined, 88
  illustrated, 89
  STS-1 signal, 89, 94, 95
  STS-3 signal, 94
Subsplit architecture, 77–78
  amplifiers, 78
  defined, 77
  illustrated, 77
  return capacity, 78
  See also Architectures
Switched virtual circuits (SVCs), 56
Switches
  ATM, 101–2
  multiservice, 350
  SONET, 102
Symmetric data communications, 35
Synchronous byte-interleaving, 94
Synchronous digital hierarchy (SDH), 84
  list of, 85
  SONET comparison, 86
Synchronous payload envelope (SPE), 88
Synchronous transfer method (STM), 238
System Network Architecture (SNA), 44
  ATM and, 116–18
  defined, 116
  layer, 117
  networks, 118
  standardization areas, 117

TCP/IP
  ATM and, 114–16
  defined, 115
  implementation, 115
Technical and office protocol (TOP), 47
Technical specifications, 307–8
  documentation, 308
  hardware and equipment, 307
  installation, 307–8
  software, 307
  testing, 308
Telecommunications Act of 1996, 24–25

competitive developments, 24–25
competitive guidelines, 25
goals, 24
implementation, 24
Telecommunications Distribution Methods Manual, 297
Telecommunications environment, 341
Telephony, 35, 37
  alternative RF broadband
    solution, 41–42
  analog and digital technology, 38–39
  ATM and, 119–21
  ATM-based cable, 42
  computer integration, 178, 243–44
  in corporate LAN, 243–47
  corporate network, future of, 41
  development of, 38–42
  WAN, future of, 40–41
  See also Cable telephony
Termination equipment, 268–71
  connectors, 270–71
  fiber distribution frames, 269–70
  manufacturers, 268
  preassembled patchcords, 271
  rack mounting and cable
    management, 270
  See also Single-mode fiber
Terms and conditions, 308–9
  bid evaluation, 309
  general terms, 309
  specific terms, 309
  vendor qualifications, 309
  warranty requirements, 309
Testing, 308
Tight buffer cables, 265–67
  defined, 265
  illustrated, 267
  indoor/outdoor, 266
  mechanical specifications, 267
  riser-rated, 265
  See also Single-mode fiber
Traffic management, 104–5
  ATM, 105
  defined, 104
  in planning multiservice LANs, 206–7

UBR service, 108

Ultrahigh frequency (UHF) band, 13
UNI signaling, 100–101
   defined, 100
   messages, 101
Unshielded twisted-pair (UTP) cable
   attenuation performance, 70
   bandwidth, 67–70
   CAT-3, 39
   CAT-4, 43
   CAT-5, 43, 52, 68, 69, 147, 177, 198,
199–200, 320–21
   CAT-6, 69, 200
   CAT-7, 69, 201
   limitations, 69
   performance concerns, 69
   standards, 198–202
   testing guidelines, 201–2
User requirements, 210–11

Variable bit rate (VBR)
   services, 7
   transmission, 76
Vendor
   capability assessments, 302–5
   qualifications, 309
Vertical risers, 192, 300–302
   cabling in, 300
   coaxial, 300–301, 302
   fiber, 300–301
   requirements, 210
   See also Active riser distribution;
      Equipment room; Passive riser
      distribution
Very high frequency (VHF) band, 13
Video
   ATM and, 119–21
   ISDN, 120
   networks, 49–50
Videoconferencing, 249–57
   analog, 250–51
   connections, 250
   costing, system, 338
   digital, 251–57
   in LANs, 255
   RF broadband and, 182–83
   standards, 251–53
   in WANs, 257

Video-on-demand (VOD), 50, 257–62
   bandwidth control, 260
   content management, 260
   control center, 21
   corporate television, 258
   costing, network, 338–39
   distance learning and, 258
   HFC-based, 338–39
   interactivity and, 260
   RF broadband and, 181–82
   servers, 259
   technology, 258–60
   transmission format, 259
   See also VOD network architectures
Virtual channel identifier (VCI), 97
Virtual LANs (VLANs), 111, 155
Virtual path identifier (VPI), 97
Virtual private networks
      (VPNs), 56–60, 348–49
   categories, 57–58
   defined, 56
   dependent, 58
   hybrid, 58
   illustrated, 59
   independent, 58
   over Internet, 57–60
   in public switched telephone
      network, 56–57
   solutions, 348
   types of, 348
Virtual private remote network
      (VPRN), 348
Virtual private trunking (VPT), 348
Virtual tributary (VT) frame structure, 92
VOD network architectures, 260–62
   ATM and IP, 261–62
   illustrated, 261
   RF broadband, 262
   See also Video-on-demand (VOD)
Voice over IP (VOIP), 11, 37
   cable telephony, 238–40
   transmission, 6

WANs
   IP tunneling in, 40–41
   low-speed ATM in, 40
   multiservice, 346–49

telephony future, 40–41
videoconferencing in, 257
voice networking in, 39–40
Warranty requirements, 309

Wavelength division multiplexing
(WDM), 26
Wireless local loop (WLL), 13

Recent Titles in the Artech House Telecommunications Library

Vinton G. Cerf, Senior Series Editor

Access Networks: Technology and V5 Interfacing, Alex Gillespie

Advanced High-Frequency Radio Communications,
Eric E. Johnson, Robert I. Desourdis, Jr., et al.

Advances in Telecommunications Networks, William S. Lee and
Derrick C. Brown

*Advances in Transport Network Technologies: Photonics Networks,
ATM, and SDH,* Ken-ichi Sato

*Asynchronous Transfer Mode Networks: Performance Issues,
Second Edition,* Raif O. Onvural

ATM Switches, Edwin R. Coover

ATM Switching Systems, Thomas M. Chen and Stephen S. Liu

Broadband Network Analysis and Design, Daniel Minoli

Broadband Networking: ATM, SDH, and SONET, Mike Sexton and
Andy Reid

Broadband Telecommunications Technology, Second Edition,
Byeong Lee, Minho Kang, and Jonghee Lee

*Client/Server Computing: Architecture, Applications, and Distributed
Systems Management,* Bruce Elbert and Bobby Martyna

Communication and Computing for Distributed Multimedia Systems,
Guojun Lu

Communications Technology Guide for Business, Richard Downey,
Seán Boland, and Phillip Walsh

Community Networks: Lessons from Blacksburg, Virginia,
Andrew Cohill and Andrea Kavanaugh, editors

Computer Mediated Communications: Multimedia Applications,
Rob Walters

Computer Telephony Integration, Second Edition, Rob Walters

Convolutional Coding: Fundamentals and Applications,
    Charles Lee

Desktop Encyclopedia of the Internet, Nathan J. Muller

*Distributed Multimedia Through Broadband Communications
    Services,* Daniel Minoli and Robert Keinath

Electronic Mail, Jacob Palme

*Enterprise Networking: Fractional T1 to SONET, Frame Relay to
    BISDN,* Daniel Minoli

FAX: Digital Facsimile Technology and Applications, Second Edition,
    Dennis Bodson, Kenneth McConnell, and Richard Schaphorst

Guide to ATM Systems and Technology, Mohammad A. Rahman

Guide to Telecommunications Transmission Systems,
    Anton A. Huurdeman

A Guide to the TCP/IP Protocol Suite, Floyd Wilder

Information Superhighways Revisited: The Economics of Multimedia,
    Bruce Egan

International Telecommunications Management, Bruce R. Elbert

Internet E-mail: Protocols, Standards, and Implementation,
    Lawrence Hughes

Internetworking LANs: Operation, Design, and Management,
    Robert Davidson and Nathan Muller

Introduction to Satellite Communication, Second Edition,
    Bruce R. Elbert

Introduction to Telecommunications Network Engineering,
    Tarmo Anttalainen

Introduction to Telephones and Telephone Systems, Third Edition,
    A. Michael Noll

LAN, ATM, and LAN Emulation Technologies, Daniel Minoli and
    Anthony Alles

The Law and Regulation of Telecommunications Carriers,
    Henk Brands and Evan T. Leo

Mutlimedia Communications Networks: Technologies and Services,
  Mallikarjun Tatipamula and Bhumip Khashnabish, Editors

Networking Strategies for Information Technology, Bruce Elbert

Packet Switching Evolution from Narrowband to Broadband ISDN,
  M. Smouts

Packet Video: Modeling and Signal Processing, Naohisa Ohta

Performance Evaluation of Communication Networks,
  Gary N. Higginbottom

Practical Computer Network Security, Mike Hendry

Practical Multiservice LANs: ATM and RF Broadband,
  Ernest O. Tunmann

Principles of Secure Communication Systems, Second Edition,
  Don J. Torrieri

Principles of Signaling for Cell Relay and Frame Relay,
  Daniel Minoli and George Dobrowski

Pulse Code Modulation Systems Design, William N. Waggener

Signaling in ATM Networks, Raif O. Onvural and Rao Cherukuri

Smart Cards, José Manuel Otón and José Luis Zoreda

Smart Card Security and Applications, Mike Hendry

SNMP-Based ATM Network Management, Heng Pan

Successful Business Strategies Using Telecommunications Services,
  Martin F. Bartholomew

Super-High-Definition Images: Beyond HDTV, Naohisa Ohta

Telecommunications Deregulation, James Shaw

Telemetry Systems Design, Frank Carden

Teletraffic Technologies in ATM Networks, Hiroshi Saito

*Understanding Modern Telecommunications and the Information
  Superhighway,* John G. Nellist and Elliott M. Gilbert

Understanding Token Ring: Protocols and Standards, James T. Carlo,
  Robert D. Love, Michael S. Siegel, and Kenneth T. Wilson

Visual Telephony, Edward A. Daly and Kathleen J. Hansell

World-Class Telecommunications Service Development,
  Ellen P. Ward

For further information on these and other Artech House titles,
including previously considered out-of-print books now available
through our In-Print-Forever® (IPF®) program, contact:

Artech House   Artech House
685 Canton Street   46 Gillingham Street
Norwood, MA 02062   London SW1V 1AH UK
Phone: 781-769-9750   Phone: +44 (0)171-973-8077
Fax: 781-769-6334   Fax: +44 (0)171-630-0166
e-mail: artech@artech-house.com   e-mail: artech-uk@artech-house.com

Find us on the World Wide Web at:
www.artechhouse.com